T0206310

CAUSALITY, MEASUREMENT THEORY AND THE DIFFERENTIABLE STRUCTURE OF SPACE-TIME

Introducing graduate students and researchers to mathematical physics, this book discusses two recent developments: the demonstration that causality can be defined on discrete space-times; and Sewell's measurement theory, in which the wave packet is reduced without recourse to the observer's conscious ego, nonlinearities or interaction with the rest of the universe.

The definition of causality on a discrete space-time assumes that space-time is made up of geometrical points. Using Sewell's measurement theory, the author concludes that the notion of geometrical points is as meaningful in quantum mechanics as it is in classical mechanics, and that it is impossible to tell whether the differential calculus is a discovery or an invention. This mathematical discourse on the relation between theoretical and experimental physics includes detailed accounts of the mathematically difficult measurement theories of von Neumann and Sewell.

R. N. Sen was a Professor in the Department of Mathematics at Ben-Gurion University, Beer-Sheva, Israel, and is now retired. His main research interests were the theory of symmetry of infinite quantum-mechanical systems and mathematical investigations into the relation between mathematics and physics, particularly the origins of the differentiable structure of space-time. He has taught a broad spectrum of courses on physics and mathematics, as well as demography. A Life Member of Clare Hall, Cambridge, he has been a Gauss Professor in Göttingen and is also a member of the International Association for Mathematical Physics and of the Israel Mathematical Union.

CAMBRIDGE MONOGRAPHS ON MATHEMATICAL PHYSICS

General Editors: P. V. Landshoff, D. R. Nelson, S Weinberg

S. J. Aarseth *Gravitational N-Body Simulations: Tools and Algorithms*
J. Ambjørn, B. Durhuus and T. Jonsson *Quantum Geometry: A Statistical Field Theory Approach*
A. M. Anile *Relativistic Fluids and Magneto-fluids: With Applications in Astrophysics and Plasma Physics*
J. A. de Azcárraga and J. M. Izquierdo *Lie Groups, Lie Algebras, Cohomology and Some Applications in Physics*†
O. Babelon, D. Bernard and M. Talon *Introduction to Classical Integrable Systems*†
F. Bastianelli and P. van Nieuwenhuizen *Path Integrals and Anomalies in Curved Space*
V. Belinski and E. Verdaguer *Gravitational Solitons*
J. Bernstein *Kinetic Theory in the Expanding Universe*
G. F. Bertsch and R. A. Broglia *Oscillations in Finite Quantum Systems*
N. D. Birrell and P. C. W. Davies *Quantum Fields in Curved Space*†
K. Bolejko, A. Krasiński, C. Hellaby and M-N. Célérier *Structures in the Universe by Exact Methods: Formation, Evolution, Interactions*
D. M. Brink *Semi-Classical Methods for Nucleus-Nucleus Scattering*†
M. Burgess *Classical Covariant Fields*
E. A. Calzetta and B.-L. B. Hu *Nonequilibrium Quantum Field Theory*
S. Carlip *Quantum Gravity in* $2+1$ *Dimensions*†
P. Cartier and C. DeWitt-Morette *Functional Integration: Action and Symmetries*
J. C. Collins *Renormalization: An Introduction to Renormalization, the Renormalization Group and the Operator-Product Expansion*†
P. D. B. Collins *An Introduction to Regge Theory and High Energy Physics*†
M. Creutz *Quarks, Gluons and Lattices*†
P. D. D'Eath *Supersymmetric Quantum Cosmology*
F. de Felice and C. J. S Clarke *Relativity on Curved Manifolds*
B. DeWitt *Supermanifolds,* 2nd edition†
P. G. O Freund *Introduction to Supersymmetry*†
Y. Frishman and J. Sonnenschein *Non-Perturbative Field Theory: From Two Dimensional Conformal Field Theory to QCD in Four Dimensions*
J. A. Fuchs *Affine Lie Algebras and Quantum Groups: An Introduction, with Applications in Conformal Field Theory*†
J. Fuchs and C. Schweigert *Symmetries, Lie Algebras and Representations: A Graduate Course for Physicists*†
Y. Fujii and K. Maeda *The Scalar-Tensor Theory of Gravitation*
J. A. H. Futterman, F. A. Handler, R. A. Matzner *Scattering from Black Holes*†
A. S. Galperin, E. A. Ivanov, V. I. Orievetsky and E. S. Sokatchev *Harmonic Superspace*
R. Gambini and J. Pullin *Loops, Knots, Gauge Theories and Quantum Gravity*†
T. Gannon *Moonshine beyond the Monster: The Bridge Connecting Algebra, Modular Forms and Physics*
M. Göckeler and T. Schücker *Differential Geometry, Gauge Theories and Gravity*†
C. Gómez, M. Ruiz-Altaba and G. Sierra *Quantum Groups in Two-Dimensional Physics*
M. B. Green, J. H. Schwarz and E. Witten *Superstring Theory* Volume 1*: Introduction*†
M. B. Green, J. H. Schwarz and E. Witten *Superstring Theory* Volume 2*: Loop Amplitudes, Anomalies and Phenomenology*†
V. N. Gribov *The Theory of Complex Angular Momenta: Gribov Lectures on Theoretical Physics*
J. B. Griffiths and J. Podolský *Exact Space-Times in Einstein's General Relativity*
S. W. Hawking and G. F. R. Ellis *The Large Scale Structure of Space-Time*†
F. Iachello and A. Arima *The Interacting Boson Model*†
F. Iachello and P. van Isacker *The Interacting Boson–Fermion Model*
C. Itzykson and J. M. Drouffe *Statistical Field Theory* Volume 1*: From Brownian Motion to Renormalization and Lattice Gauge Theory*†
C. Itzykson and J. M. Drouffe *Statistical Field Theory* Volume 2*: Strong Coupling, Monte Carlo Methods, Conformal Field Theory and Random Systems*†
C. V. Johnson *D-Branes*†
P. S. Joshi *Gravitational Collapse and Spacetime Singularities*
J. I. Kapusta and C. Gale *Finite-Temperature Field Theory: Principles and Applications,* 2nd edition
V. E. Korepin, N. M. Bogoliubov and A. G. Izergin *Quantum Inverse Scattering Method and Correlation Functions*†
M. Le Bellac *Thermal Field Theory*†
Y. Makeenko *Methods of Contemporary Gauge Theory*

N. Manton and P. Sutcliffe *Topological Solitons*†
N. H. March *Liquid Metals: Concepts and Theory*
I. Montvay and G. Münster *Quantum Fields on a Lattice*†
L. O'Raifeartaigh *Group Structure of Gauge Theories*†
T. Ortín *Gravity and Strings*†
A. M. Ozorio de Almeida *Hamiltonian Systems: Chaos and Quantization*†
L. Parker and D. J. Toms *Quantum Field Theory in Curved Spacetime: Quantized Fields and Gravity*
R. Penrose and W. Rindler *Spinors and Space-Time* Volume 1*: Two-Spinor Calculus and Relativistic Fields*†
R. Penrose and W. Rindler *Spinors and Space-Time* Volume 2*: Spinor and Twistor Methods in Space-Time Geometry*†
S. Pokorski *Gauge Field Theories,* 2nd edition†
J. Polchinski *String Theory* Volume 1*: An Introduction to the Bosonic String*
J. Polchinski *String Theory* Volume 2*: Superstring Theory and Beyond*
V. N. Popov *Functional Integrals and Collective Excitations*†
R. J. Rivers *Path Integral Methods in Quantum Field Theory*†
R. G. Roberts *The Structure of the Proton: Deep Inelastic Scattering*†
C. Rovelli *Quantum Gravity*†
W. C. Saslaw *Gravitational Physics of Stellar and Galactic Systems*†
R. N. Sen *Causality, Measurement Theory and the Differentiable Structure of Space-Time*
M. Shifman and A. Yung *Supersymmetric Solitons*
H. Stephani, D. Kramer, M. MacCallum, C. Hoenselaers and E. Herlt *Exact Solutions of Einstein's Field Equations,* 2nd edition
J. Stewart *Advanced General Relativity*†
T. Thiemann *Modern Canonical Quantum General Relativity*
D. J. Toms *The Schwinger Action Principle and Effective Action*
A. Vilenkin and E. P. S. Shellard *Cosmic Strings and Other Topological Defects*†
R. S. Ward and R. O. Wells, Jr *Twistor Geometry and Field Theory*†
J. R. Wilson and G. J. Mathews *Relativistic Numerical Hydrodynamics*

† Issued as a paperback

Causality, Measurement Theory and the Differentiable Structure of Space-Time

RATHINDRA NATH SEN

Ben-Gurion University, Beer-Sheva

CAMBRIDGE
UNIVERSITY PRESS

University Printing House, Cambridge CB2 8BS, United Kingdom

Cambridge University Press is part of the University of Cambridge.

It furthers the University's mission by disseminating knowledge in the pursuit of education, learning and research at the highest international levels of excellence.

www.cambridge.org
Information on this title: www.cambridge.org/9781107424586

First published 2010
First paperback edition 2014

A catalogue record for this publication is available from the British Library

ISBN 978-0-521-88054-1 Hardback
ISBN 978-1-107-42458-6 Paperback

To the memory of my father, mother, brother and sister

Upendra Nath Sen
Mukul Sen
Tejendra Nath Sen
Aditi Datta Chowdhury

Contents

Preface

Someone once said that a book is never finished; it is merely abandoned. As the truth of it began to sink in, I sometimes wondered what on earth made me start upon this one!

I first met Richard Eden while visiting Jacques Mandelbrojt in Marseilles, in the late 1960s or early 1970s. We were next thrown together in 2000–2001 in Clare Hall, Cambridge, where he was one of the presiding deities and I a visiting fellow. In the intervening decades we had both moved from one production line to another. Richard's questioning about what I had been doing compelled me to organize my thoughts, and this book is the result; it would not have come into being without the encouragement offered by Richard, and by Jamal Nazrul Islam.

The influence of Wigner, whom I barely knew, permeates almost every page of this book. I first met him in 1961, at the Weizmann Institute. I was then working for my Ph.D. under Professor Giulio Racah on a group-theoretical problem of nuclear spectroscopy, and in awe of Wigner. As I was introduced to him, he blew my self-confidence to smithereens by asking for my opinion about the unreasonable effectiveness of mathematics in physics – I had none – but he had the kindness to send me a typescript of his article when he returned to Princeton. That question has haunted me ever since.

It was not only Richard Eden whom I met in Marseilles; another person was Hans-Jürgen Borchers. We had rooms next to each other in the old CNRS guest house on Chemin Joseph Aiguier, which was in the middle of nowhere. As a result, we would often sit in one of our rooms late into the night over a cup of approximate coffee, engaged in what Borchers was to describe later as 'furious discussions', but which to me were long silences punctuated by the occasional remark. Out of these discussions, or whatever, grew a friendship and collaboration that has lasted till now, and the fruits of which are reflected in this book. Since then, I have been a regular visitor to Göttingen where I made lifelong friends, first and foremost Hansjörg Roos and Helmut Reeh, from whom I have learnt more than I can tell. For me the words *Extra Göttingam non est vita* have a special meaning.

My meeting with Geoffrey Sewell was also accidental. It happened in the cloakroom, not of Victoria Station, but of a student hostel in Swansea. I have not met Abner Shimony in person at all, but we have corresponded for several years by e-mail; my daughter, Mandu, was a friend of his wife, Manana Sikic, and worked for her at Yale. Bertrand Russell said that one should choose one's parents with care, but then he was a philosopher; a physicist may counter that one should choose one's children with care. It is not easy to be both a physicist and a philosopher.

The last accident that has left its imprint on this book is recent. Professor Gustav Born, the son of Max Born, gave a talk about his father in Göttingen in late October, 2008. I did not go to his talk, but Hansjörg Roos did, and found that we both had been unaware of Born's later works. I have not had enough time to absorb all of them, but I hope to have given a brief but fair indication of the relevant ones. I believe that Max Born's later works have the potential to stand physics on its head; this possibility is discussed, fittingly, in the last section of the Epilogue. These works of Born (and Pauli) mean that I can no longer make a certain claim to originality, but I could not have wished for more illustrious predecessors.

My colleagues Daniel Berend, Theodore Eisenberg and Michael Lin in Beer-Sheva have befriended me both intellectually and otherwise. Their names do not appear in the Index because they work in other areas, but Chapter 1 and the Mathematical Appendices have gained very considerably from their recensions, and that was no accident.

Shelley once wrote that 'Life, like a dome of many-coloured glass, Stains the white radiance of Eternity'. So, I believe, does the relation between theoretical and experimental physics, which is the unstated interlinear of this book. As I abandon this dome, I realize that there are many more pieces of stained glass to be put in place, and hope that it will be done by craftsmen more skilled than I.

Pardes Hanna, Israel
April 2009

R N Sen
E-mail: rsen@cs.bgu.ac.il

Acknowledgements

The contents of Chapters 2–5 have been adapted from a monograph by H-J Borchers and the author (Lecture Notes in Physics, Vol. 709), published by Springer, Berlin-Heidelberg. The treatment here is rather different, but for the convenience of readers who may wish to refer to LNP 709, the notations of the latter have been retained. Additionally, Figs. I.1, 2.1–2.7, 2.10–2.14, 3.1, 3.3, 3.5, 4.1, 4.2, 5.2 and 5.4 and the List of symbols for Part I have been reproduced, with some changes, from LNP 709. I would like to thank Springer for permission to use this material.

Many people have contributed to this book. J Avron, S K Bose, D Buchholz, D J BenDaniel, H Brown, S Chowdhury, G Derfel, G G Emch, M Goldberg, I Goldshied, M Goldstein, H Goenner, R Haag, G Hegerfeldt, A Iserles, Y-S Kim, (the late) A B Pippard, C F Roos, K Schönhammer, K B Sinha and S J Summers have answered my questions by e-mail or ordinary mail, provided me with references, reprints and photocopies, made telephone calls, pointed out errors during talks, made visits possible, and so on. In addition, H-J Borchers, H Reeh, G L Sewell, A Shimony, and N Panchapakesan have read Part I or II or both, T Eisenberg has read Chapter 1, D Berend Appendices A1–A4 and A8, and M Lin Appendices A5–A7, often more than once. They have caught slips of the pen, errors of thinking, and major omissions, and have suggested numerous improvements. H Roos has read the whole book, most of it several times, and has also enlisted the help of his son. Thanks to their help, the volume that is being offered to the reader is much improved from its original draft. The flaws and errors that remain are my own responsibility.

Maureen Storey has copy-edited this book with the kind of care that I thought had vanished from the surface of this earth. She has, inter alia, changed my British Empire English into the Queen's English, and has unearthed, and corrected, an embarassingly large number of inconsistencies. It is only proper that her contribution be acknowledged here.

Finally, I should like to thank the physicians and surgeons who have kept me in working order: Drs M Blumenthal, M Goldstein and D Rahima, but most of all our family doctor, Dr Sami Abu. It is a rare pleasure to be treated by a physician who also talks about Fermat's last theorem and Maxwell's equations.

To the reader

I have tried to make this book accessible to the average senior undergraduate or beginning postgraduate student in physics, irrespective of specialization. The mathematics used above and beyond what every physics undergraduate is taught is provided in the Mathematical appendices.

A very good idea of the contents of the book may be formed by reading the Prologue and the Introductions to Parts I and II, in that order. They are written, as far as possible, in plain English. They will also give the reader an idea of the mathematics he or she may be lacking.

A firm grasp of sets and mappings and of the mathematical structure of the real number system is basic to the whole book. Appendices A1 and A2 develop this material in sufficient detail. Part I requires an equally firm grasp of basic point-set topology, which is provided in Appendix A3. Completions, both metric and uniform, are treated in Appendix A4. Differentiable manifolds are defined in Appendix A8. There are no exercises so-called, but every now and then the reader is invited to satisfy himself or herself of some point or the other. They should not impose undue hardship, and may usefully be taken up.

Chapters 6–11 of Part II use von Neumann's definition of a separable Hilbert space over the complex numbers, and the basic results of the structure theory of self-adjoint operators, known as spectral theory, on it. This is given in Appendix A6. Students of quantum mechanics will be familiar with much of this material, but maybe not at the required level of precision. The theory of measure and integral, on which it is based, is given in Appendix A5. Chapter 10 uses the notion of conditional expectation, which is the subject of Appendix A7. Some of the material of Appendix A8 is needed in Chapter 7, and some in Chapter 12.

The accounts these appendices give are selective, and – quite deliberately – present mathematics as a part of culture, not as a part of technology. For the same reason, there are remarks on an assortment of topics that range from curiosities to watersheds, and even an occasional unsolved problem. They are there to add to the reader's enjoyment.

The appendices may also be used as a work of reference. To facilitate this, all mathematical terms being defined are set in **boldface**. Some mathematical terms are defined in the body of the main text, and they too are set in boldface. Terms that are specific to Part I are set in italics, as are terms from physics that need definition or emphasis.

The contents of Part I are built from scratch; the possibility of exploiting standard mathematical results emerges only towards the end. The treatment offered is adapted to this circumstance. In Chapter 2, almost all the results are proven; in Chapter 3, about half, and hardly any in Chapters 4 and 5. Most of the results are self-evident in Minkowski space, but require proof in our setting. Some of the proofs are difficult, but this happens only in the later stages; by then they can be omitted.

There are a large number of quotations in the text. The published source is almost always indicated, including the page number. References to these page numbers are abbreviated to p. or pp. Cross-references to pages in this book are spelled out e.g. (page 23) or (pages 161–163). The index is an index of definitions and names; it does not list all occurrences of a term. It is supplemented by a list of symbols that are specific to Part I.

Formulae and LaTeX environments (theorem, lemma, definition, remark, etc.) are numbered consecutively, within chapters, the former within brackets, the latter without. Thus, (3.17) means formula (3.17) in Chapter 3, and Lemma 10.7 means the environment (which happens to be a lemma) 10.7 in Chapter 10. The appendices are numbered A1–A8, with (A1.6) being a formula and A6.31 a definition. Formulae are referred to by number only; other environments are referred to by name and number.

Parts I and II of the book can be read independently of each other. The link between them is explained in the Prologue, and in the Introduction to Part II.

Prologue

The quantitative data obtained in any physical experiment are recorded as finite, ordered sets of rational numbers. All such sets are *discrete*. However, when a physicist sits down to make sense of such data, the tools he or she employs are generally based upon the *continuum*: analytic (or at least smooth) functions, differential equations, Lie groups, and the like. It is the view of many eminent mathematicians that '*bridging the gap between the domains of discreteness and of continuity* ... is a central, presumably even *the* central problem of the foundations of mathematics',[1] yet Fritz London did not seem to have had the slightest hesitation in writing, in the very first paragraph of his book on superfluidity,[2] 'that *new differential equations* were required to describe [the observed behaviour of]... "superfluid" helium... ' The physicist had stepped over the gap which has occupied philosophers for two millenia without even noticing that it existed![3]

This gap is but a fragment of one that separates theoretical from experimental physics. Some of the most important physicists of the first half of the twentieth century have expressed themselves on the subject, and it is instructive to compare their views. Dirac, for example, had the following to say:[4]

> The physicist, in his study of natural phenomena, has two methods of making progress: (1) the method of experiment and observation, and (2) the method of mathematical reasoning. The former is just the collection of selected data; the latter enables one to infer results about experiments that have not been performed. There is no logical reason why the second method should be possible at all, but one has found in practice that it does work and meets with reasonable success. This must be ascribed to some *mathematical quality in Nature*, a quality which the casual observer of Nature would not suspect, but which nevertheless plays an important role in Nature's scheme.

There can be no clearer acknowledgement of this gap than Dirac's remark: 'There is no logical reason why the second method should be possible at all.'

[1] See (Fraenkel, Bar-Hillel and Levy, 2001, p. 211).
[2] See (London, 1964, p. 1).
[3] The emphases in the quotations are in the originals.
[4] See (Dirac, 1938–39, first paragraph).

It is well known that Heisenberg stumbled upon matrix mechanics while attempting to express quantum theory entirely in terms of observable quantities. Many years later, he wrote an article in the literary journal *Encounter* in which he recounted the following:[5]

> It is generally believed that our science is empirical, and that we draw our concepts and our mathematical constructs from empirical data. If this was the whole truth, we should when entering a new field introduce only those quantities that can directly be observed, and formulate laws only by means of these quantities.
>
> When I was a young man I believed that this was just the philosophy which Albert Einstein had followed in his theory of Relativity. I tried, therefore, to take a corresponding and related step in Quantum theory by introducing the matrices. But when I later asked Einstein about it, he told me: 'This may have been my philosophy, but it is nonsense all the same. It is never possible to introduce only observable quantities in a theory. It is a theory which decides what can be observed...' What he meant was that...we cannot separate the empirical process of observation from the mathematical construct and concepts.

The 'mathematical quality in nature' of Dirac's description, acknowledged if not articulated by Einstein and Heisenberg, was a philosophical position that went back to the ancient Greeks – to geometry, measuring the earth, and arithmetic, the art of counting. But, in the last three decades of the nineteenth century, Georg Cantor had developed his theory of *transfinite numbers* which challenged this wisdom. Cantor introduced the notion of a *set*, and, using this notion, established several epoch-making results. One of these was a precise characterization of infinite sets.[6] Another was the proof that the set of all subsets of a given set is, in a precisely defined sense, larger than the original set. This construction, called the power-set construction, could be applied to infinite sets to yield an unending succession of infinite sets, each larger than its predecessor; a revolutionary idea in mathematics at the end of the nineteenth century. It was this freedom to pursue ideas, unfettered by constraints other than those of consistency, that – Cantor asserted – distinguished mathematics from the other sciences.[7]

It seems unlikely that Einstein, Dirac and Heisenberg were influenced in any way by Cantor's work. The same could not be said of Wigner, if only because of his friendship with von Neumann. Fifteen years before Heisenberg's *Encounter*

[5] See (Heisenberg, 1975, pp. 55–56).

[6] A brief but adequate introduction to Cantor's theory is given in Appendix A1.

[7] A summary of Cantor's position is given in the section entitled *The nature of mathematics* in (Dauben, 1990, pp. 132–133). References to original and secondary sources will also be found in this work.

article, Wigner, in his celebrated essay on *the unreasonable effectiveness of mathematics in the natural sciences*,[8] had asked the question: what is mathematics?, and answered it, paraphrasing the logician and philosopher of science Walter Dubislav, as follows: '...mathematics is the science of skillful operations with concepts and rules invented just for this purpose.' If that were the case – and many practising mathematicians today would affirm that it *is* indeed the case – the effectiveness of mathematics in the natural sciences would be difficult to understand.

Taking stock, we may discern two world-views that are diametrically opposed to each other: the pre-Cantorian view that mathematics is, in everyday speech, discovered and not invented, and the post-Cantorian one that mathematics is invented, and not discovered.[9] It turns out, however, that between these metaphysical opposites, there is room for scientific analysis.

By a scientific analysis we mean (in the present context) one that is based upon physical principles and carried out by mathematical means. The precision required for such an analysis can only be attained by narrowing the field of enquiry. We shall confine ourselves to the following question: *is the differential calculus a discovery, or an invention*? Or, in scientific language: *is the differentiable structure of space-time a* consequence *of physical principles*?[10]

In Part I of this book, we shall establish some results that suggest that, subject to a certain caveat, the answer to the last question is in the affirmative. The physical principle that has these profound mathematical consequences is *causality* in the sense of Einstein and Weyl.[11] It turns out that the notion of Einstein–Weyl causality can be defined, as a partial order, on any infinite set of *points*, totally devoid of any predefined mathematical structure. Such *causally ordered spaces* can be *completed* – i.e., densely embedded in continua – in a unique manner, and the causal order can be extended, again uniquely, to the completed space. Furthermore, when these continua are finite-dimensional, they have the (unique) local structure of a differentiable manifold. If we agree to call a countably infinite set on which Einstein–Weyl causality is defined a *discrete space-time*, then the results can be stated as follows:

(i) Any discrete space-time can be *completed*, i.e., embedded in a continuum. The discrete space-time defines this continuum uniquely.
(ii) The causal order of the discrete space-time has a unique extension to its completion.

[8] See (Wigner, 1970, p. 224).

[9] What was simplistically described above as the pre-Cantorian view is actually a vast corpus in philosophy, with a history that goes back more than two millenia.

[10] The statement that the real line $\mathbb{R}$ has a differentiable structure is equivalent to the statement that there is such a subject as the differential calculus of a single real variable. A generalization will be found in Section A8.2.

[11] For details, see (Borchers and Sen, 2006).

(iii) The completion of a discrete, finite-dimensional space-time has the local structure of a differentiable manifold.

The results described above were obtained on the assumption that the notion of geometrical points (in the sense of Euclidean geometry) may be used in physics without further analysis. This assumption was strongly controverted by Wigner. Following earlier work by Wigner himself, Araki and Yanase established, within the framework of von Neumann's measurement theory, that an observable that does not commute with a conserved quantity cannot be measured precisely.[12] Since the position operator of a point-particle would seldom commute with the Hamiltonian, its position could not be measured precisely, which led Wigner to comment to Haag that 'there are those of us who believe that there are no points'.[13]

Part II of the book is an attempt to assuage Wigner's doubts. The strategy is extremely simple: try to show that the situation is no worse in quantum mechanics than it is in classical mechanics. But the validity of this procedure is based on the assumption that there are limits to the usefulness of Francis Bacon's motto '*dissecare naturam*'. In practical terms, a concept of measurement which is untenable in classical mechanics should be treated with suspicion in quantum mechanics.

John Bell, for example, has described quantum mechanics as 'our most fundamental physical theory'.[14] If quantum mechanics is fundamental and classical mechanics a mere $\hbar \to 0$ limit of it, then it is less than obvious how a comparison with the ills of classical mechanics can cure the ills of quantum mechanics. It is true that quantum mechanics 'explains' a set of natural phenomena that classical mechanics cannot; but it is equally true that the basic 'observables' of quantum mechanics are borrowings from the dynamical variables of classical mechanics: 'Who *is* the Potter, pray, and who the Pot?'[15]

Since the aim of theoretical physics is to understand physical phenomena that are observed, a theory – I maintain – should fit a particular *observational window*.[16] For example, the theory that is appropriate for describing the behaviour of ideal gases in thermodynamic equilibrium is inappropriate for describing the

[12] See (Araki and Yanase, 1960).

[13] This comment was made by Wigner after Haag's talk at the International Colloquium on Group Theoretical Methods in Physics in Philadelphia in 1986. Wigner's own account of his doubts will be found on page 207.

[14] The quotation, and the context, will be found on page 194.

[15] The *Rubayyat of Omar Khayyam*, translated by Edward FitzGerald.

[16] The notion of an observational window is arrived at by attempting to understand Einstein's maxim, 'it is a theory which decides what can be observed'. First, the observer decides what he or she wants to observe, and devises a theory to account for the observed regularities. A 'description of physical phenomena' is a description of the temporal evolution of the *state* of a physical system. The observational window determines the variables of state. The latter are required to be *complete*, i.e., temporal evolution is required to map the space of states into itself. Finally, this requirement constrains what can, or cannot, be observed. The term 'observational window' was first used in (Roos and Sen, 1994).

scattering of alpha-particles by thin metallic foils, and vice versa. We assume that the theories we are working with are not 'theories of everything'. Their function is to permit logical deductions from well-defined premises. It is therefore reasonable to demand that each theory be internally consistent.[17] To sum up, I do not see the question: which is more fundamental – quantum mechanics or classical mechanics – as one that advances scientific enquiry.[18] Classical mechanics has provided us with a body of concepts in terms of which equations of motion can be precisely framed for several classes of state spaces. Quantum mechanics has not changed these concepts; it has added a single new concept, but the result has been a revolutionary change in the space of states, which is the same as its observational window. I therefore believe that the strategy mentioned earlier is well conceived.

After this explicit statement of the assumptions that underlie our endeavour, we may turn to the essential point. We want to show that, as far as limitations on the accuracy of a measurement are concerned, they are no worse in quantum mechanics than they are in classical mechanics. But what are the factors that limit the accuracy of measurements in classical mechanics?

In the theory called classical mechanics, there are no physical principles that limit the accuracy of measurements. Measurements are assumed to be instantaneous, and therefore even the position of a moving point-particle can be measured precisely at any instant of time. What, then, is the source of limitations on the accuracy of classical measurements on which we are trying to build our case? We begin with a few historical remarks, some of which are common knowledge while others have hardly entered into the consciousness of the scientific community.

Although Einstein's contributions were decisive in establishing the particle aspect of light, Einstein himself remained a lifelong sceptic of quantum mechanics. His exchanges with Bohr are well known;[19] Einstein remained unconvinced. However, Einstein also carried on a lifelong correspondence with Max Born on the subject. Born too failed to convince Einstein, but, in the process – sometime before 1954 – he came to a crucial realization: the reason why an exact determination of the state of a physical system – be it classical or quantum-mechanical – was impossible lay in the *mathematical structure of the real number system.* He

[17] Unfortunately, this demand cannot always be met. As far as agreement between theory and experiment is concerned, quantum electrodynamics (QED) is by far the most successful physical theory that we have, but it is known to be logically inconsistent. The logical inconsistency of QED was pointed out, from two different directions, by Dyson and Haag (Dyson, 1952; Haag, 1955). The explanation of this puzzle remains – or so the present author contends – the most important unsolved problem in theoretical physics. By *explanation* we mean a logical deduction from accepted premises, and not a *belief*, as articulated by Weinberg (Weinberg, 1995, p. 499, last paragraph).

[18] It may be that in making this assertion I am, as Keynes would have it, being driven by the ghosts of defunct philosophers.

[19] See, for example, (Wheeler and Zurek, 1983), which contains 47 pages on the Bohr–Einstein dialogue, and reprint of the EPR paper (Einstein, Podolsky and Rosen, 1935) and Bohr's reply to it.

pointed out a fact to which no physicist, before him, seems to have paid the slightest attention.[20] This fact is the following: the rational numbers are countable, and therefore form a set of Lebesgue measure zero on the real line. That is, almost every real number is irrational. Now the explicit decimal representation of an irrational number is nonrecurrent, and requires an infinite number of digits. It is therefore absurd to assert that the position of a point-particle on a real line can be measured precisely. At least ten years before the paper by Lorenz that set off the chaos revolution,[21] Born noticed what has since become known as the 'sensitive dependence on initial conditions' of nonlinear classical mechanics, and used these facts to make the following assertions about classical point-particle mechanics:

(i) From the viewpoint of the experimentalist, it makes little sense to talk about the position of a point-particle. What makes sense is the notion of a *probability distribution* about its position.
(ii) In view of the sensitive dependence on initial conditions, a second determination of the position of a point-particle – if successful – could be interpreted as *a reduction of the probability distribution*; it would effect a drastic change in the probability distribution.

We may now hone our strategy to the following. Since the measurement of a continuous variable in classical physics is possible only within an error ε, where ε is an arbitrarily small *but positive* number, what we have to show is that, given any $\varepsilon > 0$, the corresponding quantum-mechanical observable can be measured within this error.

To sum up: the gap between the domains of discreteness and of continuity in mathematics is equally *a gap between experimental and theoretical physics.* Here classical mechanics and nonrelativistic quantum mechanics are on a par with each other, as they both rely on the same local topological–geometrical structure of space-time.

As is well known, von Neumann's measurement theory, the source of Wigner's doubts, requires the intervention of the observer's 'conscious ego'. The mathematical part of the theory cannot account for the reduction of the wave packet; it is, as we shall find, a theory of entanglement (the term was coined by Schrödinger three years after the appearance of von Neumann's book) rather than a theory of measurement. A resolution of Wigner's doubts requires, first and foremost, a resolution of the quantum measurement problem in a *mathematical* manner: namely, a theory that accounts for the reduction of the wave packet in which appeal to the observer's conscious ego is replaced by a significantly weaker mathematical hypothesis.

[20] In (Sen, 2008) I referred to this fact as 'known to all but honoured by none'. I was wrong; Max Born had seen its implications more than 50 years earlier.

[21] The reference is to (Lorenz, 1963).

Such a theory has been proposed by Sewell, and extended by the present author to continuous spectra.[22] This theory may aptly be described as a bridge over the Heisenberg cut, with movement across it being controlled by the Schrödinger–von Neumann equations. In this theory observables with discrete, rational spectra can be measured precisely, and therefore observables with continuous spectra can be measured with an error $\varepsilon > 0$, where ε can be made arbitrarily small. This is not very different from the measurement of a continuous variable in classical mechanics, *if the result of measurement is constrained to be a rational number.* Therefore one is tempted to claim that quantum mechanics does not make the situation any worse than it already is in classical physics.

However, Sewell's theory assumes that the state spaces of object and apparatus are finite-dimensional. As is well known, the canonical commutation relation $qp - pq = i\hbar$ cannot be realized on finite-dimensional vector spaces; it runs afoul of the identity $\mathrm{Tr}\,(qp - pq) = 0$. This causes no trouble in measurement theory, as quantum-mechanical uncertainties are negligibly small by comparison with errors of observation.[23] Nevertheless, the assumption may be at variance with the general principles of quantum mechanics as they are commonly understood, and that is why the above claim should be tempered with caution.

Is it possible to lift the assumption of finite-dimensionalities from Sewell's measurement theory? The answer is not known, but to do so it will almost certainly be necessary to devise a framework (within nonrelativistic physics) for describing interactions of microscopic quantum systems with macroscopic systems *considered as a whole.* In the opinion of the present author, this is a key unsolved problem in nonrelativistic quantum mechanics.

The development of Parts I and II is mathematically rigorous. The results that are quoted without proof – and there are many – *have been proven.* The phrase 'it may be shown that...' (or something similar) invariably means 'it has been shown that...'. I use the former phrase because it sounds better to my ears. Again, mindful of the intended readership, many concepts defined in the appendices are recapitulated in footnotes to the text, or else a reference is given to the page on which it is defined. The word 'page' (or 'pages') is spelled out when it refers to a page in this book, and abbreviated to p. (or pp.) when it refers to some other source.

In a book such as the present one, it is neither possible nor desirable – or so the present author contends – to avoid expression of the opinions and beliefs that guide the endeavours of physicists. My personal opinions may diverge from the consensus (or, when there is no consensus, from commonly held views), and I have tried to keep the two separate. My personal beliefs and opinions are

[22] See (Sewell, 2005; Sen, 2008).

[23] The observable consequences of quantum mechanics derive mainly from the superposition principle, which holds on any linear space, irrespective of its dimensionality.

expressed either in the first person singular, or marked by a qualifier as in the first sentence of this paragraph.

Many of the references cited have been reprinted in various collections. Many articles originally published in German have been translated into English. I have referred to reprint volumes and English translations wherever I have had access to them, but I have omitted the names of the translators (which were not always available). Some of the references to books have been dictated by the desire to provide a historical perspective, but without pretensions to historical scholarship; others, by what I own, or have access to. Few of them are of recent vintage; I have not referred to later editions or reprints that I have not been able to consult.

Part I of this book is based entirely on the special theory of relativity; Part II, entirely on *nonrelativistic* quantum mechanics. Since I do not consider the unity of physics to be a good working hypothesis for a mathematical treatment,[24] I think that nonrelativistic quantum mechanics should stand as an autonomous, logically consistent edifice, despite its inadequacy as a physical theory at high energies. But I should add that nonrelativistic mechanics, both classical and quantum, assumes mathematical structures on space and time that appear to have their origins on Einstein–Weyl causality.

As stated earlier in different words, we have fallen short of our goal of deciphering whether the differential calculus is a discovery or an invention. But the search has revealed some new questions of interest in mathematics, theoretical physics and possibly in experimental physics, and these are discussed, or speculated upon, in the Epilogue.

[24] This is only one possible viewpoint among many. Theorists have long been attempting to integrate a larger class of problems as one unit, and some of the more recent attempts, such as string and superstring theories, have led to phenomenal advances in mathematics. If these endeavours succeed in reaching even a few of their goals, I will have to change my opinion.

Part I

Causality and differentiable structure

Introduction to Part I

In the special theory of relativity, the principle that no signal can propagate faster than light (which we shall call *Einstein–Weyl causality*) determines a partial order (the past–future order) on Minkowski space $\mathbb{M}^4$, which is just the topological space $\mathbb{R}^4$ with the Minkowski inner product defined on it. In 1959, A D Aleksandrov observed that this partial order defines a topology on $\mathbb{M}^4$ which is the same as the usual (product) topology of $\mathbb{R}^4$ (Aleksandrov, 1959). Let $p \in \mathbb{M}^4$ and $C_p^{\pm}$ be the forward (future) and backward (past) cones at p. Finally, let $q \in C_p^+$. Denote by $I[p,q]$ the set $C_p^+ \cap C_q^-$. The interior of $I[p,q]$ will be denoted by $I(p,q)$ and called an open order interval. It is nonempty if q does not lie on the mantle of C_p^+ (Fig. I.1(b)). One sees immediately that the family of open order intervals forms a base for the usual topology on $\mathbb{R}^4$. Note that this topology is Hausdorff; if a, b are two distinct points in $\mathbb{M}^4$, then one can find open order intervals I_a, I_b such that $a \in I_a$, $b \in I_b$ and $I_a \cap I_b = \emptyset$, and this is true for any $\mathbb{M}^n$, $n \geq 2$.

What happens in nonrelativistic physics? Here too one can claim to have a principle of causality (which we shall call Newton causality), which asserts that there are no limits on signal velocities: the 'velocity of light' is infinite. Irrespective of the physical content of this principle, it is mathematically nontrivial. For simplicity, consider a two-dimensional space-time $\mathbb{R}^2$, space and time being represented on the X- and Y-axes respectively. The 'forward cone' at $p = (x_0, t_0)$ is the half-plane $\{(x,t) | \text{all } x, t \geq t_0\}$. Let $q = (x_1, t_1)$ with $t_1 > t_0$. Then

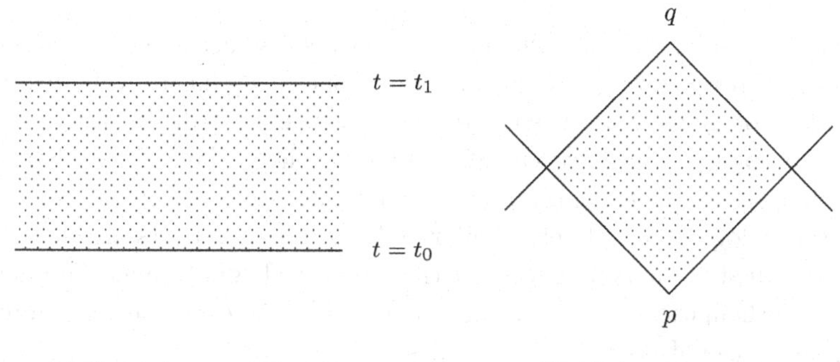

Fig. I.1. Order intervals in two-dimensional space-time

$I(p,q) = \{(x,t)|\text{ all } x, t_0 < t < t_1\}$ is an open order interval (Fig. I.1(a)). These order intervals define a topology, which is not the standard topology of $\mathbb{R}^2$. It is not Hausdorff; distinct points with the same value of t *cannot* be separated by open order intervals.

Since causality principles are physical principles, the fact that they have *mathematical* consequences deserves notice. However, Minkowski space and Newtonian space-time come with so many predefined mathematical structures[1] that the precise contribution of causality principles becomes difficult to separate. Can one define causal structures on point-sets that have *no* predefined mathematical structures on them?

The answer should be in the affirmative if one could define the analogues of light cones on them. Light cones in Minkowski space are families of light rays with very specific properties. Light rays themselves have a total order defined on them, the order in the direction of propagation. In the following, we shall use the term *light rays* to denote the space-time paths of light rays, as one pictures them in Minkowski spaces. In Minkowski spaces, two light rays intersect only once, or not at all. However, in the formation of an image through a lens – which takes us out of Minkowski space – two light rays intersect twice, but at well-separated points; the local behaviour of families of light rays may not be the same as their global behaviour. The causal properties of space-time are completely determined, both locally and globally, by light rays *and their intersections.* The example of image formation by a lens shows that the local and global properties of the intersections of light rays may be very different, and one has to remain sensitive to this fact.

The order properties of light rays and their intersections can be abstracted. A key property is that between any two distinct points on a light ray lies a third, which means that a light ray must have infinitely many points. That rules out finite sets. It turns out that the abstracted properties of light rays and their intersections – which may legitimately be called *causal structures* – can be defined on *countably infinite sets.* Such sets are sharply distinguished from continuua, and cannot carry any differential structure in the usual sense.

We start by making precise the notion of a causal structure on a structureless set. This requires a certain amount of care, as one really has to walk a tightrope between the too general and the too restrictive. As expected, the order structure defines a topology, which we call the order topology. The essential part is a careful analysis of this topology, and the crowning result of this analysis is that the order topology is completely regular and that one-point sets are closed. Such spaces are called Tychonoff spaces. At this point we begin to make contact with standard mathematics – so far we have not had use of even the real numbers – and can make use of the following results:

[1] Topological, linear, differential, metric, invariance groups...

(i) A completely regular space is uniformizable.
(ii) A uniform space admits a uniform completion.

But much work remains; the whole exercise would be pointless if the order structure of the original space could not be extended to the completed space. It can, but only after a substantial effort. The original ordered space is densely embedded in the new ordered space so obtained. The new ordered space is also completely regular – in fact, Tychonoff – and therefore uniformizable. However, it is already complete in this uniformity.[2] Such spaces are called *completely uniformizable.* Now a remarkable theorem due to Shirota states that a topological space is completely uniformizable if and only if (i) it is Tychonoff, (ii) *it can be embedded as a closed subspace of* $\mathbb{R}^J$ *for some* J, and (iii) it has an arcane property which is briefly discussed on page 286, and which is possessed by all known sets. So, if our completed space is finite-dimensional, it is a closed subspace of some $\mathbb{R}^n$, and will surely have, locally, a differentiable structure. At this point we can make no statement about a global differential structure.

That is, a causally ordered space which cannot have a differentiable structure can be densely embedded in a continuum. If the latter is finite-dimensional, it has the local structure of a differentiable manifold. Since it is impossible, experimentally, to distinguish between the original space and the manifold in which it is embedded, it becomes difficult to tell – or so the present author contends – whether the differentiable structure of space-time is a consequence of the physical principle of Einstein–Weyl causality or a mathematical invention.

This part is organized as follows. Chapter 1 is aimed at readers who are not familiar with 'abstract' mathematics. The point is to decouple the idea of mathematical structures from the underlying sets by giving examples of different mathematical structures that can be defined on the same set of points (which may be finite or infinite), and are of interest to the physicist. Chapter 2 defines the notion of causality as a partial order on sets that are devoid of predefined mathematical structures. The next chapter studies the topology of these ordered sets in some detail. It establishes that these spaces have the Tychonoff property, as well as some global homogeneity properties. Chapter 4 studies the completion of these spaces and introduces the notion of order completion, which is marginally different from the corresponding mathematical notion. The extension of order to the completed space is also studied in this chapter. Our final result, that locally compact order-complete spaces have the local structure of differentiable manifolds, is established in Chapter 5.

Until Chapter 5, no use is made of 'established mathematics'. Everything has to be built 'from the ground up'. Intuition is lacking, and as a result it

[2] There is a small but physically significant difference between uniform completion and what we shall eventually define as *order completion* that is being disregarded here. At this stage, it is a harmless simplification.

is almost impossible to absorb definitions and results without proofs. At the beginning, quite a few proofs are given in full, but their density thins out as the work progresses. Fortunately, the long and difficult constructions and proofs come towards the end, by which stage the reader is prepared enough to see the wood without counting the trees. Wherever possible, formal statements and arguments are supplemented by verbal explanations and examples that only presume familiarity with the special theory of relativity.

1
Mathematical structures on sets of points

Every student of physics is familiar with the Gibbs paradox, and its resolution by 'correct Boltzmann counting'. The paradox arises because identical particles in classical physics are assumed distinguishable; the question of how they are to be distinguished is not asked. The resolution, correct Boltzmann counting, is equivalent to the assumption that even in classical mechanics, two point-particles with the same mass *cannot* be distinguished from each other.

The opposite is true in set theory. Take the assembly of geometrical points that constitute an open interval on a straight line, and recall that in Euclidean geometry a point has no structure. Now perform a thought-experiment in which two points are pulled out at random from the interval and shown to an observer in the next room. There is no way in which the observer can tell one from the other. However, admitting two identical objects in a set is a recipe for disaster (see the example on page 242); if one does, then it becomes impossible to define the notion of a function in a sensible manner. A set, in mathematics, has to be a collection of *distinct* objects, considered as a single entity.[1]

How, then, is one supposed to understand 'the set of points that constitute the real line' or the two-dimensional plane? The short answer is: exactly as Gibbs understood an assembly of n classical point-particles – distinguishable. Familiarity with quantum mechanics may have made it counterintuitive to today's physicist, but it was clearly not counterintuitive to Gibbs, and does not appear to be counterintuitive to mathematicians.

If S is a set containing n elements, then the **Cartesian product** $S^2 = S \times S = \{(a, b)|a, b \in S\}$ contains exactly n^2 elements; S^2 has more elements than S. Until 1888, mathematicians believed that something similar also held true for infinite sets (although *infinity* was not a very well-defined concept); for example, there were believed to be many more points on the plane than on the line.

In the 1870s, Georg Cantor succeeded in placing the notion of infinity – in the mathematical context – on a precise logical footing (pages 243–244). Using his newly developed tools, he showed in 1888 that one could map the unit interval $I = [0, 1]$ on the real line $\mathbb{R}$ **bijectively** (i.e., one-to-one) onto I^2, the unit square. Two years later, Giuseppe Peano constructed a map $f : I \to I^2$ which

[1] One should note carefully that not every collection of distinct objects is a set; see pages 248–249.

was bijective and *continuous.*[2] Peano's construction was rapidly followed by others, and such maps are now known as *space-filling* or *Peano* curves. These examples demonstrated that differences between infinite sets like I and I^2 were to be sought, not in the 'number of points' they contained, but elsewhere.

The 'elsewhere' are the mathematical structures that are placed on a set of points. Shorn of all such structures, a set of points X has only one attribute, namely its cardinality (Section A1.5.1, page 245), which is denoted by $|X|$. The sets I^m and $\mathbb{R}^m$, which are defined for positive integers m, have the same cardinality for all such m; the sets $\mathbb{R}^m$ and $\mathbb{R}^n$ have the same cardinality for all positive integers m and n. The common cardinality of all these sets is $\aleph$ (Section A1.5.1).[3] In other words, each of the sets I^m and $\mathbb{R}^n$ is obtained by defining different mathematical structures on the same underlying set of cardinality $\aleph$.

In this chapter we shall review some of the mathematical structures that can be defined on finite sets, and infinite sets of cardinalities $\aleph_0$ and $\aleph$. We shall have no occasion to use sets of higher cardinalities. Our aim is to demonstrate that a vast array of distinct mathematical structures can be defined on the same underlying set.

1.1 Mathematical structures on finite sets

There are relatively few classes of mathematical structures that one can define on a finite set. The most important of them, at least for the physicist, are certain finite groups like the permutation groups and the crystallographic point groups, which are finite subgroups of the rotation group in three dimensions.

Recall that, in mathematics, a **field** is a set on which two operations, both commutative, are defined. These operations are called addition and multiplication, and indeed the field one encounters most often is the field of real numbers. The cardinality of the underlying set is $\aleph$. Another example of a field is the field of rational numbers, of cardinality $\aleph_0$. However, there exist finite fields as well. It can be shown that, for any prime p and any positive integer n, there exists a field F_{p^n} with cardinality p^n (Jacobson, 1974). Finite fields, once mere mathematical curiosities, are now used extensively in cryptography and the theory of coding.

1.2 Mathematical structures on countably infinite sets

In this section our *tabula rasa* will be a countably infinite set, also known as a set of cardinality $\aleph_0$ (see page 244), which we shall denote by Ξ. The sets $\mathbf{N}$,

[2] The inverse map $f^{-1} : I^2 \to I$ was well defined, but it was *not* continuous; it could not be, because I is not homeomorphic to I^2. The proof that I^m and I^n (or for that matter $\mathbb{R}^m$ and $\mathbb{R}^n$; $\mathbb{R}$ denotes the set of real numbers in their natural order) are not homeomorphic to each other for $m \neq n$, where m and n are positive integers, came much later, and required the development of the new subject of algebraic topology.

[3] Some authors prefer to use the letter $\mathfrak{c}$ rather than $\aleph$; However, everyone – as far as the present author knows – is happy with $\aleph_0$, the cardinality of the set of integers.

of **natural numbers** (i.e., of positive integers), $\mathbb{N}$, of **nonnegative integers**, $\mathbb{Z}$, of **integers** and $\mathbb{Q}$, of the **rationals** are all constructed from Ξ. They have an **order** defined on them, the familiar $<$ or $>$ of arithmetic. To specify the algebraic structures on them, it is convenient to begin by defining the notion of a **semigroup**, which is a group without the inverse:

Definition 1.1 (Semigroup) A **semigroup** $(S, \cdot)$ is a set S on which a binary operation $\cdot : S \times S \to S$ is defined. The image $\cdot(a, b)$ of (a, b) under this map is written $a \cdot b$, or simply ab. The operation is **associative**, i.e., $a(bc) = (ab)c$.

If the operation is commutative, i.e., if $ab = ba$, it is denoted by the plus sign and written $a + b$ $(= b + a)$ rather than ab, and called *addition*. In this case the semigroup is called **Abelian**, or **commutative**. Unlike a group, a semigroup need not have an identity, and therefore we shall state explicitly whether or not it has one. The reader is invited to prove that the identity, if it exists, is unique.

If an Abelian semigroup has an identity, then the latter is generally denoted by 0; $a + 0 = a$ for all $a \in S$.

We now turn to the order and algebraic structures[4] on Ξ that turn it into $\mathbf{N}$, $\mathbb{N}$, $\mathbb{Z}$ and $\mathbb{Q}$.

(i) $\mathbf{N}$, the set of natural numbers. This is obtained from Ξ by defining on it the (arithmetical) order $<$, and the operation of addition $+$. With respect to the order $<$, $\mathbf{N}$ has a smallest element, namely 1, and each element has a unique successor. With respect to addition, $\mathbf{N}$ is a semigroup without identity.

Note that one can also define the operation of *ordinary multiplication* on $\mathbf{N}$. Addition and multiplication obey the distributive law $a(b+c) = ab + bc$. $\mathbf{N}$ is a multiplicative semigroup with identity, the multiplicative identity being 1.

(ii) $\mathbb{N}$, the set of nonnegative integers. $\mathbb{N}$ differs from $\mathbf{N}$ in having the additive identity 0, which now becomes the smallest element with respect to the order $<$.

(iii) $\mathbb{Z}$, the set of integers. This set is also totally ordered with respect to $<$, but it has no smallest (or largest) element. Each element has a unique successor. $\mathbb{Z}$ is a *group* under addition, but only a semigroup under multiplication.

(iv) $\mathbb{Q}$, the set of rational numbers. This set is also totally ordered by $<$, but *no element has a unique successor*. For suppose that r_2 is the unique successor of r_1; then $r_1 < (r_1 + r_2)/2 < r_2$, a contradiction, because $(r_1 + r_2)/2$ is also a rational number. $\mathbb{Q}$ is a group under addition, and $\mathbb{Q} \setminus \{0\}$ is a group under multiplication; in mathematical terms, $\mathbb{Q}$ is an **ordered field**.

[4] For the reader who is particular, we should add that we are not going into the monoid, ring, vector space or module structures that these sets may have.

The *order topology* (page 259) can be defined on each of the above sets by choosing as a basis the family of open intervals (a, b), augmented by the intervals $[1, b)$ for $\mathbf{N}$ and $[0, b)$ for $\mathbb{N}$. In this topology, one-point sets are open in $\mathbf{N}$, $\mathbb{N}$ and $\mathbb{Z}$, but not in $\mathbb{Q}$. Therefore the topology thus defined on $\mathbf{N}$, $\mathbb{N}$ and $\mathbb{Z}$ is discrete, but that defined on $\mathbb{Q}$ is not. One can also define a notion of distance d on each of the above sets: $d(a, b) = |a - b|$. This satisfies the **triangle inequality**

$$d(x, z) \leq d(x, y) + d(y, z),$$

where x, y, z are distinct points, equality holding if either $x < y < z$ or $x > y > z$. The *metric topology* (page 269) defined by d is the same as the order topology on each set.

1.3 Mathematical structures on uncountable sets

As we noted earlier, the sets $\mathbb{R}^n$ have the same cardinality. They are distinguished, not by the 'number of points' they contain, but by the mathematical structures that are defined on them. The cases $n = 1$ and $n > 1$ have to be considered separately.

For mathematical analysis,[5] the big difference between $\mathbb{Q}$ and $\mathbb{R}$ is that all Cauchy sequences converge in $\mathbb{R}$, but not in $\mathbb{Q}$. As a result, the limit that defines the derivative (say of a function f) and exists in $\mathbb{R}$ may fail to exist upon restriction to $\mathbb{Q}$. The space $\mathbb{Q}$ cannot be endowed with a differentiable structure in the usual sense.[6] The differentiable (or differential) structure, in turn, serves as the substratum for further geometrical structures, such as the Riemannian structure.

The real line $\mathbb{R}$ has the following structures defined on it:

(i) The order structure. $\mathbb{R}$ is a totally (or linearly) ordered set, ordered by the relation $<$.
(ii) The topological structure. The order on $\mathbb{R}$ defines the order topology on it.
(iii) The algebraic structure. $\mathbb{R}$ is a field.
(iv) The linear structure. $\mathbb{R}$ is a one-dimensional vector space over itself (the reals).
(v) The metric structure. $\mathbb{R}$ has a metric defined on it. This metric defines a topology on it, and the metric topology coincides with the order topology.
(vi) The differentiable structure. $\mathbb{R}$ is a one-dimensional differentiable manifold.
(vii) The Borel structure (page 301).

The structures enumerated above are interrelated, and are compatible with each other.

[5] *Mathematical analysis* is also the name for a branch of mathematics that has grown out of the *theory of functions* of the nineteenth century.

[6] The concept of a differentiable structure will be defined in Section A8.2.1.

It is interesting to compare the mathematical structures on $\mathbb{R}$ with those on $\mathbb{Q}$. On $\mathbb{Q}$, one may define the order, topological algebraic and metric structures just as one does on $\mathbb{R}$. Furthermore, $\mathbb{Q}$ is a one-dimensional linear space over the rationals. When one *completes* $\mathbb{Q}$ to $\mathbb{R}$, these structures may be extended to the completion,[7] and these extensions coincide with the usual order, topological, algebraic and metric structures on $\mathbb{R}$, which is also a one-dimensional vector space over the rationals.

1.4 Global geometrical structures on $\mathbb{R}^n$

We shall denote by $\mathbb{R}^n$ the Cartesian product of n copies of $\mathbb{R}$, topologized by the product topology (page 263). The following structures are defined *globally* on $\mathbb{R}^n$:

(i) **The Euclidean structure.** Let $x = (x_1, \ldots, x_n), y = (y_1, \ldots, y_n)$ be points in $\mathbb{R}^n$. Define on $\mathbb{R}^n$ the **Euclidean metric**

$$d(x, y) = \{(x_1 - y_1)^2 + \cdots + (x_n - y_n)^2\}^{1/2}. \tag{1.1}$$

The topology induced by this metric on $\mathbb{R}^n$ coincides with the product topology.[8] The group of transformations (i.e., maps of $\mathbb{R}^n$ onto itself) that leave the metric (1.1) invariant is called the **Euclidean group**.

(ii) **The Minkowski structure.** This structure is defined by the indefinite **Minkowski form**

$$s^2(x, y) = (x_0 - y_0)^2 - (x_1 - y_1)^2 - \cdots - (x_{n-1} - y_{n-1})^2. \tag{1.2}$$

In (1.2), we have denoted the components of x by $(x_0, x_1, \ldots, x_{n-1})$, in keeping with the practice in relativity theory. One may use the Minkowski form to define a topology on $\mathbb{R}^n$ via **light cones.** Define

$$\begin{aligned} \operatorname{int} C_a^+ &= \{x | s^2(a, x) > 0, a_0 - x_0 < 0\}, \\ \operatorname{int} C_a^- &= \{x | s^2(a, x) > 0, a_0 - x_0 > 0\}. \end{aligned} \tag{1.3}$$

The reader will notice that $\operatorname{int} C_a^{\pm}$ are respectively the interiors of the forward and backward cones at a. Let now

$$I(a, b) = \operatorname{int} C_a^+ \cap \operatorname{int} C_b^-.$$

Then $I(a, b) = \emptyset$ unless b is in the interior of the forward cone at a. The collection of all $I(a, b)$ is a basis for a topology on $\mathbb{R}^n$, and this topology

[7] The subject of *completions* is treated in detail in Appendix A4.

[8] Recall that there exist different, but topologically equivalent, metrics on $\mathbb{R}^n$, as well as topologically inequivalent ones.

coincides with the product topology on $\mathbb{R}^n$. The group of transformations that leave the Minkowski form invariant is known as the **Poincaré group**, or the **inhomogeneous Lorentz group**.

(iii) **The affine structure**. The rotation and Lorentz groups are subgroups of the group $GL(n,\mathbb{R})$ of **general linear transformations** of $\mathbb{R}^n$. Likewise, the Euclidean and Poincaré groups are subgroups of the group of transformations

$$x_i' = \sum_{j=1}^{n} A_{ij}x_j + b_i, \tag{1.4}$$

where the matrix $[A_{ij}]$ is invertible. Such transformations were called **affine** or **linear** by Weyl (Weyl, 1950, bottom of p. 21). This usage is no longer current. The term *linear* is reserved for the homogeneous part of the transformations (1.4), and the term affine survives mainly in *affine connections* in differential geometry. Weyl's group of affine transformations is the inhomogeneous general linear group, which is sometimes denoted by $IGL(n,\mathbb{R})$.

1.5 Local geometrical structures on n-manifolds

We shall now define some local geometrical structures on differentiable manifolds. (The reader who is unfamiliar with this notion will find a brief account in Appendix A8.) We remind the reader that the mathematician's conception of the local-global dichotomy is based on the notion of *neighbourhood of a point*, which can be defined independently of the notion of distance between points. This is discussed in greater detail in Appendices A3 and A8.

1.5.1 Riemannian and Lorentz structures

A **Riemannian structure** on an n-manifold is a smooth assignment of a positive-definite quadratic form (in local coordinates)

$$\mathrm{d}s^2 = \sum_{i,j=1}^{n} g_{ij}(x)\mathrm{d}x^i\mathrm{d}x^j \tag{1.5}$$

to each point of the manifold. It is a fundamental theorem in differential geometry that *every differentiable manifold admits a Riemannian metric* (see Appendix A8). A Riemannian metric defines a metric topology on the manifold which is identical with its original topology.

A **Lorentz structure** on an n-manifold is a smooth assignment of an **indefinite** quadratic form of signature

$$+ - \cdots -$$

at every point of the manifold. Formula (1.5) continues to hold, but the signature of $\mathrm{d}s^2$ has changed. As opposed to the Riemannian structure, not every manifold admits a Lorentz structure. The topology of a Lorentzian manifold may be thought of as determined by its Riemannian metric. It is a pleasing fact of differential geometry that most of the results that hold on Riemannian manifolds hold on Lorentzian manifolds as well (Eisenhart, 1964).

1.5.2 The conformal structure

In an attempt to unify general relativity and electrodynamics, Weyl investigated a class of transformations on Riemannian and Lorentzian manifolds that did not leave $\mathrm{d}s^2$ invariant, but multiplied it by a local scaling factor $\Omega(x)$. These transformations left the angle between two directions, which is a dimensionless quantity, unchanged, and therefore he called them **conformal** transformations. Conformal transformations of a Lorentz manifold therefore map light cones,

$$\mathrm{d}s^2 = 0,$$

to light cones. Conformal transformations on a four-dimensional Minkowski space form a 15-parameter group consisting of the Lorentz transformations, dilatations and the (nonlinear) transformations of reciprocal radii $x_i \to x_i/x^2$. Weyl showed that conformal transformations left the tensor

$$\begin{aligned} C^{\lambda}_{\mu\nu\sigma} &= R^{\lambda}_{\mu\nu\sigma} + \frac{1}{n-2}\left(\delta^{\lambda}_{\nu}R_{\mu\sigma} - \delta^{\lambda}_{\sigma}R_{\mu\nu} + g_{\mu\sigma}R^{\lambda}_{\mu} - g_{\mu\nu}R^{\lambda}_{\sigma}\right) \\ &\quad + \frac{R}{(n-1)(n-2)}\left(\delta^{\lambda}_{\sigma}g_{\mu\nu} - \delta^{\lambda}_{\nu}g_{\mu\sigma}\right) \end{aligned} \tag{1.6}$$

invariant. In (1.6) $R^{\lambda}_{\mu\nu\sigma}$ is the Riemann tensor, $R_{\mu\nu}$ the Ricci tensor and R the scalar curvature. The tensor $C^{\lambda}_{\mu\nu\sigma}$ was called the *conformal curvature tensor* by Weyl, but nowadays it is usually called the *Weyl tensor* or the *conformal tensor.*

1.5.3 The Weyl projective structure

In general relativity, paths of freely falling particles are timelike geodesics; paths of light rays are null-geodesics, as in special relativity. Transformations of a Lorentzian manifold that map geodesics to geodesics were studied by Weyl, who demonstrated that they leave the following tensor invariant:

$$W^{\lambda}_{\mu\nu\sigma} = R^{\lambda}_{\mu\nu\sigma} + \frac{1}{n-1}\left(\delta^{\lambda}_{\nu}R_{\mu\sigma} - \delta^{\lambda}_{\sigma}R_{\mu\nu}\right). \tag{1.7}$$

This tensor was called the *projective curvature tensor* by Weyl, and the structure associated with it the *projective structure*. Note, however, that the Weyl projective structure on a Lorentz manifold does not necessarily map null-geodesics to null-geodesics, or light cones to light cones.

The reader is referred to (Eisenhart, 1964) for more information on the Weyl conformal and projective structures.

2

Definition of causality on a structureless set

In the following we shall assume that the space, time and space-time of physics are made up of 'points' in the sense of Euclidean geometry. This assumption has been questioned by Wigner since 1952, and his reservations will be the subject of Part II of this book.

We shall begin by defining the notion of causality (which we shall call *Einstein–Weyl causality*) on a set of points devoid of any mathematical structure. It will be defined as a partial order on the set. Our point of departure is the observation that the propagation of a light ray determines a total order (the past–future order) on its path. We shall try to build up the causal structure by abstraction from the intersection properties of these paths in Minkowski space, and in real life.

2.1 Light rays

The fundamental objects in our scheme will be:

(i) A nonempty set of *points* M.
(ii) A distinguished family of subsets of M, called *light rays.*

The fundamental relation in our scheme will be a *total order* defined on the light rays.

We shall use the term *light ray* as a shorthand for the space-time path of a light ray; more precisely, for a mathematical abstraction from the corresponding physical concept. Light rays will generally be denoted by the letter l. A light ray through the point x will be denoted by l_x, one through the points x, y by $l_{x,y}$. Distinct light rays through x will be differentiated, if necessary, by superscripts, thus: l_x, l'_x, l^1_x, l^2_x, etc. Every light ray l will be a totally ordered subset of M, and the total order on every ray l will be denoted by $<^l$ or (equivalently) by ${}^l\!>$. The statements $x <^l y$ and $y \,{}^l\!> x$ will be read, respectively, as 'x precedes y on l' and 'y follows x on l', and will mean exactly the same thing.

We do not exclude the possibility that both $x <^l y$ and $x \,{}^l\!> y$ hold simultaneously; however, they can hold simultaneously if and only if $x = y$. If $x <^l y$ and $x \neq y$, we shall write $x <^{ll} y$.

2.2 The order axiom

We shall demand that a light ray have the property that between any two distinct points lies a third, that is, if $x, z \in l$, $x <^{ll} z$, then there exists $y \in l$ such that $x <^{ll} y <^{ll} z$. This requirement can only be met if there are infinitely many distinct points on a light ray. We shall also demand that a light ray have no beginning and no end.[1] To sum up, a light ray will be assumed to satisfy the following:

Axiom 2.1 (The order axiom)

(a) If $x, y \in l$ and $x \neq y$, then $x <^l y$ or[2] $y <^l x$. If $x, y \in l$ and both $x <^l y$ and $y <^l x$, then $x = y$.

(b) If $x, z \in l$ and $x <^{ll} z$, then there exists[3] $y \in l$ such that $x <^{ll} y <^{ll} z$.

(c) For every $y \in l$, there exist $x, z \in l$ such that $x <^{ll} y <^{ll} z$.

(d) If l^1, l^2 are two distinct light rays and $x, y \in l^1 \cap l^2$, then

$$x <^{l^1} y \Leftrightarrow x <^{l^2} y.$$

The last condition means that no light ray propagates 'backward in time' (see Fig. 2.1).

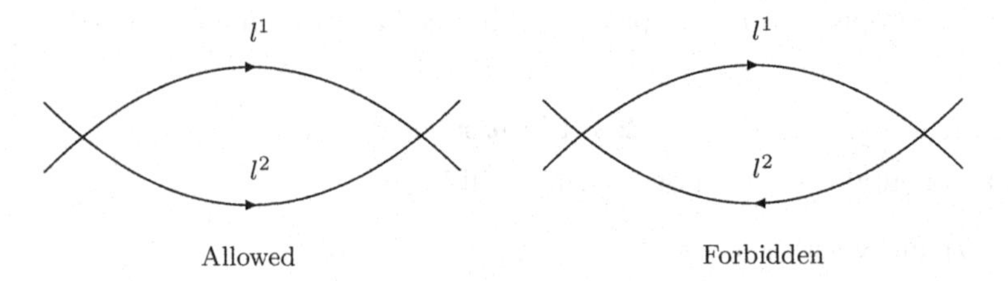

Fig. 2.1. Illustrating the order axiom, part (d)

Notations 2.2 l_x will denote a light ray through the point x, and $l_{x,y}$ will denote a light ray through x which passes through the point y.

We now define *light ray segments* (not *intervals*; the term interval will be reserved for a different use, to be defined later) as follows. For any light ray l

[1] We should reassure the reader that this is essentially a technical assumption that facilitates the subsequent mathematical development; it does not prevent one from introducing singularities, if desired, at the end of the process.

[2] In everyday speech, the phrase 'A or B' may mean one of two things: (i) 'A (is true), or B (is true), *or* both (are true)'; (ii) 'A, or B, *but not* both'. One has to decide from the context which of the two is meant, which may be difficult if the writing is sloppy. In mathematics, this particular ambiguity is avoided by convention; **or** always means the *inclusive* or, i.e., case (i). However, the inclusive or would often be unacceptable in physics; for example, two distinct points may be spacelike or timelike to each other, but cannot be both! These contradictory usages are firmly established. We shall try to avoid the problem as much as possible by using the phrase **either** A **or** B, which will mean 'either A or B, but not both'.

[3] Again, in mathematics, the phrase *there exists* means that there exists *at least one* object in question; there can be, and often are, many distinct objects satisfying the condition in question.

and $x, y \in l, x <^l y$, define

$$\begin{aligned} l(x,y) &= \{a | a \in l, x <^{ll} a <^{ll} y\}, \\ l[x,y] &= \{a | a \in l, x <^l a <^l y\}. \end{aligned} \tag{2.1}$$

$l(x,y)$ and $l[x,y]$ will be called *open* and *closed* light-ray segments respectively. Note that $l(x,x) = \emptyset$ and $l[x,x] = \{x\}$ for any $x \in l$.

Remark 2.3 Although, according to the above definition, $l[x,x] = \{x\}$ is a closed light-ray segment, when we talk about closed light-ray segments $l[x,y]$ in this and the following chapters, we shall generally assume that $x <^{ll} y$, i.e., $l(x,y)$ is nonempty. If the possibility $x = y$ is to be included, it will be stated explicitly.

In view of the centrality of light rays in our scheme, it is not surprising that we shall often have to decide whether or not two given points x, y are joined by a light ray. We shall denote by $\lambda(x,y)$ the statement that there is a light ray through x, y, and by $\sim\lambda(x,y)$ its negation: there is no light ray that passes through both x and y.

Let $\{x_0, x_1, \ldots, x_n\}$ be a *finite* set of points such that $\lambda(x_k, x_{k+1})$ for $k = 0, 1, \ldots, n-1$. Define $l^*[x,y] = l[x,y]$ if $x <^{ll} y$ and $l^*[x,y] = l[y,x]$ if $y <^{ll} x$. We shall call $l^*[x,y]$ an *unoriented segment*. The concatenation (symbol: $\bowtie$) of unoriented segments

$$l^*[x_0, x_1] \bowtie l^*[x_1, x_2] \bowtie \cdots \bowtie l^*[x_{n-1}, x_n] \tag{2.2}$$

will be called an *l-polygon*, and denoted by

$$P(x_0, x_1, \ldots x_n).$$

Examples of l-polygons are shown in Fig. 2.2. Note that, although three successive points x_{k-1}, x_k, x_{k+1} in $P(x_0, x_1, \ldots x_n)$ will not generally lie on the same light ray, i.e., $l^*_{x_{k-1},x_k} \neq l^*_{x_k,x_{k+1}}$, this is not a requirement; degenerate cases will be admissible, including the case in which the points $x_0, \ldots, x_n$ lie on the same light ray, as well as the case $x_0 = x_1$ in $P(x_0, x_1)$, i.e., $P(x_0, x_1) = l[x_0, x_0] = \{x_0\}$. The points $x_0, x_1, \ldots, x_n$ will be called the *nodes* of the l-polygon P.

We now make the following definition:

Definition 2.4 A space M will be called *l-connected* if there is an l-polygon connecting any two points x and y in M.

The essential part of this definition is the fact that an l-polygon consists of a *finite* number of light-ray segments.

From now on, we shall make the following nontriviality assumptions:

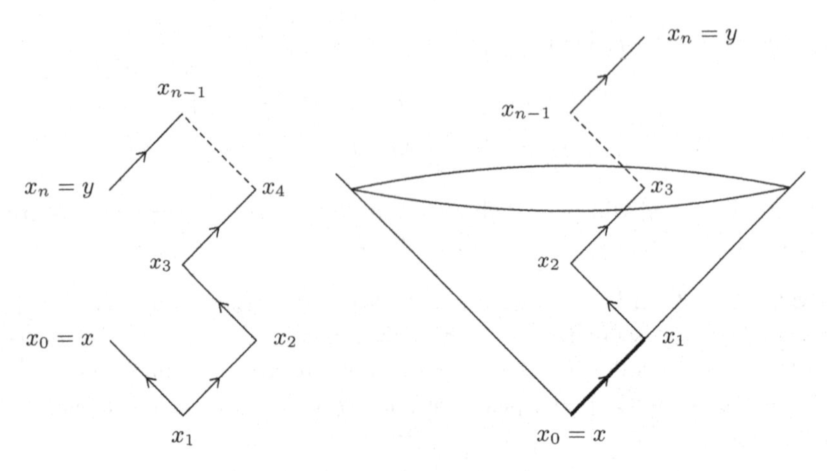

Fig. 2.2. Examples of l-polygons

Assumption 2.5 (Nontriviality assumptions)

(i) M does not consist of a single point.
(ii) M does not consist of a single light ray.
(iii) M is l-connected.

2.3 The identification axiom

We now turn to the intersections of light rays; the structure of the theory will be decided by the intersection properties of light rays.

The defining physical property of Minkowski space is that the velocity of light is constant, so that paths of light rays are straight lines. Consequently two light rays intersect only once, if they intersect at all. Consider now an optical device that forms a real image. To do so, actual light rays must intersect twice, which means that the paths of light rays in the optical device cannot be described in Minkowski space. However, the object point and the image point are well separated, which suggests that the space-times containing optical imaging devices may be locally Minkowski, like the space-time of general relativity.

We shall therefore assume that if two light rays intersect more than once, the points of intersection are well separated, in a precise sense that will be defined below using only the notion of order. We shall also assume that two light rays do not merge, and a light ray does not split into two (or more). These conditions, which are mainly aimed at avoiding excessive generality, are illustrated in Fig. 2.3. We formulate these assumptions mathematically as follows:

Axiom 2.6 (The identification axiom) If l, l' are two distinct light rays, $S = l \cap l'$ and $a \in S$, then there exist $p, q \in l$ such that $p <^{ll} a <^{ll} q$ and $l(p,q) \cap S = \{a\}$. The same situation holds on l'.

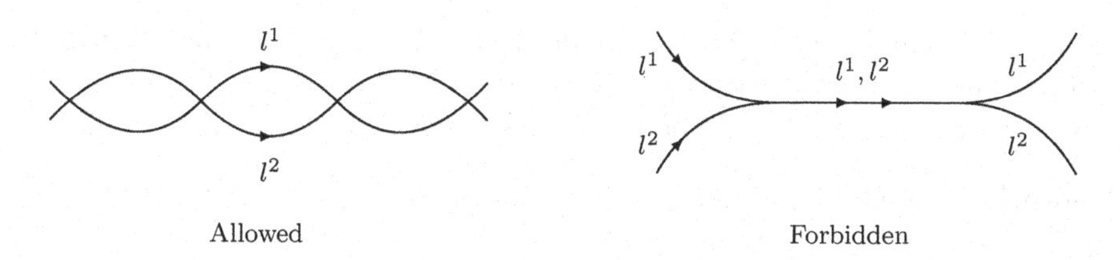

Fig. 2.3. Illustrating the identification axiom

It follows immediately from this axiom that if l is any light ray and $l(x, y)$ any open segment on it, then $l(x, y) \subset l' \Rightarrow l' = l$; in trying to avoid excessive generality, we have arrived at a far-reaching 'analyticity condition'. This condition will eventually allow us to extend a light-ray segment indefinitely and unambiguously in both directions, and this extension property will prove to be an essential tool in later chapters.

2.4 Light cones

An l-polygon $P(x_0, x_1, \ldots x_n)$ will be called *ascending* if $x_k <^{ll} x_{k+1}$ for $k = 0, 1, \ldots, n-1$, and denoted by $P^{\uparrow}(x_0, x_1, \ldots x_n)$. It will be called *descending* if $x_{k+1} <^{ll} x_k$ for $k = 0, 1, \ldots, n-1$, and denoted by $P^{\downarrow}(x_0, x_1, \ldots x_n)$.

We now define *semi-infinite* light-ray segments as follows:

$$\begin{aligned} l^+_{b,c} &= \{x | x \in l_{b,c}, b <^{ll} x; b <^{ll} a\}, \\ l^-_{b,a} &= \{x | x \in l_{b,a}, x <^{l} a; a <^{ll} b\}. \end{aligned} \tag{2.3}$$

The second argument in $l^+_{b,c}$ or $l^-_{b,a}$ is one that picks a unique light ray out of the many that may pass through b; the implicit assumption is that such a point exists; on any nonempty light-ray segment one can find two points that are traversed by only one light ray. This assumption is inessential; it may be avoided, but then the discussion leading to the definitions of cones becomes much longer. This discussion may be found in (Borchers and Sen, 2006).

Using semi-infinite segments, we define an extension of the notion of ascending and descending polygons by allowing the last segment to extend indefinitely. In the forward direction, with $x_0 <^{ll} x_1 <^{ll} \cdots <^{ll} x_n$, this will be the concatenation

$$l[x_0, x_1] \bowtie \cdots \bowtie l[x_{n-2}, x_{n-1}] \bowtie l^+_{x_{n-1}, x_n}. \tag{2.4}$$

The object defined by (2.4) will be called an *unbounded* ascending l-polygon. Unbounded descending l-polygons can be defined analogously. We shall denote by $P^{\uparrow}_x$ an ascending l-polygon from x, either bounded or unbounded, but still with a finite number of nodes. The nodes other than x are left unspecified. $P_x^{\downarrow}$ will be defined similarly, with order reversed.

We are now in a position to define light cones at x.

Definition 2.7 (Light cones at x)

$$\begin{aligned} C_x^+ &= \bigcup_{\text{all } P_x^{\uparrow}} P_x^{\uparrow}, \\ C_x^- &= \bigcup_{\text{all } P_x^{\downarrow}} P_x^{\downarrow}. \end{aligned} \tag{2.5}$$

C_x^+ and C_x^- are called, respectively, the *forward* and the *backward* cone at x. The forward cone is also called the *future* or *positive* cone; the backward cone is also called the *past*, or *negative* cone. The union

$$C_x = C_x^+ \cup C_x^- \tag{2.6}$$

is called the *cone* at x. We now define

$$l_x^{\pm} = l_x \cap C_x^{\pm}. \tag{2.7}$$

l_x^+ and l_x^- will be called, respectively, forward and backward rays from x. The notations $l_{x,a}^{\pm}$ will mean, respectively, a forward and a backward ray from x through a. In the first case, one will have $x <^{ll} a$; in the second, $a <^{ll} x$. These notations have been anticipated in (2.3).

2.4.1 Timelike points

Let $P(x_0, x_1, \ldots, x_n)$ be an l-polygon. The unoriented segment $l^*[x_0, x_1]$, with $x_0 \neq x_1$, will be called the *initial* of P.

Definition 2.8 (Timelike points)

(a) A point $y \in C_x^+, y \neq x$ will be called a *timelike point* of C_x^+ if, *for any forward ray l_x^+ through x*, there is an ascending l-polygon P from x to y such that the initial of P is a subset of l_x^+. The set of all timelike points of C_x^+ will be denoted by τC_x^+, and called the τ-*interior* of C_x^+.

(b) The set τC_x^- is defined analogously, by reversal of order.

Equivalently, one could say that y is a timelike point of C_x^+ if, for any l_x^+, there is $a \in l_x^+, a \neq x$ such that $y \in C_a^+$.

The significant part of this definition is the italicized phrase *for any forward ray l_x^+ through x*, which cannot be omitted; it is not enough that there be *an* ascending polygon from x to y.

2.4.2 Extension of order

We shall now extend the total order on the light rays to a partial order on all of M.

Definition 2.9 (Order on M) Define

$$x < y \quad (\text{or } y > x) \quad \text{iff} \quad y \in C_x^+,$$

equivalently,

$$x < y \quad (\text{or } y > x) \quad \text{iff} \quad x \in C_y^-.$$

The equivalence of $x < y$ and $y > x$ follows from the fact that an ascending polygon from x to y is equally a descending polygon from y to x. Note also that if $y \in l_x^+$, then $x < y \Leftrightarrow x <^l y$.

Theorem 2.10 $<$ *(or* $>$*) defines a reflexive* (page 263) *partial order on* M.

Proof Reflexivity means $x < x$, which holds because, by definition, $x \in C_x^+$. To prove transitivity, i.e., $x < y$ and $y < z \Rightarrow x < z$, it suffices to note that the concatenation of an ascending polygon from x to y and one from y to z is an ascending polygon from x to z. □

The reader is invited to verify that the following lemma is an immediate consequence of the definitions:

Lemma 2.11 *If* $y \in \tau C_x^+$ *and* $z > y$, *then* $z \in \tau C_x^+$, *and the same for order reversed.*

We now define

Definition 2.12 (β-boundary)

$$\beta C_x^+ = C_x^+ \setminus \tau C_x^+, \qquad \beta C_x^- = C_x^- \setminus \tau C_x^-.$$

βC_x^+ and βC_x^- will be called the β-boundaries of C_x^+ and C_x^- respectively.

Proposition 2.13 *Let* $y > x$. *Then*

(1) $y \in \tau C_x^+$ *is equivalent to* $\beta C_x^+ \cap C_y^+ = \emptyset$.
(2) $x \in \tau C_y^-$ *is equivalent to* $\beta C_y^- \cap C_x^- = \emptyset$.

Proof The first assertion is proven as follows. Let $y \in \tau C_x^+$ and suppose that there exists a point $z \in \beta C_x^+ \cap C_y^+$. Then $z > y$. Hence, by Lemma 2.11, $z \in \tau C_x^+$, a contradiction. In the opposite direction, since $y \in C_y^+$, $\beta C_x^+ \cap C_y^+ = \emptyset$ implies $y \notin \beta C_x^+$. But $y \in C_x^+$, from which it follows that $y \in \tau C_x^+$. The second assertion follows by reversal of order. □

There are two reasons why we cannot use the topological terms *interior* and *boundary* in the present context. One is that we do not yet have a topology! The other is that when we do have a topology, the β-boundary of a cone may fail to agree with its topological boundary. An example is provided by the two-dimensional Minkowski space from which the strip $x_1 \geq x_0$, $2 \leq x_1 \leq 3$ has been excised, as shown in Fig. 2.4. Of the two forward rays from the point $(1, 3)$, only one, marked l, is shown in the figure. It does not belong to the cone C_x^+ (shaded region), but does belong to the topological boundary ∂C_x^+ of C_x^+.

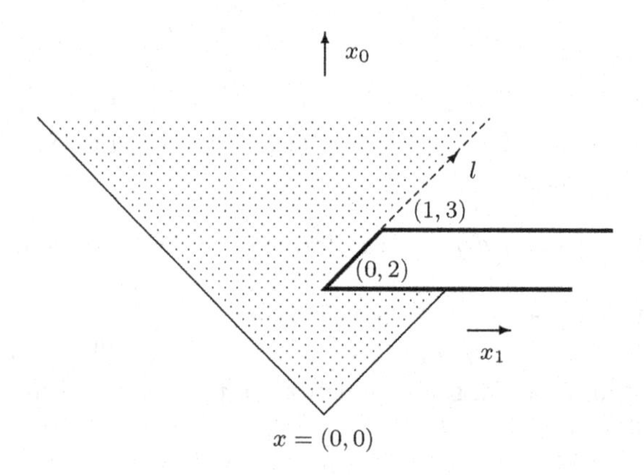

Fig. 2.4. Illustrating that βC_x^+ and ∂C_x^+ may be different

2.5 The cone axiom

The order defined by x is *not antisymmetric*, i.e., $x < y$ *and* $x > y$ together do *not* imply that $x = y$. By definition, $x \in C_x^+ \cap C_x^-$; antisymmetry fails when $C_x^+ \cap C_x^-$ contains points other than x. Consider the physics. Owing to part (a) of the order axiom (page 24), light rays cannot form a closed loop; but there is nothing in our scheme so far to prevent timelike curves[4] from forming closed loops. Indeed, the anti-de-Sitter space of Fig. 2.5, which is a one-sheeted hyperboloid (light rays are the two families of straight lines which generate the hyperboloid), permits closed timelike loops, one of which is shown in the figure. This must be regarded as a physical pathology which has to be eliminated. We do this via the following axiom:

Axiom 2.14 (The cone axiom)

$$C_x^+ \cap C_x^- = \{x\} \quad \text{for all } x \in M.$$

[4] A precise definition of timelike curves will be given in Section 5.1. The definition will not contradict physical intuition, which is what we are using here.

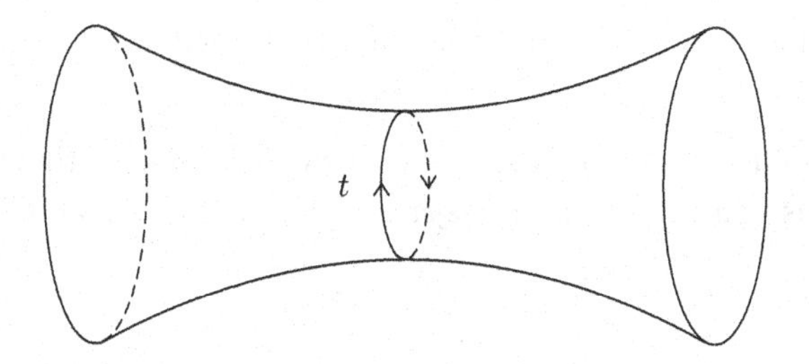

Fig. 2.5. This anti-de-Sitter space violates the cone axiom

We shall now make the following notational conventions:

Notations 2.15 If $y \in \tau C_x^+$, we shall write $x \ll y$. If $x \in \tau C_y^-$, we shall write $y \gg x$. The symbols $\not\ll$ and $\not\gg$ will denote the negations of $\ll$ and $\gg$ respectively.

As a mnemonic, one could note that the past–future orientation of all three symbols $<^l$, $<$ and $\ll$ is the same.

The odd thing is that $x \ll y$ does not imply $y \gg x$ (which is why we have called the above notational conventions rather than definitions); our scheme is still too general. This is shown by the following example (Fig. 2.6).

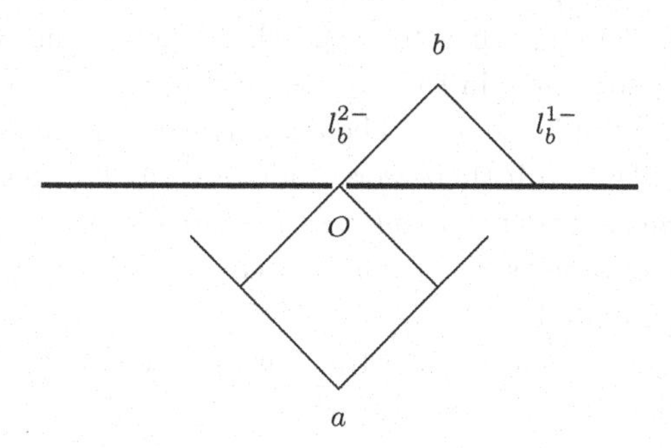

Fig. 2.6. $a \ll b$ does not imply $b \gg a$

Example 2.16 All points on the X-axis except the origin O are excised from two-dimensional Minkowski space. This leaves the upper and lower half-planes (without boundary), connected by the single point O. One sees from Fig. 2.6 that there are descending l-polygons from b that meet each of the two forward rays from a at points above a. However, *no* ascending l-polygon from a can meet the ray l_b^- at a point below b. Therefore $a \ll b$, but $b \not\gg a$.

The problem – and it is definitely a problem – lies not in the definitions of $\ll$ and $\gg$ but in the fact that the two half-planes are joined by a single point. It is easy to see that it disappears if the two half-planes are joined but by an open interval, no matter how small. The necessary restrictions will be placed in

Axiom 2.43 (which requires some preparation). Anticipating this, we make the following definition:

Definition 2.17 (S-spaces) A subset of M satisfying the order, identification and cone axioms will be called an *S-space* (S from symmetry) if

$$x \ll y \Leftrightarrow y \gg x.$$

It will be seen later that we shall only be concerned with S-spaces, for which we have the following

Theorem 2.18 *In an S-space, $\ll$ (equivalently, $\gg$) defines a nonreflexive, nonsymmetric* (page 263) *partial order.*

To avoid confusion, we shall not use the symbol $\gg$ until we have established Theorem 2.33 (page 40).

Terminology 2.19 In keeping with standard practice in physics, we shall say that two distinct points $x, y \in M$ are *spacelike*, *timelike* or *lightlike* to each other according as (i) $x \notin C_y$ (equivalently, $y \notin C_x$), (ii) $x \ll y$ (or $y \ll x$), or (iii) $\lambda(x, y)$.

If x, y are two distinct points in Minkowski space, then they would be either spacelike, or timelike, or lightlike to each other; there is no fourth possibility. The situation is not so simple in the present setting.

Let $a \in M$ and $x \in \beta C_a^+$, $x \neq a$. Then x is neither spacelike nor timelike to a. It would be lightlike to a if there were a forward ray l_a^+ which passes through x. However, *the axioms we have adopted so far provide no assurance that this is true.* It will turn out to be true, *locally*, as a consequence of Axiom 2.43, which we have not yet formulated.

This fact should serve as a warning that, in the present setting, results which are obvious in Minkowski space need proof.

2.6 D-sets

Of the three axioms that we have adopted so far, one (the order axiom) introduces structure while the other two (the identification and cone axioms) rule out excessive generality. The order axiom itself is basically global, without any explicitly local content. For example, it allows two distinct light rays to intersect more than once. This is essential for admitting optical image-forming devices, but one would not want two light rays to intersect more than once in a small neighbourhood of a point. We need an axiom that defines *small neighbourhoods*, i.e. the local structure of the theory.

As a guide to intuition, consider a closed ball B around the origin O in Euclidean n-space. One of the important properties of B is that it can be contracted upon itself to the point O; i.e., if $x = (x_1, x_2, \ldots, x_n) \in B$ and

$O = (0, 0, \ldots, 0)$, then the maps $x \mapsto \lambda x = (\lambda x_1, \lambda x_2, \ldots, \lambda x_n)$, $0 < \lambda \leq 1$ are homeomorphisms, and the limit $\lambda \to 0$ of these homeomorphisms is a continuous map. One hardly needs any reflection to realize that a vast amount of mathematical structure is involved in expressing this seemingly simple idea, none of which is available to us yet.

The objects that *are* available to us, as substitutes for open and closed balls, are open and closed *order intervals* $I(a, b)$ and $I[a, b]$, which we now define. In the following, a, b are any two points in M.

Definition 2.20 (Order intervals)

$$I[a, b] = C_a^+ \bigcap C_b^-,$$

$$I(a, b) = \tau C_a^+ \bigcap \tau C_b^-.$$

If $a \ll b$, then the order intervals $I[a, b]$ and $I(a, b)$ will also be called *double cones*. We shall write $\beta I[a, b] = (\beta C_a^+ \cup \beta C_b^-) \cap I[a, b]$.

Remarks 2.21 It follows from these definitions that if a and b are spacelike to each other, then $I(a, b) = I[a, b] = \emptyset$. If $a = b$, then $I(a, b) = \emptyset$ and $I[a, b] = \{a\}$. If $\lambda(x, b)$ and $b <^{ll} x$, then $I[x, b] = \emptyset$; similarly, if $b \ll x$, then $I[x, b] = \emptyset$. But, if $y \in \beta C_x^+$ and $y \neq x$, we cannot exclude the possibility that $\sim\lambda(x, y)$; we know very little about $I[x, y]$ in this situation.

What are the properties that we would like the order intervals to have? Our arsenal is limited: we have at our disposal only the light rays (which are our substitutes for straight lines), and we can impose conditions only on their intersections. While avoiding global commitments, we should like order intervals $I[a, b]$ that are small enough to have the following properties:

(i) A light ray from the τ-interior of $I[a, b]$ should not exit $I[a, b]$ without intersecting its β-boundary (see Example 2.24).
(ii) However, if a light ray l traverses $I(a, b)$, then no segment $l[x, y]$ of it (with $x \neq y$) should lie on $\beta I[a, b]$.
(iii) There should be at most one light ray through two distinct points of $I[a, b]$. (This cannot be required of all $I[x, y] \subset M$.)
(iv) There should be no 'holes' inside $I[a, b]$; this has to be expressed precisely in terms of light rays and their intersections.
(v) If $x \in \beta C_a^+ \cap I[a, b]$, then there should be a light ray through a and x. Similarly, if $y \in \beta C_b^- \cap I[a, b]$, then there should be a light ray through y and b.
(vi) $I[a, b]$ should be an S-space.

The above requirements would seem to be obvious; i.e., they should be incorporated in the theory unless they turn out to have unacceptable (and, at the present stage, unforeseen) consequences. However, they leave one important question unanswered.

In Minkowski space, one cannot form a triangle by intersecting three light rays. That is, if l_x^{1+} and l_x^{2+} are two forward rays from x, $a_1 \in l_x^{1+}$, $x <^{ll} a_1$ and $a_2 \in l_x^{2+}$, $x <^{ll} a_2$, then there is no light ray that passes through a_1 and a_2. It is not clear a priori whether or not this is a desirable feature in the present setting. Locally, the conditions stipulated above would prevent a light ray through a_1 and a_2 from traversing the τ-interior of $I[x, y]$, so that such a ray has to lie entirely on βC_x^+. It turns out, somewhat surprisingly (we shall not prove it), that this situation is inconsistent with the conditions stipulated above, and that one has to rule it out explicitly. (This will be achieved by Definition 2.22(d).)

With this preparation, we proceed to define, formally, the notion of a *D-set*,[5] preparatory to stating our final axiom. The definition will be formulated as six separate conditions, which will be assigned names for easier recall.

Definition 2.22 (***D*-sets**) A subset U of M will be called a D-set iff it fulfils the following conditions:

(a) **The order-convexity condition.** U contains the entire closed order interval between any two of its points: $x, y \in U \Rightarrow I[x, y] \subset U$ (Fig. 2.7(a)).
A body in Euclidean n-space is called **convex** if a line segment joining any two points in the body lies wholly within the body. A D-set, while being order-convex, need not be convex (see Fig. 2.8).

(b) **The openness condition.**[6] For every $x \in U$ and every light ray l through x, there are points $p, q \in l \cap U$ such that $p <^{ll} x <^{ll} q$ (Fig. 2.7(a)).
That is, the intersection of a light ray with a D-set does not have a minimum or a maximum with respect to the order $<^{ll}$.

(c) **The intersection condition.** If $y \in U$, $r \in \tau C_y^- \cap U$ and l_r is a ray through r, then (Fig. 2.7(b), upper part)

$$l_r^+ \cap \{\beta C_y^- \setminus \{y\}\} \cap U \neq \emptyset,$$

and the same with order reversed.
In words, a forward ray from the backward τ-interior of a cone in a D-set intersects its backward β-boundary below the apex of the cone, and the same with order reversed.

[5] From the German *Durchschnittseigenschaft*, roughly translatable here as *standard* or *normal*. However, these two words are so over-used in the mathematical literature that it seems better to avoid them.

[6] When the topology is introduced, D-sets will turn out to be open sets.

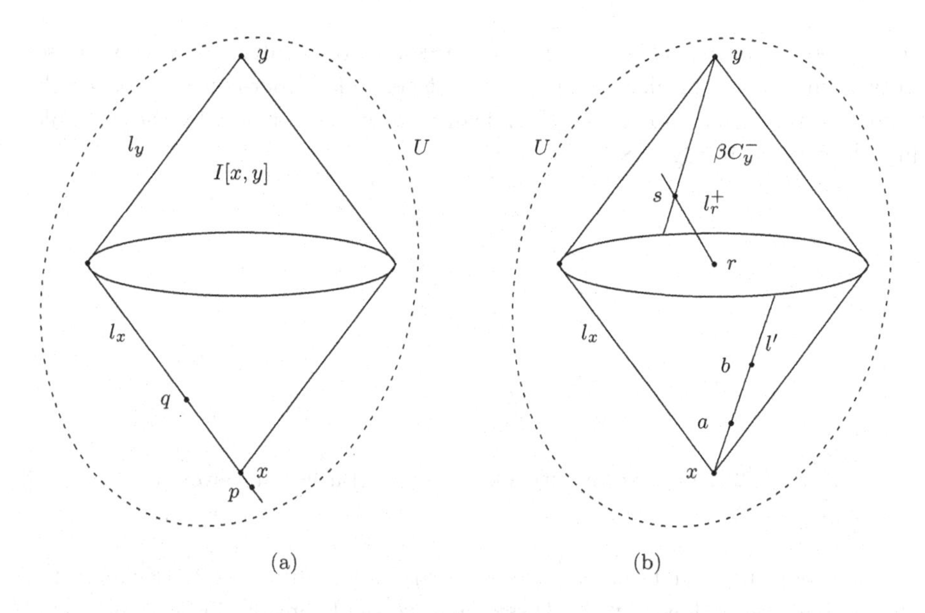

Fig. 2.7. Some defining properties of D-sets

(d) **The convexity condition.** If $x \in U$ and $l' \cap \beta C_x^+ \cap U$ contains two distinct points a, b, then (Fig. 2.7(b), lower part)[7]

$$x \in l'_{a,b} \cap C_x^+ \cap U \subset \beta C_x^+ \cap U,$$

and the same with order reversed.
That is, in a D-set U, a light ray that passes through two distinct points on the β-boundary of a forward (backward) cone lies wholly on the β-boundary of that forward (backward) cone and passes through its vertex.

(e) **The l-uniqueness condition.** If $a, b \in U$ and $\lambda(a,b)$, then the ray $l_{a,b}$ is unique.
That is, through any two distinct points in a D-set there passes at most one light ray. (It follows that two distinct light rays cannot intersect more than once in a D-set.) As we observed earlier, this condition will not hold globally except in particularly simple spaces.

(f) **The l-dimension condition.** Through any point of U, there are at least two distinct light rays.

Remarks 2.23 (About the figures) In the figures, some D-sets are generally indicated by dashed lines enclosing circular or elliptical regions, which are

[7] We shall use the term *convexity* only in this sense. On the rare occasion when we have to refer to the property called *convexity* of bodies in Euclidean spaces, we shall state it explicitly.

convex bodies in $\mathbb{R}^2$. This should not be taken too literally; circles and ellipses in two-dimensional Minkowski space are not generally order-convex. Conversely, order-convexity does not imply Euclidean convexity, as shown by the example, in $\mathbb{M}^3$, pictured in Fig. 2.8.

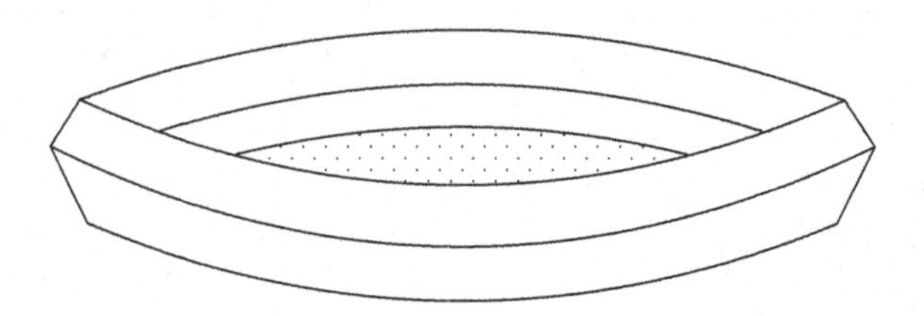

Fig. 2.8. Order-convexity does not imply Euclidean convexity

One should note that the empty set $\emptyset$ satisfies Conditions 2.22 trivially; it is therefore a D-set. However, the D-sets that we shall consider in Section 2.7 will all be assumed nonempty, unless the contrary is stated explicitly. Note that we have not required a D-set to be an S-space; this property will be established later as a theorem.

2.7 Properties of D-sets

We shall now study the basic properties of D-sets. In Minkowski space these properties are either obvious or trivial to prove. The reason is that light rays are homeomorphic to the real line $\mathbb{R}$, and there are no holes or gaps in $\mathbb{R}^n$. By contrast, $\mathbb{Q}$ may be described as being 'full of gaps', and the same is true of the real subfield $\mathbb{F}$ of the field of algebraic numbers (which is countable; see page 250). One cannot rule out a priori that two light rays cross each other without intersecting; consider the following example:

Example 2.24 Let $M = \mathbb{Q}\times\mathbb{R}$. This is a subset of $\mathbb{R}^2$ in which the X-coordinate is restricted to be rational. Let α be any irrational number. The lines $x = y$ and $x + y = \alpha$ in M cross each other without intersecting.

This is one reason why the results of this chapter require proof.

We begin by defining yet another convexity property:

Definition 2.25 A closed order interval $I[a, b]$ will be called *l-convex* if any light ray through any point $x \in I(a, b)$ intersects $\beta I[a, b]$ at exactly two points, one lying on βC_a^+ and the other on βC_b^-.

Not every closed order interval is l-convex. For example, the one (in two dimensions) shown in Fig. 2.9 is not; it has a hole in its middle. The topological

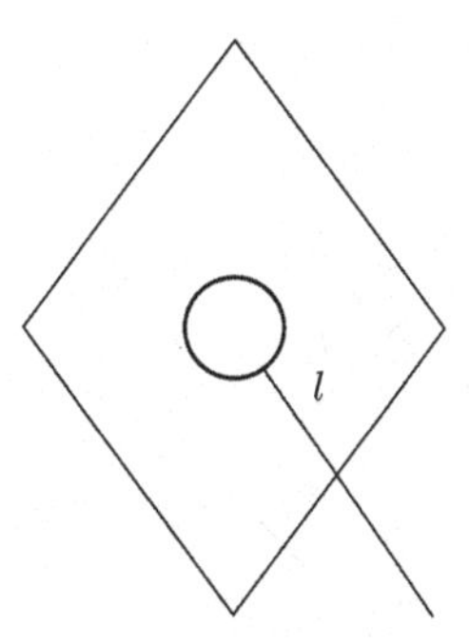

Fig. 2.9. An order interval that is not l-convex

boundary of the hole cannot belong to M; if it did, the light ray l shown in the figure will have a point that is maximal with respect to the order $<^l$.

Lemma 2.26 *Let U be a D-set and $I[a,b] \subset U$. Then $I[a,b]$ is l-convex.*

Proof This follows immediately from the intersection and the convexity conditions. □

It is the property of l-convexity that allows us to distinguish between order intervals that have holes in them and those that do not. When we have defined the topology (and are therefore allowed to use the terms interior and boundary), we shall find that a light ray from the interior of a D-set intersects its boundary at exactly two points. We shall then be able to say that every D-set is l-convex.

2.7.1 D-sets and timelike order

In Section 5.1 we shall establish the existence, in D-sets, of *timelike curves*, which will be subsets of M that are totally ordered by $\ll$ and are homeomorphic to light-ray segments. However, for the moment we cannot take for granted that there are enough mutually timelike points in a D-set. In this section we shall establish that there are enough such points.

Theorem 2.27 *Let U be a D-set and let $y \in U$. Then there exist points $x, z \in U$ such that $x \ll y \ll z$.*

In words, every point in a D-set has a timelike predecessor and a timelike successor.

Proof Let l_y be a ray through y. By the openness condition (b) of Definition 2.22, there exists a point $p \in l_y^+ \cap U$ such that $y <^{ll} p$ (see Fig. 2.10). By the l-dimension condition 2.22(f) there is at least one light ray l_p other than l_y through p. Again, the openness condition implies that there exists a point $z \in l_p^+$ such that $p <^{ll} z$. Then $\sim\lambda(y,z)$, so that $y <^{ll} p, p <^{ll} z$ imply that $y < z$. It follows that either

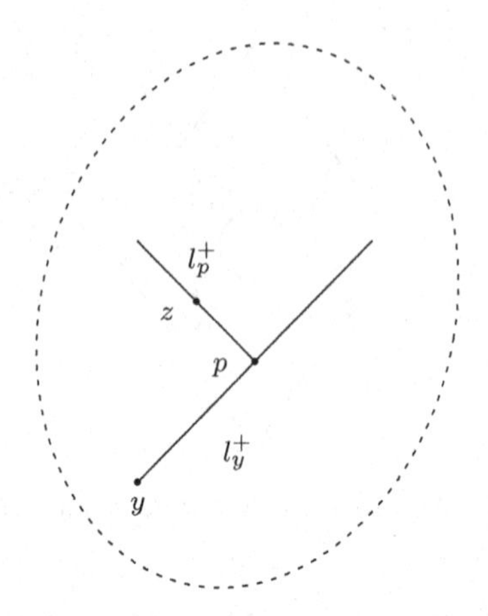

Fig. 2.10. Every point in a D-set has a timelike successor

$y \ll z$ or $z \in \beta C_y^+$. If $z \in \beta C_y^+$, then, since $p \in \beta C_y^+$, the intersection $l_p \cap \beta C_y^+ \cap U$ contains two distinct points p and z. Then from the convexity condition 2.22(d) it follows that $z \in l_p \cap C_y^+ \cap U \subset \beta C_y^+ \cap U$, i.e., the ray l_p passes through y, a contradiction. This proves that $y \ll z$.

The existence of the point x, $x \ll y$ is established by the same argument, with order reversed. □

The next result shows that, in a D-set, between any two distinct timelike points lies a third.

Lemma 2.28 *Let U be a D-set and $x, z \in U$ such that $x \ll z$. Then there exists $y \in U$ such that $x \ll y \ll z$. The same holds with order reversed.*

Proof We have to prove that $I(x, z)$ is nonempty. The spirit of the proof is the same as the proof of Theorem 2.27. The details may be found in (Borchers and Sen, 2006), but may also be filled in by the reader without much difficulty. □

The following result is an immediate corollary of Theorem 2.27:

Corollary 2.29 *Let U be a D-set, and let $x_0 \in U$. Then there exists an infinity of points $x_n \in U$, $n \in \mathbb{Z}$ such that $x_n \ll x_{n+1}$ for all $n \in \mathbb{Z}$. The same holds with order reversed.*

2.7.2 Light rays from τ-interiors of D-sets

The intersection condition 2.22(c) stipulates that, in a D-set, the intersection of a light ray from the τ-interior of a cone with its β-boundary be nonempty. We prove below that this intersection consists of a single point.

Theorem 2.30 *Let U be a D-set, and let $x, y \in U$ such that $y \in \tau C_x^+$. Let l_y be a light ray through y. Then*

$$l_y^- \cap \beta C_x^+$$

consists of a single point, and the same holds with order reversed.

Proof By Proposition 2.13, $y \in \tau C_x^+$ is equivalent to $\beta C_x^+ \cap C_y^+ \cap U = \emptyset$. By the intersection condition, if $l_y \cap \beta C_x^+ \cap U$ is nonempty, then

$$l_y \cap \beta C_x^+ = l_y^- \cap \beta C_x^+ \cap U \subset I[x, y] \subset U.$$

Then, if $l_y^- \cap \beta C_x^+$ contained two distinct points, it would follow from the convexity condition that

$$l_y \cap C_x^+ \cap U \subset \beta C_x^+,$$

contradicting the assumption $y \in \tau C_x^+$. The same argument holds with order reversed. □

2.7.3 Incidence of light rays on cone boundaries

In this section we shall establish two results on the incidence of light rays on cone boundaries in D-sets. The first is that in a D-set, every point on the β-boundary of a cone is connected to its vertex by a light ray. From this it would follow that in a D-set, two distinct points are either spacelike, or timelike, or lightlike to each other (see the paragraphs that follow Terminology 2.19 on page 32). The second is that, in a D-set, every light ray through the vertex of a cone lies wholly on its β-boundary.

Lemma 2.31 *Let U be a D-set and let $x \in U$. If a is any point on βC_x^+ such that $a \neq x$, then there exists a light ray l_x through x such that $a \in l_x^+ \cap \beta C_x^+ \cap U$ and $x <^{ll} a$. The same holds with order reversed.*

Proof Since $a \in \beta C_x^+$, there is an ascending l-polygon $P^\uparrow(x_0, \ldots, x_{n-1}, x_n)$ from $x = x_0$ to $a = x_n$, where the points $x_0, \ldots, x_n$ are distinct. Suppose that $x_{n-1} \in \beta C_x^+$. Then, by the convexity condition, $x \in l_{x_{n-1},a}$. That is, there exists a light ray l_x through x and a. (By the l-uniqueness condition, there is only one light ray through the points x and a.)

Suppose now that $x_{n-1} \notin \beta C_x^+$. Then $x_{n-1} \in \tau C_x^+$. Now $a = x_n > x_{n-1}$, and therefore it follows from Lemma 2.11 that $a \in \tau C_x^+$, a contradiction. This proves the stated result. The same argument holds with order reversed. □

We define an *inner* light ray at $x \in M$ to be a ray that joins x with a point $y \in \tau C_x^+$ (or in τC_x^-).

Lemma 2.32 *There are no* inner *light rays in a D-set.*

Proof Let $y \in \tau C_x^+$ and suppose that there exists a light ray $l_{x,y}$ that joins x with y. By Theorem 2.30, a backward ray from y intersects βC_x^+ at only one point. By the intersection condition, this point cannot be x itself. The same argument holds for $a \in \tau C_x^-$. □

2.7.4 A D-set is an S-space

We now have the tools to prove the following fundamental result:

Theorem 2.33 *Every D-set is an S-space.*

Proof Let U be a D-set and $x, y \in U$ such that $y \in \tau C_x^+$. Let l_y be a light ray through y. Then, by Theorem 2.30, the intersection $l_y^- \cap \beta C_x^+$ is a unique point, say u. If $u \neq x$ then it follows that $x \in \tau C_y^-$. The case $u = x$ is ruled out by Lemma 2.32. □

2.7.5 New D-sets from old

In this section we shall state two results on generating new D-sets from given ones. The proof of the first is straightforward, and that of the second is lengthy. We shall omit the proofs, and refer the interested reader to (Borchers and Sen, 2006).

Proposition 2.34 *The intersection of two D-sets is a D-set.*

Proposition 2.35 *Let U be a D-set and $x, y \in U$ such that $x \ll y$. Then $I(x, y)$ is an l-connected D-set.*

It should be stressed that a D-set need not be l-connected. In two-dimensional Minkowski space, the sets U and V shown in Fig. 2.11 are D-sets. In each

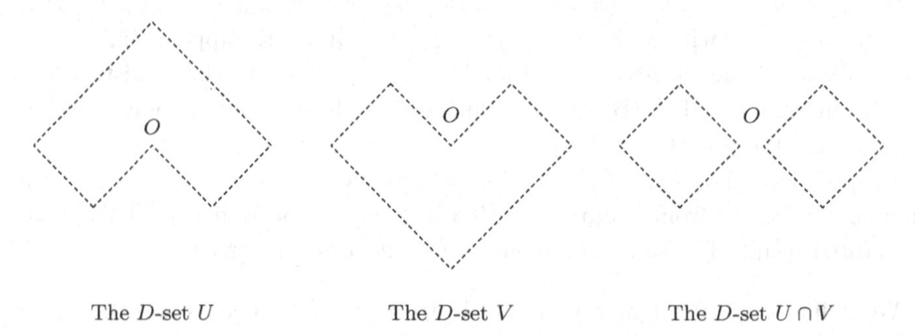

Fig. 2.11. D-sets need not be l-connected

diagram, the origin of coordinates is marked O. Their intersection $U \cap V$ consists of two disjoint pieces, i.e., $U \cap V = \emptyset$. For clarity, the two rectangles constituting $U \cap V$ have been shifted slightly with respect to each other.

The sets $I(x, y)$ that are D-sets will be used so frequently that it is useful to give them a name:

Definition 2.36 An order interval $I(x, y)$ will be called a *D-interval* if $I[x, y] \subset U$, where U is a D-set.

If a D-set contains two disjoint pieces U and V, then the two must be spacelike to each other; no point of U can be either timelike or lightlike to any point of V.

2.7.6 An incidence theorem for β-boundaries

In this section, we shall establish a variant of Proposition 2.13 that holds in D-sets. We begin with a preliminary lemma.

Lemma 2.37 *Let U be a D-set and $x = z_0$, $y = z_n \in U$ such that $x < y$. Let $P^{\uparrow}(z_0, z_1, \ldots, z_n)$ be an ascending polygon from z_0 to z_n. If the light ray l_{z_1,z_2} is distinct from the ray l_{z_0,z_1}, then $y = z_n \in \tau C_x^+$. The same statement holds with order reversed.*

Proof We shall assume that $z_2 \in \beta C_x^+$ and obtain a contradiction. If $z_2 \in \beta C_x^+$, then, by the convexity condition, $l_{z_1,z_2} \cap U \subset \beta C_x^+ \cap U$ and $x \in l_{z_1,z_2}$. Since z_0 and z_1 belong to both light rays, the two rays (restricted to U) must be the same, which is the desired contradiction. Hence $z_2 \in \tau C_x^+$. The same argument applies with order reversed. □

Theorem 2.38 *Let U be a D-set, $x \in U$ and $p \in \beta C_x^+ \cap U$, $p \neq x$. Then*

(1) $\beta C_x^+ \cap \beta C_p^+ \cap U = l_{x,p}^+ \cap \beta C_p^+ \cap U$;

(2) $\beta C_x^+ \cap \beta C_p^- \cap U = l[x, p]$.

Proof Let $a \in \beta C_p^+$, $p \neq a$. Then $P(x, p, a)$ is an ascending l-polygon from x to a (Fig. 2.12). By Lemma 2.37, if $l_{x,p} \neq l_{p,a}$, then $a \in \tau C_x^+$. If $a \notin \tau C_x^+ \cap U$, then necessarily $a \in \beta C_x^+ \cap U$, i.e., $a, p \in \beta C_x^+ \cap U$. Therefore, by the convexity condition, $x \in l_{p,a}$. Then, by the l-uniqueness condition, $l_{x,p} \cap U = l_{p,a} \cap U$, which proves the first assertion.

To prove the second assertion, note simply that owing to the convexity condition any point in the intersection $\beta C_x^+ \cap \beta C_p^-$ has to lie on the unique ray that passes through both x and p. □

2.7.7 Spacelike separation in D-sets

So far we have concerned ourselves mainly with pairs of points that were timelike or lightlike to each other. We now turn to spacelike separations in D-sets.

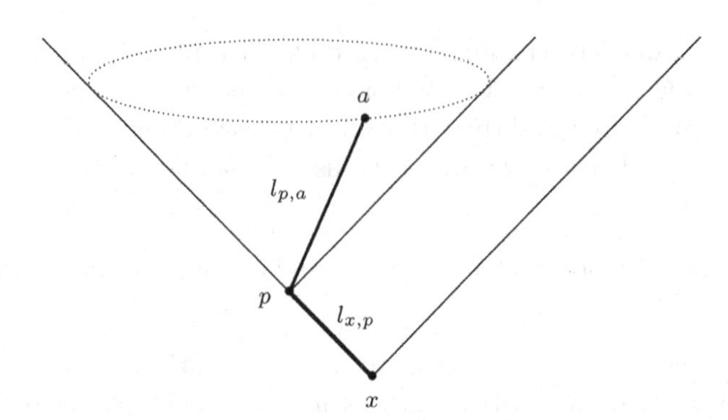

Fig. 2.12. Illustrating Theorem 2.38

Theorem 2.39 *Let U be a D-set, $x \in U$ and l_x^1, l_x^2 two distinct light rays through x. Pick a point $a \in l_x^{1-} \cap U$, $a \neq x$ and a point $b \in l_x^{2+} \cap U$, $b \neq x$. Then*

(1) *$b \gg a$, and*

(2) *$I(a, b) \cap C_x = \emptyset$,*

i.e., every point in $I(a, b)$ is spacelike to x.

Proof The fact that $b \gg a$ follows immediately from Lemma 2.37 and Proposition 2.34. Next, Theorem 2.27 ensures that $I(a, b)$ is nonempty. Finally, Theorem 2.38 shows that $C_x^+ \cap \tau C_b^- = C_x^- \cap \tau C_a^+ = \emptyset$. Hence $C_x \cap I(a, b) = \emptyset$. □

If $r, s \in U$ and $r \ll s$, then, by definition, $C_r^- \cap U \subset C_s^- \cap U$ and the inclusion is such that $\beta C_r^- \cap \beta C_s^- \cap U = \emptyset$; the situation shown in Fig. 2.13(a) cannot arise. If $r \in \beta C_s^-$ and $r \neq s$, then, from Theorem 2.38(1), we have $\beta C_s^- \cap \beta C_r^- \cap U = l_{s,r}^- \cap U$. Therefore if $\beta C_r^- \cap \beta C_s^- \cap U \neq \emptyset$ and the condition $r <^{ll} s$ does not hold, then r and s must be spacelike to each other (Fig. 2.13(b)). Conversely:

Lemma 2.40 *Let U be a D-set, $x, r, s \in U$ such that $x \ll r$, $x \ll s$ and r, s are mutually spacelike. Then $\beta C_r^- \cap \beta C_s^- \cap C_x^+$ is nonempty.*

Proof Let l_x be a light ray through x. From the definition of D-sets, $l_x \cap \beta C_r^- \cap U$ is a single point. Call it q_r. Similarly, let $\{q_s\} = l_x \cap \beta C_s^- \cap U$. Since q_r and q_s lie on the same light ray, there are three possibilities:

(1) $q_r <^{ll} q_s$. Then Lemma 2.37 implies that $q_r \ll s$, so that the ray $l_{q_r,r}$ intersects βC_s^- at a unique point p, and $p \in \beta C_r^- \cap \beta C_s^- \cap C_x^+$.

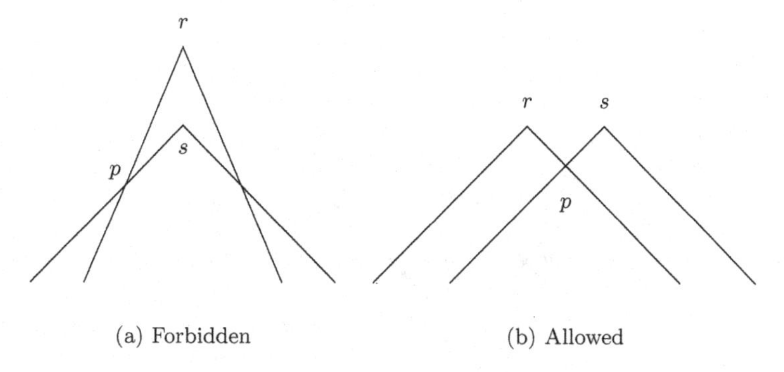

(a) Forbidden (b) Allowed

Fig. 2.13. Difference between pairs of timelike and spacelike points

(2) $q_r = q_s$. Then this point lies, by definition, on $\beta C_r^- \cap \beta C_s^- \cap C_x^+$.

(3) $q_s <^{ll} q_r$. This is the same as case (1) above, with r and s interchanged.

□

2.7.8 Timelike order and D-subsets

We have established, in Theorem 2.27, that every point in a nonempty D-set has timelike predecessors and successors. We shall now establish that similar results hold with respect to D-subsets.

Proposition 2.41 *Let U be a D-set and V a D-subset of U. Let $x \in V$ and $z \in U \setminus V$ such that $x \ll z$. Then there exists $y \in V$ such that $x \ll y \ll z$. The same holds with order reversed.*

Remark: The existence of points $y \in V$ such that $x \ll y$ has been established in Theorem 2.27. What remains to be proved is that among these there are points which also satisfy $y \ll z$.

Proof Owing to the order-convexity condition, $I[x, z] \in U$. If $I(x, z) \subset V$, then every point y in $I(x, z)$ satisfies $x \ll y \ll z$. If $I(x, z) \not\subset V$, then for any forward ray l_x^+ from x, there is a backward ray $l_z^{(1)-}$ from z such that $\{q\} = l_x^+ \cap l_z^{(1)-} \in U$ (see Fig. 2.14). Next, let $p \in l_x^+ \cap V$ such that $x \ll p <^l q$. Finally, let $l_z^{(2)-}$ be a second backward ray from z, and let $\{s\} = l_z^{(2)-} \cap \beta C_p^+$. According to the openness condition, there exist points $y \in l(p, s)$, $y \neq s$ such that $y \in V$. These points fulfil the requirement $x \ll y \ll z$.

The same argument holds, with order reversed. □

The main result of this section follows easily from the above.

Theorem 2.42 *Let U be a D-set and V a D-subset of U. Let $x \in V$ and $z \in U \setminus V$ such that $z \gg x$. Then there exist points $u, y \in V$ and $w \in U$ such*

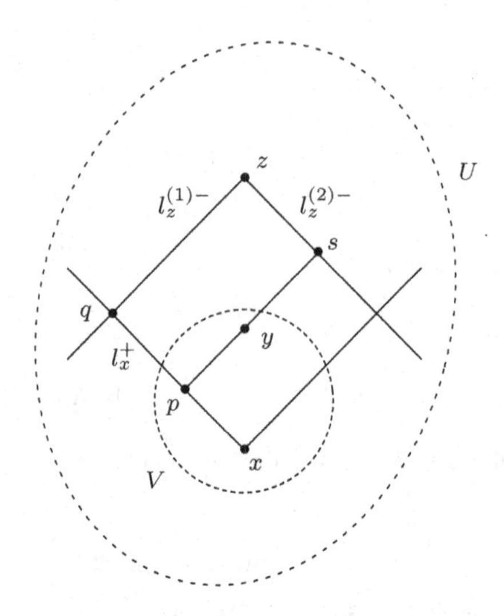

Fig. 2.14. Illustrating Proposition 2.41

that

$$u \ll x \ll y \ll z \ll w.$$

Proof

(1) Use Theorem 2.27 to obtain from $x \in V$ a point $u \in V$ such that $u \ll x$.

(2) Use Theorem 2.27 to obtain from $z \in V$ a point $w \in U$ such that $z \ll w$.

(3) Use Proposition 2.41 to obtain from $x \in V$ and $z \in U$ (as $V \subset U$) a point $y \in V$ such that $x \ll y \ll z$.

Then $u \ll x \ll y \ll z \ll w$. □

2.8 The local structure axiom

We conclude this chapter with our last axiom, the *local structure axiom*:

Axiom 2.43 (The local structure axiom) *For each $x \in M$ there is a D-set U_x such that $x \in U_x \subset M$.*

This axiom ensures that the order structure of the space M has, locally, many of the properties of a Minkowski space, always bearing in mind that M need not be a continuum; its cardinality could well be $\aleph_0$.

2.9 Ordered spaces

We are now ready to make the following definition:

Definition 2.44 (Ordered space) A space M which satisfies the order, identification, cone and local structure axioms will be called an *ordered space.*

The reader may wonder, legitimately, whether the system of axioms we have adopted is free of internal inconsistencies. A simple counterexample would prove that a system of axioms is inconsistent, but there is no way of proving consistency in mathematics. The best one can hope for is to exhibit a nontrivial example that satisfies the axioms. n-dimensional Minkowski space $\mathbb{M}^n$ satifies all our axioms.

We end this chapter with the definition of order-preserving maps:

Definition 2.45 (Order preserving maps) Let M, M' be ordered spaces and $\varphi : M \to M'$ a map from M to M'. Let $x, y \in M$ and $x' = \varphi(x), y' = \varphi(y)$. The map φ will be called *order preserving* if the following conditions are satisfied:

(a) $x = y \Rightarrow x' = y'$; $x \neq y \Rightarrow x' \neq y'$.

(b) If l is a light ray in M, then its image $l' = \varphi(l)$ is a light ray in M'.

(c) $\lambda(x, y) \Rightarrow \lambda(x', y')$; $\sim\!\lambda(x, y) \Rightarrow \sim\!\lambda(x', y')$.

(d) $x <^l y \Rightarrow x' <^l y'$.

(e) $x < y \Rightarrow x' < y'$.

(f) $x \in \tau C_y^{\pm} \Rightarrow x' \in \tau C_{y'}^{\pm}$.

An order-preserving map is thus required to be injective (condition (a)), but not bijective. The injectivity condition is rather strong; it means, among other things, that the projections of three-dimensional Minkowski space onto its two-dimensional subspaces will not be order-preserving.

3

The topology of ordered spaces

The order structure on M defines a topology on it, which will be called the order topology. This topology makes M into a Tychonoff space which, additionally, has some strong homogeneity properties. We shall define the order topology and study its basic properties in this chapter.

3.1 The order topology

We begin with two preliminary results on separating points by D-sets.

Theorem 3.1 *Let U be a D-set, $y \in U$ and $b \in U \setminus C_y^-$. Then there exists a point $a \in U \setminus C_y^-$ such that $b \gg a$.*

Proof There are three possibilities, depending on the situation of b with respect to y. They are:

(1) $b \gg y$.

(2) $b \in \beta C_y^+$.

(3) $b \notin C_y^+$.

We shall establish the existence of the point a case-by-case.

(1) By Proposition 2.35, if $b \gg y$, then $I(b, y)$ is a nonempty l-connected D-set. The condition $a \ll b$ will be satisfied by any point $a \in I(y, b)$.

(2) If $b \in \beta C_y^+$, then, by the convexity condition, there is a light ray $l_{y,b}$ through y and b. Pick a point $p \in l_{y,b}$ such that $y <^{ll} p <^{ll} b$. From p, choose a backward ray l_p^- different from $l_{y,b}$, and on it a point $a \in U$ such that $a <^{ll} p$. Then Lemma 2.37 implies that $a \ll b$. Since $C_y^- \subset C_p^-$ and $\beta C_y^- \cap \beta C_p^- = l_{y,b} \cap C_y^-$, it follows that $a \notin C_y^-$ (see Fig. 3.1(a)).

(3) Finally, let $b \notin C_y^+$. By Theorem 2.27 there exists $z \in U$ such that $z \ll b$. (i) If $z \notin C_y^-$, then there is nothing to prove. (ii) If $z \in \beta C_y^-$, then, by Lemma 2.28, there exists $a \in U$ such that $z \ll a \ll b$. Then, by Theorem 2.39(2), $a \notin C_y^-$. (iii) If $z \in \tau C_y^-$ then, by Lemma 2.40, $\beta C_y^- \cap \beta C_b^-$ is nonempty. Let $p \in \beta C_y^- \cap \beta C_b^-$ and choose a point $z' \in l_{p,y}$ $z' <^{ll} p$ (Fig. 3.1(b)). Then, by Lemma 2.37, $z' \ll b$ and we are back to situation (b).

□

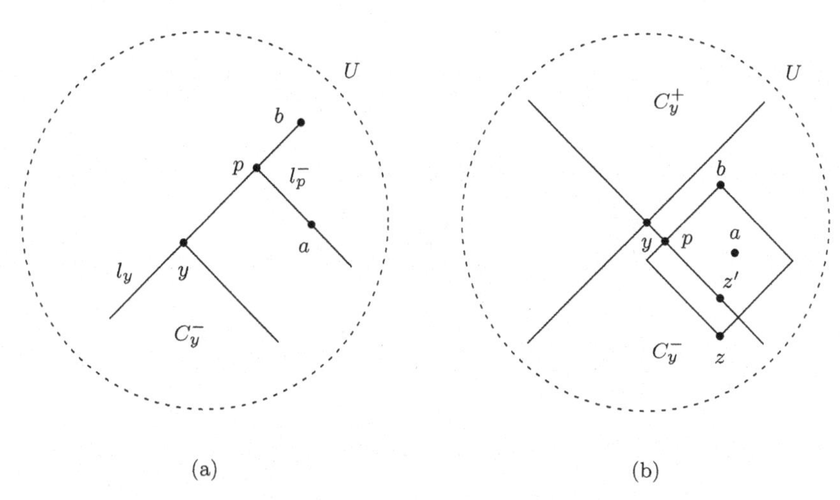

Fig. 3.1. Separation by forward and backward cones

The above separation allows one to construct the desired separation by D-sets.

Theorem 3.2 *If x and y are two distinct points in a D-set U, then there exist D-subsets U_x, U_y of U such that $x \in U_x$, $y \in U_y$ and $U_x \cap U_y = \emptyset$.*

Proof There are three possibilities, according to whether x and y are timelike, lightlike or spacelike to each other. We take them in turn.

(1) x and y are timelike to each other. We may assume, without loss of generality, that $x \ll y$. Then, from Corollary 2.29, there exist points $a, b, c \in U$ such that $a \ll x \ll b \ll y \ll c$. The sets $U_x = I(a, b)$ and $U_y = I(b, c)$ fulfil the requirements.

(2) x and y are lightlike to each other. We may assume, without loss of generality, that $x <^{ll} y$. By Theorem 2.27, there exist points $a, b \in U$ such that $a \ll x$ and $b \gg y$. Let $q \in l(x, y)$ (see Fig. 3.2). Let l_q be a second ray through q, $l_q \neq l_{x,y}$, and choose $p, r \in l_q \cap U$ such that $p <^{ll} q <^{ll} r$. Let $a \in \tau C_x^- \cap U$ and $b \in \tau C_y^+ \cap U$. The sets $U_x = I(a, r)$ and $U_y = I(p, b)$ fulfil the requirements. Details are left to the reader.

(3) x and y are spacelike to each other, i.e., $y \notin C_x \cap U$. Then $y \notin C_x^- \cap U$, and therefore, from Theorem 3.1, there exists a point $p \in U$ such that $p \ll y$ and $p \notin C_x^- \cap U$. Then $x \notin C_p^+$. We may therefore apply Theorem 3.1 with order reversed to x and C_p^+ to obtain a point $s \in U$ such that $s \gg x$ and $s \notin C_p^+ \cap U$ (Fig. 3.3). Then $C_s^- \cap C_p^+ = \emptyset$. Proposition 2.41 now tells us that there exist points $r, q \in U$ such that $p \ll y \ll q$, $r \ll x \ll s$. The sets $U_x = I(r, s)$ and $U_y = I(p, q)$ fulfil the requirements.

□

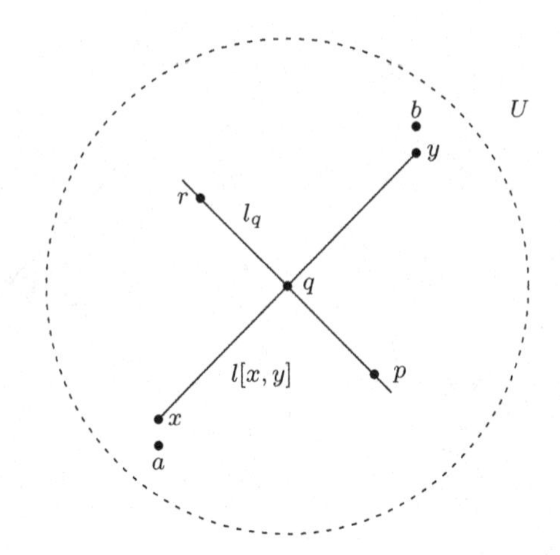

Fig. 3.2. Proving Theorem 3.2(2)

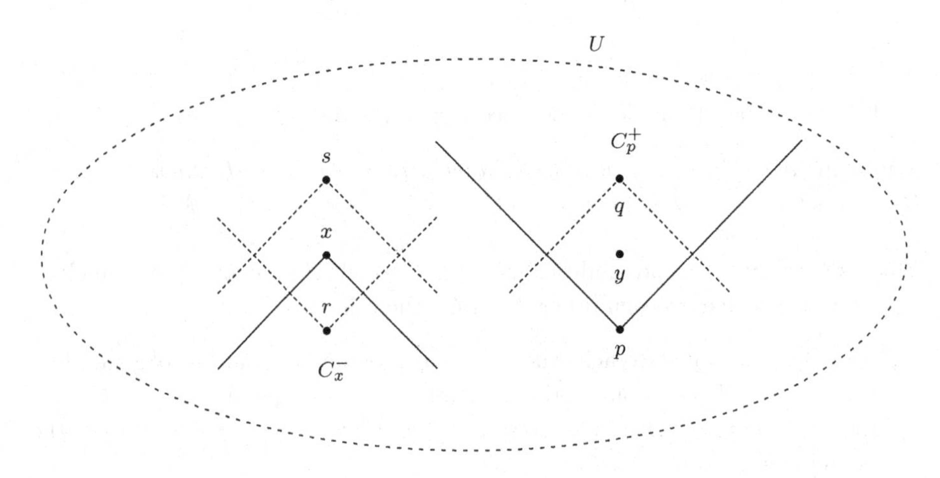

Fig. 3.3. Proving Theorem 3.2(3)

We shall now prove the following crucial lemma:

Lemma 3.3 *The family of D-intervals in M is a base* (page 258) *for a topology on M.*

Proof Let us temporarily denote D-intervals by I, distinguishing different ones by subscripts. We have only to prove that for any two I_1, I_2 such that $I_1 \cap I_2 \neq \emptyset$, there exists I_3 such that $I_3 \subset I_1 \cap I_2$.

Note first that every point is contained in a D-interval; by the local structure axiom, for any point $x \in M$ there is a D-set U_x such that $x \in U_x \subset M$. Then, by Theorem 2.27, there is an open order interval I such that $x \in I \subset U_x$.

By Proposition 2.34, $W = I_1 \cap I_2$ is a D-set. If it is nonempty, then for $w \in W$ there exist, again by Theorem 2.27, points $a, b \in W$ such that $a \ll w \ll b$. Then, by the order-convexity condition, $I_3 = I(a, b) \subset W$. □

Definition 3.4 (Order topology) The topology on M that has the family of D-intervals as a base will be called the *order topology* on M.

It follows immediately from this definition and Theorem 3.2 that:

Theorem 3.5 *The order topology on M is Hausdorff* (page 266).

It follows that the order topology is T_1 (page 266), i.e., *one-point sets are closed* in an ordered space.

Having defined a topology, we are now in possession of the topological notions of interior and boundary. We omit the simple proof of the following:

Theorem 3.6 *In a D-set U, $\beta C_x^+ \cap U = \partial C_x^+ \cap U$ (the boundary of C_x^+) and $\tau C_x^+ \cap U = \operatorname{int} C_x^+ \cap U$ (the interior of C_x^+), and the same with order reversed. It follows that if $x, y \in U$, $x \ll y$, then $I[x, y]$ is the topological closure of $I(x, y)$.*

The family of D-sets is also a base for a topology on M, and it is easily seen that the topology defined by this base is again the order topology. It follows that every D-set is an open set. The converse is false. If $a \ll b \ll c \ll d$, then $U = I(a, b) \cup I(c, d)$ is clearly an open set; however, it is not order-convex, and therefore is not a D-set. When is an open set a D-set? This question is answered by the following lemma:

Lemma 3.7 *An open subset U of an ordered space M is a D-set if and only if it satisfies the following conditions:*

(1) *U is order-convex.*

(2) *If $I[x, y] \subset U$, then every light ray that traverses the interior of $I[x, y]$ intersects $\beta I[x, y]$ at exactly two distinct points.*

(3) *If $a, b \in U$ and $\lambda(a, b)$, then the ray $l_{a,b}$ is unique.*

If U is a D-set, then the above conditions are satisfied by definition. To prove that if U satisfies these conditions then it is a D-set one has to verify the six defining conditions of D-sets. These verifications are straightforward, and will be omitted. They may be found in (Borchers and Sen, 2006).

3.1.1 The Tychonoff property

Ordered spaces are Tychonoff spaces (page 267), as we shall prove below. A Tychonoff space is a *completely regular* space (page 267) which is also a T_1 space. Ordered spaces are Hausdorff (Theorem 3.5) and therefore T_1, so that we have only to prove that they are completely regular.

The separation property called complete regularity was first proved by Urysohn, in one of the deepest results of point-set topology which has become known as *Urysohn's lemma.* Urysohn proved his result for normal spaces (which need not be T_1) by a certain construction which, as we shall see below, can be transferred almost word-for-word to ordered spaces without the assumption of normality.

Theorem 3.8 (Complete regularity) *Let M be an ordered space, $A \subset M$ a closed subset and $b \in M$ a point such that $b \notin A$. Then there exists a continuous real-valued function $f : M \to [0,1]$ such that $f(b) = 0$ and $f(x) = 1$ for $x \in A$.*

Proof The reader who is familiar with Urysohn's lemma will recognize that the proof is exactly the same as that of the latter, except that, instead of normality, one uses Lemma 2.28 to obtain a family of open sets with the required nesting property. The reader who is unfamiliar with the proof of Urysohn's lemma will find that the proof uses a strikingly original idea.

We first set up the notations specific to the purpose. In the rest of this proof, I will denote a D-interval $I(.,.)$, and $\overline{I}$ its closure $I[.,.]$. Write $\mathbb{P} = \mathbb{Q} \cap [0,1]$ and let $p_0, p_1, \ldots, p_n, \ldots$ be an enumeration of $\mathbb{P}$ such that $p_0 = 0$ and $p_1 = 1$. Set $\mathbb{P}_n = \{p_0, \ldots, p_n\}$.

Since $b \in M \setminus A$ and $M \setminus A$ is open, there exists a D-interval I_1 such that $b \in I_1 \subset M \setminus A$. By Lemma 2.28, there exist points $x, y \in I_1$ such that $x \ll b \ll y$, i.e., $b \in I(x,y) \subset I[x,y] \subset I_1$. Set $I_0 = I(x,y)$ (see Fig. 3.4).

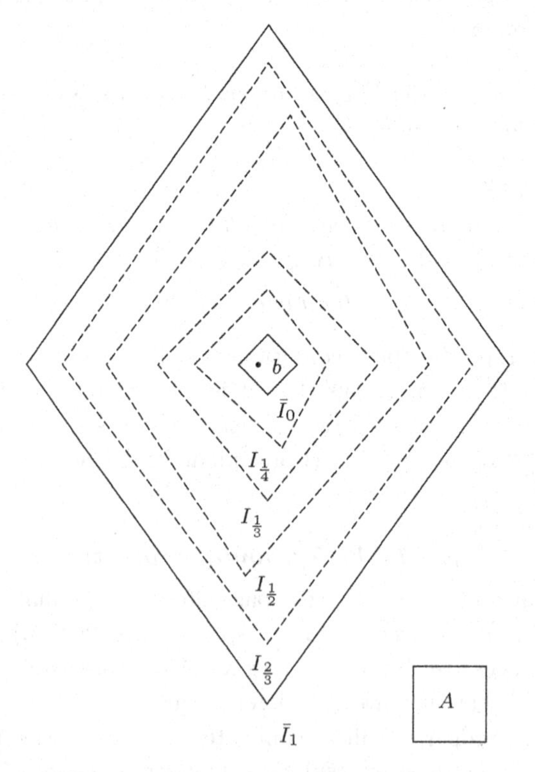

Fig. 3.4. Proof of complete regularity

Suppose that, for $r \in \mathbb{P}_n$, D-intervals I_r that have the property

$$r < s \Rightarrow \overline{I}_r \subset I_s \ \text{ for } \ r, s \in \mathbb{P}_n \tag{3.1}$$

have already been defined. We shall define a D-interval I_{n+1} such that property (3.1) holds for $r, s \in \mathbb{P}_{n+1}$ (Fig. 3.4).

Since $0, 1 \in \mathbb{P}_n$ and $0 < p_{n+1} < 1$, the number p_{n+1} partitions the set $\mathbb{P}_n$ into two disjoint subsets L and R such that $r \in L \Rightarrow r < p_{n+1}$ and $r \in R \Rightarrow r > p_{n+1}$. Since $\mathbb{P}_n$ is finite, L has a largest member w and R has a smallest member u, and

$$w < p_{n+1} < u.$$

Since $u, w \in \mathbb{P}_n$, the D-intervals I_u, I_w are already defined. Let now

$$u_1, u_2, w_1, w_2 \in I_1$$

such that $I_u = I(u_1, u_2)$ and $I_w = I(w_1, w_2)$. Then

$$u_1 \lll w_1 \lll w_2 \lll u_2.$$

By Lemma 2.28, there exist $v_1, v_2 \in I_1$ such that

$$u_1 \lll v_1 \lll w_1 \lll w_2 \lll v_2 \lll u_2.$$

Then $\overline{I}_v \subset I_u$ and $\overline{I}_w \subset I_v$.

The above procedure defines, recursively, a set of D-intervals I_r, indexed by $r \in \mathbb{P}$, that have the properties

(1) $b \in I_0$;
(2) $r < s \Rightarrow \overline{I}_r \subset I_s$ for all $r, s \in \mathbb{P}$;
(3) $I_1 \cap A = \emptyset$.

Define now the function

$$f(x) = \begin{cases} 1, & \text{if } x \notin \text{any } I_r; \\ \inf\{r | x \in I_r\}, & \text{otherwise.} \end{cases} \tag{3.2}$$

Clearly, $f(b) = 0$ and $f(x) = 1$ for all $x \in A$. It remains to prove that f is continuous. This proof is identical with that in the proof of Urysohn's lemma, and may be found in any textbook on the subject, for example (Munkres, 1975). □

Since one-point sets are closed in a Hausdorff space, the result we want follows immediately:

Corollary 3.9 *The ordered space M is a Tychonoff space.*

3.1.2 Order equivalence

Let M, M' be two ordered spaces and $f : M \to M'$ a map. We shall denote by x' and U' the images in M', under f, of a point x and a subset U of M. We shall, temporarily, distinguish different light rays in M by subscripts, e.g., l_1 and l_2, and denote their images in M' by l'_1 and l'_2. Using these notations, we define the notion of *order equivalence* as follows:

Definition 3.10 (Order equivalence) Two ordered spaces M and M' will be said to be *order equivalent* if there exists a map $f : M \to M'$ that satisfies the following conditions:

(a) f is bijective.

(b) If l is a light ray in M, its image l' is a light ray in M', and $a <^{ll} b$ implies $a' <^{ll} b'$.

(c) If $l_1 \cap l_2 = \{a_n\}$, then $l'_1 \cap l'_2 = \{a'_n\}$, and $a_n <^{ll} a_{n+1} \Rightarrow a'_n <^{ll} a'_{n+1}$.

(d) If U is a D-set in M, then U' is a D-set in M'.

It is easy to check that, according to the Definition 2.45 of order-preserving maps, the map f is order-preserving. Order equivalence, as the reader is invited to verify, is an equivalence relation in the mathematical sense. Definition 3.10(d) implies that two order-equivalent spaces M and M' are homeomorphic with each other. However, the converse is not true, as the following examples show.

Examples 3.11

(i) The de Sitter space in $1+1$ dimensions is a one-sheeted hyperboloid. It has two families of generators, which are straight lines, and two generators from the two different families intersect exactly once. Identifying the generators with light rays furnishes the $1+1$ de Sitter space with an order structure.

Consider now the cylinder $S_1 \times \mathbb{R}$ with the circle S_1 as base. Define the light rays to be curves that are inclined at 45° with the circular sections of the cylinder parallel to the base. This makes the cylinder into an ordered space. Two light rays that intersect each other do so infinitely many times.

The cylinder $S_1 \times \mathbb{R}$ is homeomorphic with the $1+1$ de Sitter space. However, the order structures defined on the two are manifestly different; they are not order-equivalent.

These examples suggest that the phenomena of gravitational lensing may be independent of the global topological structure of space-time.

(ii) Start with two copies of two-dimensional Minkowski space. (a) Excise the origin from one, and (b) excise the closed disc of unit radius and centre at the origin from the other. The resulting spaces are homeomorphic with each other. However, they are not order-equivalent. In case (a), only the two light rays through the origin are cut into two; in case (b), uncountably many light rays are cut into two.

We end this section with the following proposition:

Proposition 3.12 *The subspace topology on a light ray l induced by the order topology on M is the same as the topology of the order $<^l$ on l.*

The straightforward proof is omitted; it may be found in (Borchers and Sen, 2006). (Had this proposition been false, further development of the theory would have ground to a halt.) The result will be used in the following without attribution.

3.2 Homogeneity properties

In this section we shall establish two homogeneity properties of D-sets. The first is that any two closed light-ray segments that lie entirely in D-sets are homeomorphic with each other, as are any two open segments. To state the second, we need a notation. If U is a D-set, $a, b \in U$, $a \ll b$, we define

$$S(a,b) = \beta C_a^+ \cap \beta C_b^-. \tag{3.3}$$

The definition ensures that $S(a,b)$ is nonempty, does not consist of a single point, and lies in a D-set. The homogeneity property we wish to express is that if U, V are D-sets, $S(a,b) \subset U$ and $S(p,q) \subset V$, then $S(a,b)$ and $S(p,q)$ are homeomorphic with each other. In imitation of Minkowski space, the $S(a,b)$ will be called *spacelike hyperspheres.*

We shall establish plausibility, but shall not give the proofs of any of the results in this section. With one exception, they are all straightforward, and all of them may be found in (Borchers and Sen, 2006). The results will be based on certain maps of light-ray segments onto each other in a D-set, which we shall now define.

3.2.1 The standard maps

Definition 3.13 Let U be a D-set, $x, y \in U$ such that $x \ll y$, and l_x a light ray through x. Set (Fig. 3.5)

$$\{p\} = l_x \cap \beta C_y^-.$$

Then, by order-convexity, $p \in U$. Let l_y be a light ray through y such that $l_y \neq l_{p,y}$, and set

$$\{q\} = l_y \cap \beta C_x^+.$$

Then $q \in U$. Next, let r, s be any two points in U that satisfy

$$r \in l_x, \quad x <^l r <^l p,$$
$$s \in l_y, \quad q <^l s <^l y.$$

Define now the maps ρ and σ,

$$\rho : l_x[x, p] \to l_y,$$
$$\sigma : l_y[q, y] \to l_x,$$

as follows:

$$\begin{aligned} \rho(r) &= l_y \cap \beta C_r^+, \\ \sigma(s) &= l_x \cap \beta C_s^-, \end{aligned} \tag{3.4}$$

respectively. Since the right-hand sides of (3.4) are unique points, the maps ρ and σ are well defined.

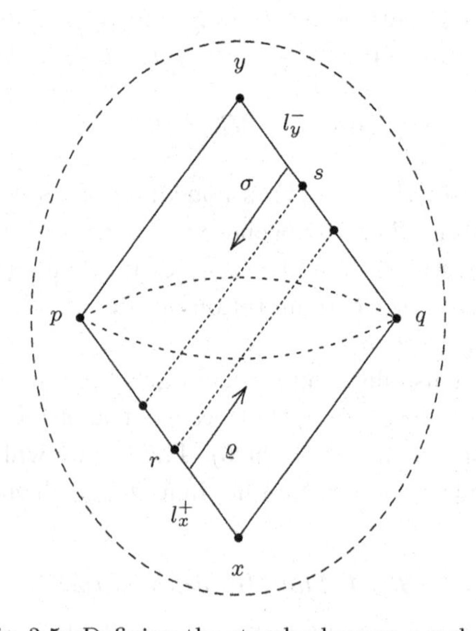

Fig. 3.5. Defining the standard maps ρ and σ

The maps ρ and σ are *natural maps*, in the mathematical sense of the term; one does not need a coordinate system to define them. Their key properties are given in the following theorem:

Theorem 3.14 *The maps ρ and σ defined by* (3.4) *are injective and order-preserving. Furthermore, they are homeomorphisms onto their ranges.*

The range of ρ is $l[q, y]$ and that of σ is $l[x, p]$. When ρ and σ are restricted to their ranges, we shall denote their inverses by ρ^{-1} and σ^{-1} respectively. We have, in this case,

$$\rho = \sigma^{-1} \text{ and } \sigma = \rho^{-1}.$$

The segments $l[x, p]$ and $l[q, y]$ are homeomorphic;[1] we shall write this as

$$l[x, p] \overset{\text{hom}}{=} l[q, y].$$

We shall call the maps ρ and σ the *standard maps.*

3.2.2 Three or more dimensions

We begin by remarking straightaway that the term *dimension* has not yet been defined. We are allowing ourselves to use it because, in this case, intuition will not lead us astray.

By arguments identical with the above, one can prove that

$$l[x, q] \overset{\text{hom}}{=} l[p, y].$$

But what about the possibility that (for example)

$$l[x, p] \overset{\text{hom}}{=} l[x, q]\,?$$

To answer this question, one has to distinguish between the following cases: (i) There are exactly two light rays through any point in the D-set U. (ii) There are more than two light rays through any point $x \in U$. In case (i), $S(x, y) = \{p, q\}$ and there are not enough natural maps to answer the question (Fig. 3.6). However, there is a simplifying factor that comes into play, and it is the following. Let $a \in I(x, y)$. Then, of the two backward rays l_a^{1-} and l_a^{2-} from a, one intersects $l(x, p)$ and the other intersects $l(x, q)$ (Fig. 3.6). Therefore, *when it becomes possible to introduce coordinates on* $l[x, p]$ *and* $l[x, q]$, one can assign coordinates (a_1, a_2) to any point $a \in I[a, b]$. This, in turn, enables the results of interest to be deduced quickly and easily. This case, which will be called the *two-dimensional case*, will not be pursued any further.

Case (ii) is more interesting. From now on we shall deal exclusively with this case, which we state as follows:

Assumption 3.15 There are infinitely many light rays through any point of M.

[1] In a D-set, a segment $l[u, v]$ will determine, uniquely, the ray l on which it lies.

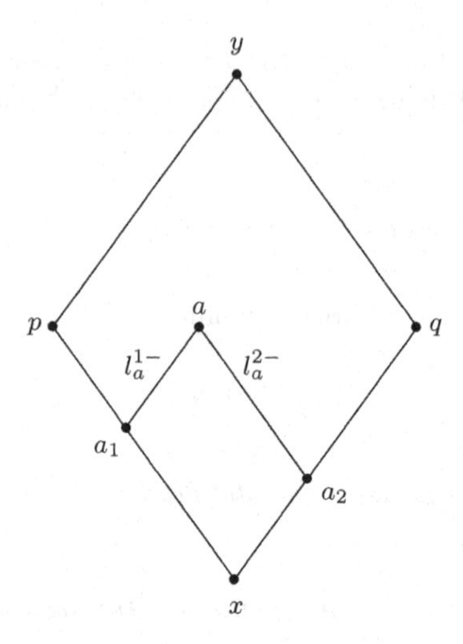

Fig. 3.6. The two-dimensional case

A word of explanation may be required here. We know that in two-dimensional Minkowski space there are exactly two light rays through each point, but in higher-dimensional Minkowski spaces there are infinitely many light rays through every point. In our present setting, known proofs of the results that will be stated below require the existence of three or more light rays through each point. However, it can be shown at a later stage of development that if there are more than two light rays through each point, there are infinitely many. Assumption 3.15 is therefore a shortcut, and not a loss of generality.

Let $I(x, y)$ be a D-interval, and $p \in S(x, y)$. We shall call the segments $l[x, p]$ and $l[p, y]$ *boundary segments* of $I(x, y)$. Figure 3.7 shows a D-interval $I(x, y)$ with three points $p_1, p_2, p_3 \in S(x, y)$. Clearly,

$$l[x, p_1] \overset{\text{hom}}{=} l[p_3, y]$$

and

$$l[x, p_2] \overset{\text{hom}}{=} l[p_3, y],$$

from which it follows that

$$l[x, p_1] \overset{\text{hom}}{=} l[x, p_2].$$

The following result then becomes almost self-evident:

Theorem 3.16 *In a D-set, any two boundary segments of an order interval are homeomorphic to each other.*

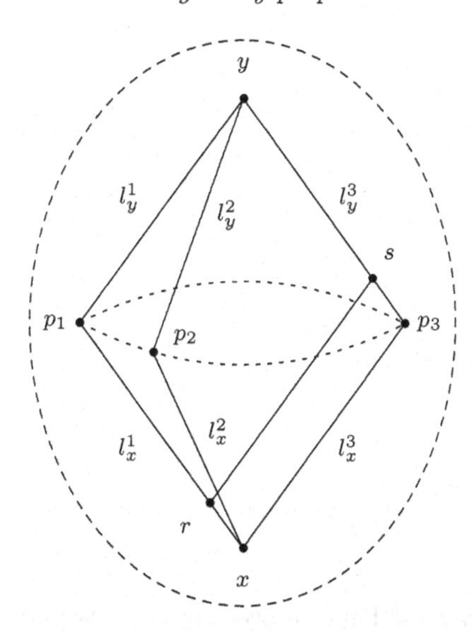

Fig. 3.7. Homeomorphisms of light ray segments in D-sets

The points s and x in Fig. 3.7 are timelike to each other, with $x \ll s$. Therefore $l[x,r] \overset{\text{hom}}{=} l[x,p_3]$, from which it follows that $l[x,r] \overset{\text{hom}}{=} l[x,p_1]$. We state this result as a theorem:

Theorem 3.17 *Let U be a D-set and $l[a,p] \subset U$ any light ray segment in it. Then any closed subsegment of $l[a,p]$ is homeomorphic with $l[a,p]$.*

Using l-polygons that lie entirely within U, Theorem 3.17 may be combined with Theorem 3.16 to yield the following result:

Theorem 3.18 *In a D-set, any two closed light-ray segments are homeomorphic to each other.*

It is natural to ask if it is possible to extend these results to all of M. Clearly, the answer would be in the affirmative if it were possible to extend Theorem 3.17 to all of M. This depends on the existence of *overlapping covers*, which we shall now define.

Definition 3.19 (Overlapping cover) Let M be an ordered space, l a light ray in M and $\mathcal{U} = \{U_\alpha | \alpha \in A\}$ a cover of l by D-sets. The cover $\mathcal{U}$ is called an *overlapping cover* if, for any $x, y \in l$ such that $x <^{ll} y$, there exist points $x = x_0, x_1, \ldots, x_n = y$ on l satisfying $x_k <^{ll} x_{k+1}$ and D-sets $U_1, \ldots, U_n$ such that $l[x_k, x_{k+1}] \subset U_{k+1}$. Then $U_k \cap U_{k+1} \neq \emptyset$ for $k = 0, \ldots, n-1$.

Although the definition is long, it is essentially simple, as illustrated in Fig. 3.8. The figure shows the ray l, the points $x_{k-2}, x_{k-1}, x_k, x_{k+1}$ on it and the D-sets U_{k-1}, U_k and U_{k+1}.

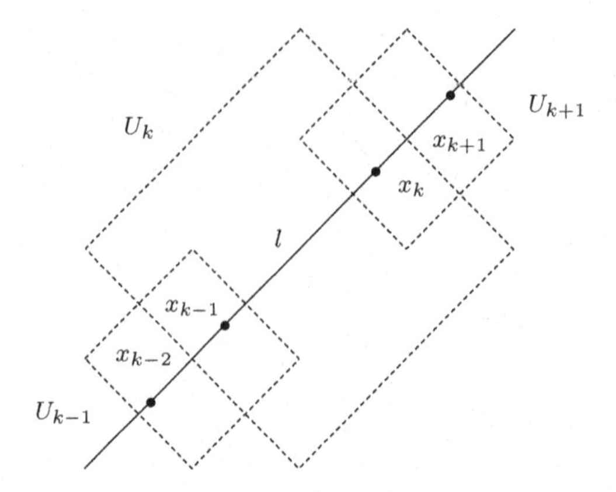

Fig. 3.8. Part of an overlapping cover

When does a light ray *not* have an overlapping cover? If l is a continuum, i.e., if it is locally homeomorphic with $\mathbb{R}$, then every open cover of it contains an overlapping subcover. The question arises only if l is a discontinuum, e.g. if it is locally homeomorphic with $\mathbb{Q}$. In that case it may happen that two ordered spaces are joined together not by a point – which would contradict the local structure axiom – but by a gap! In that case light rays could pass from one to the other, but such light rays would not have overlapping covers. An example of this kind may be found in (Borchers and Sen, 2006).

We shall regard such examples as pathological, and eliminate them by the following assumption:

Assumption 3.20 (Overlapping cover assumption) Henceforth, every light ray will be assumed to have an overlapping cover.

The following result is a direct consequence of Assumption 3.20. The proof may be found in (Borchers and Sen, 2006).

Theorem 3.21 (First homogeneity property) *Let M be an ordered space in which every light ray has an overlapping cover. Let U_1, U_2 be D-sets and $l^1[a,b] \subset U_1$, $l^2[p,q] \subset U_2$ be closed light ray segments such that $a <^{ll} b$ and $p <^{ll} q$. Then*

$$l^1[a,b] \overset{\text{hom}}{=} l^2[p,q].$$

The following result, like the one above, is also proven step-by-step, but the proof of the very first step (proving that it holds in a D-set) is complicated. We shall content ourselves with the remark that the proof does not require the introduction of artificial coordinate systems, and refer the interested reader to (Borchers and Sen, 2006).

Theorem 3.22 (Second homogeneity property) *Let M be an ordered space in which every light ray has an overlapping cover. Let $U, V \subset M$ be D-sets, $a, b \in U$, $a \ll b$ and $p, q \in V$, $p \ll q$. Then*

$$S(a,b) \overset{\text{hom}}{=} S(p,q).$$

By now, it would undoubtedly have occurred to the reader to wonder whether or not any two D-intervals in M are homeomorphic to each other. The answer is not known; it has not yet been found possible to address this question by coordinate-free methods. We shall see later that the answer is in the affirmative in order-complete spaces, which are the subject of the chapter that follows, but the proof that has been found requires the use of a coordinate system.

4
Completion of ordered spaces

The ordered spaces M that we have defined are not complete, i.e, they are not necessarily complete. Strictly speaking, we cannot yet speak of the completeness of ordered spaces. The only mathematical structure defined so far on an ordered space is the topological structure, and completeness is not a topological notion.[1]

Let us therefore start by considering light rays, which, by definition, are totally ordered sets possessing the property that between any two points lies a third. This property is shared by the set of rational numbers in their natural order. The *Dedekind completion* (page 253) of such sets invokes only the order property, and results in a set that has the *least upper bound property* (page 255). Such sets are locally homeomorphic[2] with $\mathbb{R}$. One may therefore talk about light rays that are complete, meaning thereby that they are Dedekind-complete.

In the strictly technical sense of the term, no proofs are given in this chapter, with a few exceptions in Section 4.7. However, results that require complicated proofs are broken down into smaller lemmas and propositions, and bare statements are often accompanied by an explanation of what the result is driving at. The reader who is not mathematically inclined may not be able to reconstruct the proofs independently, but should be able to follow how the argument develops. Sometimes, if the result seems obvious but the proof is not, the warning 'requires proof' is added in parenthesis.

4.1 Spaces in which light rays are complete

It turns out that there are infinite-dimensional ordered metric spaces in which light rays are complete, but the space itself is not metrically complete; there are Cauchy sequences in it that do not converge. (It should be added that we know of no finite-dimensional examples of this kind; this phenomenon may be confined to infinite-dimensional spaces.) The results that are quoted below hold in ordered spaces in which light rays are complete. In this case every light ray has an overlapping cover by D-sets. The reason is as follows. From Proposition 3.12 we know that the order topology on a light ray l is the same as the subspace topology on it induced by the order topology on M. Therefore open segments in

[1] The reader may occasionally encounter the term *topological completeness*; it is defined in Section A4.1, page 277.

[2] If we assume that light rays are *countable* subsets of M, then their Dedekind completions will be globally homeomorphic with $\mathbb{R}$.

l are intersections of l with open sets in M. It is a standard result in point-set topology that a totally ordered set is Dedekind-complete iff every cover of it by open intervals has an overlapping subcover. It follows that if l is complete, then l has an overlapping cover by open sets in M, from which it follows easily that it has an overlapping D-cover.

Proposition 4.1 *Let M be an ordered space in which light rays are complete, and let x and y be two distinct points in M. Then there exists an l-polygon $P(x_0, \ldots, x_n)$ with $x = x_0$ and $y = x_n$ such that any two successive vertices x_k, x_{k-1}, $k = 0, \ldots, n-1$ lie in a D-set.*

Using this, we may prove several interesting results:

Theorem 4.2 *Let M be an ordered space in which light rays are complete. Then M is an S-space, i.e., $x \ll y \Leftrightarrow y \gg x$.*

Using similar methods, we may prove that the following results hold globally:

Theorem 4.3 *Let M be an ordered space in which light rays are complete, and let $x \in M$. Then $\tau C_x^+ = \operatorname{int} C_x^+$, and the same with order reversed.*

Note, however, that Theorem 4.3 does *not* imply that $\beta C_x = \partial C_x$; in the example shown in Fig. 2.4, light rays are homeomorphic with $\mathbb{R}$.

Theorem 4.4 *Let M be an ordered space in which light rays are complete, let $x \in M$ and $y \in \beta C_x^+$. Then there is a light ray through x which passes through y.*

Proofs of these results do not involve any significantly new ideas, and may safely be omitted. The interested reader is referred to (Borchers and Sen, 2006). The results themselves are promising enough to warrant an investigation of the notion of completeness in ordered spaces.

4.2 Metric and uniform completions

If an ordered space were to be metrizable, then of course it would admit a metric completion. Ordered spaces are not necessarily metrizable; but, being completely regular (Theorem 3.8), they are uniformizable (Theorem A4.9). Now, according to Theorem A4.10, a necessary and sufficient condition for a uniform space to be metrizable is that it have a countable base. However, we shall show, by an example, that an ordered space need not have a countable base.

For this we need a digression into set theory. A set S is called **well-ordered** if (i) it is totally ordered by an order relation $\prec$, and (ii) every nonempty subset of S has a smallest element in the ordering $\prec$. The set of nonnegative integers $\mathbb{N}$ is well-ordered in the natural order $<$, but the sets of integers $\mathbb{Z}$ and of real numbers $\mathbb{R}$ are not. The **well-ordering theorem** of set theory asserts that every set can be well-ordered, although noone has the slighest idea of how to well-order a set like $\mathbb{R}$. This theorem, announced be Zermelo in 1904, caused consternation among mathematicians. Its proof used the axiom of choice. The well-ordering theorem was eventually proven to be equivalent to the axiom of choice, and the controversy largely subsided when the latter was shown by Kelley, in 1950, to be equivalent to the Tychonoff theorem.

Let $(X, \prec)$ be a well-ordered set, and let $\alpha \in X$. Given α, the set

$$S_\alpha = \{x | x \in X \text{ and } x \prec \alpha\}$$

is called a **section of X by α**. It can be proved that

Theorem 4.5 *There exists an uncountable well-ordered set, every section of which is either finite or countable.*

For a proof, see (Munkres, 1975, p. 66). This set is called the **minimal uncountable well-ordered set**, and is denoted by S_Ω. Using S_Ω, we define a set as follows:

Definition 4.6 Let $\mathbb{L}$ be the set $S_\Omega \times [0, 1)$ in the *dictionary order* $<_\text{d}$, with the smallest element removed.[3] The set $\mathbb{L}$ with the order $<_\text{d}$ is called the **long line**.

If $\alpha \in S_\Omega$ and $u \in [0, 1)$, the coordinates of a point on $\mathbb{L}$ are (α, u); the point $(o, 0)$ is excluded, where o is the first member of S_Ω.

Traditionally, light rays in two-dimensional Minkowski space $\mathbb{M}^2$ are parallel to the lines $x + y = 0$ and $x - y = 0$. If one rotates the coordinate axes by $\pi/4$ while keeping the light rays fixed, the light rays become the lines $x = \text{const}$ and $y = \text{const}$. By analogy, we define the two-dimensional *big Minkowski space* as the product $\mathbb{L} \times \mathbb{L}$ in which light rays are the long lines $\xi = \text{const}$ and $\eta = \text{const}$, where (ξ, η) is a point of $\mathbb{L} \times \mathbb{L}$, with $\xi = (\alpha, y)$ and $\eta = (\beta, y)$. This makes $\mathbb{L} \times \mathbb{L}$ into an ordered space. This space does not have a countable base, and therefore is not metrizable.

4.2.1 Uniformizability of ordered spaces

Such examples may seem to be of little physical relevance, but we do not wish to exclude them yet. That is, we are consciously barring the path to completions via metrization. This option is available to us because completions may also be defined via *uniformities*.[4] Unlike metrics – which are defined with the help of real numbers – uniformities do not need any external concept, and in this sense are more in the spirit of our endeavour.

[3] **Dictionary order** $<_\text{d}$ on $S_\Omega \times [0, 1)$ is defined as follows. Let $\prec$ be the order on S_Ω, $\alpha, \beta \in S_\Omega$ and $x, y \in [0, 1)$. Then $(\alpha, x), (\beta, y) \in S_\Omega \times [0, 1)$. One defines $(\alpha, x) <_\text{d} (\beta, y)$ iff either $\alpha \prec \beta$ or $\alpha = \beta$ and $x < y$. It is easily seen that $<_\text{d}$ is a nonreflexive total order.

[4] Uniformities are discussed in sufficient detail for our purposes in Sections A4.2 and A4.3.

A uniform structure on a point-set X is stronger than a topological structure, but weaker than a metric structure, in the following sense. A uniform structure on X defines a unique topology on it, which is called the *topology of the uniformity*, or the *uniform topology*. However, different uniform structures on X may define the same topology on it. In precisely the same way, a metric structure on X is stronger than a uniform structure on it. A metric on X defines a unique uniformity on it which is called the *uniformity of the metric*, or the *metric uniformity*, but different metrics on X may induce the same uniformity on it. And, just as a topological space is called metrizable if it admits a metric which induces the given topology on it, it is called *uniformizable* if it admits a uniformity which induces the given topology on it. Uniformizability, like metrizability, is a topological concept.

The question we should ask is the following: under what conditions is the order topology uniformizable? The answer is provided by Theorem A4.9: *A topological space is uniformizable if and only if it is completely regular.* As we have seen, ordered spaces are not only completely regular; they are Tychonoff (Theorem 3.8 and Corollary 3.9). We therefore conclude that:

Theorem 4.7 *Every ordered space is uniformizable.*

However, we should also pause to enquire *why* we have asked the question answered by Theorem 4.7. The property we are interested in is completeness, because complete spaces (like $\mathbb{R}^n$) admit differentiable structures in the ordinary sense, which incomplete spaces (like $\mathbb{Q}^n$) do not. Uniformizability, for us, is not an end in itself but only a means to the end of completeness.

One could visualize the situation as follows. Incomplete spaces like $\mathbb{Q}^n$ contain gaps which prevent limiting processes essential to the differentiable structure from being carried out; the process of completion fills these gaps. In incomplete metric spaces these gaps prevent some Cauchy sequences from converging. The elegant solution to this problem is to define a new space *consisting of Cauchy sequences in the old space.*[5] The metric here is merely the enabling device which makes it possible to define Cauchy sequences. In spaces which are not metrizable, one would need an alternative to the theory of convergence in metric spaces before one can try to define a completion process.

Two major alternatives – essentially equivalent – have been developed by mathematicians: nets and filters. The concept of a net involves an auxiliary notion of a 'direction' which is external to set theory (and topology); a filter, by contrast, is defined by using only the concepts of sets and subsets (Section A3.8). To the purist, this may be seen as an advantage, but it is probably correct to say that the concept of nets is closer to the intuition of those who are used to

[5] Strictly speaking, the new space consists of *equivalemce classes* of Cauchy sequences in the old; since convergence properties are determined entirely by the tail, two Cauchy sequences are defined to be equivalent iff they share the same tail.

sequences. Despite this, we have chosen the filter theory of convergence; the notion of a direction bears a certain resemblence to the notion of order, which may be a source of confusion.

A **filter** on a set X is a family $\mathcal{F}$ of nonempty subsets of X such that if $F_1, F_2 \in \mathcal{F}$, then $F_1 \cap F_2 \in \mathcal{F}$, and $F \in \mathcal{F}$, $F' \supset F$ implies that $F' \in \mathcal{F}$. Appendix A4, devoted to metric and uniform completions, also covers the essentials of the theory of convergence in the language of filters. Cauchy filters in uniform spaces are defined (Definition A4.16), a complete uniform space is defined as one in which every Cauchy filter converges (Definition A4.18), and the completion of a uniform space is discussed. The main result is Theorem A4.20, which may be stated concisely as follows:

Theorem 4.8 *Every uniform space is densely and uniformly embedded in a complete uniform space, which is called its* uniform completion. *The completion is Hausdorff if the original space is Hausdorff.*

The *topology* of a first-countable space (page 274) may be described in terms of convergent sequences. It is possible to define Cauchy sequences in a uniform space (Definition A4.13) without involving the notion of real numbers. Unsurprisingly, every Cauchy sequence converges in a complete uniform space; one says that every uniformly complete space is *sequentially complete* (Theorem A4.22). Later we shall assume first countability, which will allow us to exploit sequential completeness to obtain the required topological results – and will render these results more transparent.

4.2.2 Complete uniformizability

We now come to a concept of considerable theoretical interest. A topological space is called *completely uniformizable* if there exists a uniformity in which it is complete, and if the topology of that uniformity is identical with the original topology of the space. We shall give a few examples to clarify this concept.

Examples 4.9

(i) Consider the space $\mathbb{Q}$ in the discrete topology (in which every subset is open). This topology can be induced by the metric

$$d(x,y) = \begin{cases} 1, & \text{if} \quad x \neq y, \\ 0, & \text{if} \quad x = y. \end{cases}$$

This metric also induces a uniformity on $\mathbb{Q}$, which is the discrete uniformity. With these structures, the space $\mathbb{Q}$ is both metrically and uniformly complete, as the only Cauchy sequences are those which are ultimately constant. $\mathbb{Q}$, with the discrete topology, is completely uniformizable.

(ii) Consider $\mathbb{Q}$ again, but this time with its usual topology, which is induced by the metric $d(x, y) = |x - y|$. This space is uniformizable, but not completely uniformizable. The metric completion of $\mathbb{Q}$ is $\mathbb{R}$, as is its uniform completion.

(iii) The space $\mathbb{R}$, with its usual topology, is completely uniformizable. The open interval $(0, 1)$, which is homeomorphic with $\mathbb{R}$, is uniformizable but not completely uniformizable; its completion (both metric and uniform) is the closed interval $[0, 1]$, which is completely uniformizable.

(iv) Exactly the same is true for the n-dimensional analogues of the above spaces.

The question of *when a topological space is completely uniformizable* turns out to be more subtle than it looks. It has been discussed quite adequately, for our purposes, in Appendix A4, and the result expressed as Shirota's theorem (page 286). In view of this theorem, we may assert that the uniform completion of an ordered space is a closed subspace of some $\mathbb{R}^J$. Ignoring trivial examples like the $\mathbb{Q}^J$ (and other pathlogical ones that will surely be concocted), we may modify the earlier statement to read that completions of ordered spaces will be products of $\mathbb{R}^J$, or closed connected subsets of $\mathbb{R}^J$ for some J.

This is where the real line enters, unavoidably, into the picture.

In Part I of this book we shall use the term **continuum** to denote a nonempty open subset of $\mathbb{R}^J$ for any $J \neq \emptyset$.

4.3 The concept of order completion

Having discussed the theoretical situation at length, we now come to practical matters; the pitfalls to be avoided, and how to avoid them. The following examples illustrate both the wanted and the unwanted features of uniform completion (which may at times be the same as metric completion). Recall that we denote n-dimensional Minkowski space by $\mathbb{M}^n$.

Examples 4.10

(i) Let M be the punctured plane $\mathbb{M}^2 \setminus \{O\}$, where O is the origin. M is an ordered space according to Definition 2.44; excision of a point from $\mathbb{M}^2$ cuts each of the two light rays through it into two, but this cannot be regarded as a failing. Completion of this space restores the origin, which has the effect of joining two pairs of distinct light rays in $\mathbb{M}^2 \setminus \{O\}$ into two continua. We would want the completion process to 'complete' existing light rays, and possibly to add new light rays, but not to join two distinct light rays in M into one.

(ii) Let $M = \mathbb{M}^n \setminus \bar{B}$, where $\bar{B}$ is a closed ball of finite radius 'somewhere' in $\mathbb{M}^n$. Again, M is an ordered space. Its completion is the space $\mathbb{M}^2 \setminus B$, where B is $\bar{B}$ with its boundary removed; it is an *open* ball. The completion process adds the boundary $\partial\bar{B}$ of the hole, which become end-points of light rays in $\mathbb{M}^2 \setminus B$. The completed space violates the order axiom.

(iii) Let $M = \mathbb{Q}^2$, with the light rays being the lines $x + y =$ const and $x - y =$ const, where $x, y \in \mathbb{Q}$. Its completion is the space $\mathbb{R}^2$. To make it into an ordered space, we have to define the light rays. Some light rays define themselves, so to speak; these are the Dedekind completions of the light rays in $\mathbb{Q}^2$. However, the completion process introduces new points (the irrationals) on the X-axis, and wholly new light rays have to be defined through them. Finally, one has to show that the completed space with these new light rays is indeed an ordered space. The solution is obvious in the present example, but requires considerable effort in the general case.

(iv) Let $a, b \in$ ordered $\mathbb{Q}^2$, $a \ll b$ and let M be the open order interval $I_{\mathbb{Q}}(a, b) \subset \mathbb{Q}^2$. Then M is an ordered space. Its completion is the closed order interval $I_{\mathbb{M}}[a, b]$ in the Minkowski plane. It would be an ordered space but for the boundary. Excising the boundary, one obtains the open order interval $I_{\mathbb{M}}(a, b)$ in the Minkowski plane, which is indeed an ordered space.

In examples (i) and (ii) above, the spaces M are already complete enough, and the new points added by the completion process are merely an embarassment; in case (i), they force a change in the order structure; in case (ii), they destroy the order structure. Example (iii) illustrates what the completion process can and cannot do; the completed space is a continuum, but is no longer an ordered space; for example, the irrational points on the X-axis introduced by the completion process have no light rays defined through them. Example (iv) provides a pointer to how to proceed. Suppose that $I_{\mathbb{Q}}(a, b)$ is a D-interval in $\mathbb{Q}^2$. Its completion $I_{\mathbb{M}}[a, b]$ is a closed order interval in $\mathbb{M}^2$, but its interior $I_{\mathbb{M}}(a, b)$ is a D-interval in $\mathbb{M}^2$.

Let us temporarily denote by X^{u} the uniform completion of X in the uniformity under consideration. X is assumed Hausdorff, and therefore so is X^{u}. Let U be an open subset of X. Then U^{u} is *closed* in X^{u}; the analogue of $U \in M$ will not be $U^{\mathsf{u}} \in X^{\mathsf{u}}$, but rather its interior $\operatorname{int} U^{\mathsf{u}} \subset X^{\mathsf{u}}$.

What we really want is the following conjecture to be true:

Conjecture 4.11 Let M be an ordered space. There exists an ordered space $\check{M}$ that satisfies the following conditions:

(i) $\check{M}$ is locally a continuum.

(ii) It is possible to extend the order on M to $\check{M}$.

(iii) There exists an embedding $\iota : M \to \check{M}$ that is order-preserving (see Definition 2.45).

If such a space $\check{M}$ exists and is unique, then it may be called the *order completion* of M.

4.3.1 The order uniformity on D-sets

We shall establish the existence of the order completion $\check{M}$ of M, subject to a condition that will be stated a little later, by a constructive procedure. We begin with the definition of *the order uniformity on D-sets.*

Let $U \subset M$ be a D-set in the ordered space M, and denote by D^α a D-interval that is a subset of U. Let $\mathcal{B}_A = \{D^\alpha | \alpha \in A\}$ be a base for a topology on U. Such bases clearly exist, and the topology they define on U is clearly the order topology on U, which coincides with the subspace topology on U inherited from the order topology on M. We shall call $\mathcal{B}_A$ a D-base for the topology of U.

Lemma 4.12 *Let $\{\mathcal{B}_A | A \in \mathcal{A}\}$ be the family of D-bases on U. For each $A \in \mathcal{A}$, define*

$$E^A = \bigcup_{\alpha \in A} D^\alpha \times D^\alpha.$$

Then[6] *the family $\{E^A | A \in \mathcal{A}\}$ is a base for a filter $\mathcal{E}_U$ on $U \times U$.*

Theorem 4.13 *The filter $\mathcal{E}_U$ of Lemma* 4.12 *defines a uniformity on U.*

The uniformity defined by $\mathcal{E}_U$ will be called the *order uniformity* on U, and also denoted by $\mathcal{E}_U$. We have:

Theorem 4.14 *The order uniformity $\mathcal{E}_U$ on U is Hausdorff.*

Finally, we have:

Theorem 4.15 *The topology of the order uniformity $\mathcal{E}_U$ on U is the same as the order topology on U.*

Recall that a complete subspace of a Hausdorff uniform space is closed (Theorem A4.24). It follows that the completion of a D-interval in U is a *closed* subset of the completion of U.

We now make the following fundamental assumption:

Assumption 4.16 Let M be an ordered space and U a D-set in M. Then the uniform subspace $(U, \mathcal{E}_U)$ is totally bounded (see pages 279 and 286) in the order uniformity $\mathcal{E}_U$ on U.

We shall state Assumption 4.16 succinctly as follows: *The ordered space M is locally precompact.*

The significance of this assumption is as follows. Although we know that M is uniformizable, we do not know whether or not it admits a unique uniformization. The order uniformity on a D-set is, by definition, unique. If M is locally

[6] The proof of this lemma is not trivial. The interested reader is referred to (Borchers and Sen, 2006).

precompact, then the completion of any D-subset U of M in its order uniformity is compact and Hausdorff (in the uniform topology of the completion). On a compact Hausdorff space, there is only one uniformity that is compatible with its topology. Therefore this uniformity is bound to be the order uniformity on the completion of U. If U_1 and U_2 are two D-sets in M, then so is $U_1 \cap U_2$, and the completion of this intersection has a unique uniformity which is necessarily the subspace uniformity induced by the completions of either U_1 or U_2. This fact makes it possible to extend the order on M to its completion D-set by D-set, the results remaining valid independently of the uniqueness, or otherwise, of the uniformity on M.

Before we can proceed, we have to set up a whole array of notations and definitions. In the following, and in Section 4.4, D will always denote a D-set in M:

Notations and definitions 4.17

(i) The uniform completion of $(D, \mathcal{E}_D)$ will be denoted by $\widetilde{D}$. The topology of $\widetilde{D}$ will be the topology of its uniformity, which will be the only topology which we shall ever consider on $\widetilde{D}$.

(ii) The interior of $\widetilde{D}$ will be denoted by $\check{D}$:

$$\check{D} = \operatorname{int} \widetilde{D}. \tag{4.1}$$

(iii) Let $\{D^\alpha | \alpha \in A\}$ be a D-cover of M. The space $\check{M}$ will be defined as follows:

$$\check{M} = \bigcup_{\alpha \in A} \check{D}^\alpha. \tag{4.2}$$

The reader is invited to verify that the family of sets $\{\check{D}^\alpha | \alpha \in A\}$ is a base for a topology on $\check{M}$. This is the only topology we shall ever consider on $\check{M}$. The sets $\check{D}$ will be called $\check{D}$-sets in $\check{M}$. For the moment the terminology is only suggestive; it will be justified later.

(iv) Finally, define

$$\widetilde{M} = \bigcup_{\alpha \in A} \widetilde{D}^\alpha. \tag{4.3}$$

Define a topology $\mathcal{T}$ on $\widetilde{M}$ by taking finite unions and arbitrary intersections of the family $\{\widetilde{D}^\alpha | \alpha \in A\}$ as its closed sets. The sets $\{\widetilde{D}^\alpha\}$ cover $\widetilde{M}$, and therefore $\mathcal{T}$ is indeed a topology on $\widetilde{M}$. Clearly $\check{M}$ is a subspace of $\widetilde{M}$, and the topology of $\check{M}$ which was defined above is the subspace topology it inherits from $\mathcal{T}$. The sets $\check{D}^\alpha$ are open in both $\widetilde{M}$ and $\check{M}$. The only difference between the two spaces is that

$$\check{M} = \widetilde{M} \setminus \partial \widetilde{M}.$$

We are now in a position to define the concept of order completion.

Definition 4.18 Let M be an ordered space which is locally precompact in its order uniformity. The *order completion* of M will be defined to be the space $\check{M}$ of (4.2), with the topology which has the family of $\check{D}$-sets defined by (4.1) as a base.

4.4 Extension of order: notations and definitions

The term *order completion* defined above is suggestive, but remains to be justified. The justification will consist of showing that the order on M can be extended to $\check{M}$. That is, $\check{M}$ can be made into an ordered space in such a way that M becomes an ordered subspace of $\check{M}$. This is a long process, and it is important to devise a set of symbols and terms that are suggestive, not misleading, and can be remembered from one page to the next by the reader who is not interested in the details of the many long and elaborate proofs.

We begin by extending some of the notations and definitions introduced above to larger classes of subsets of $\check{M}$.

More notations and definitions 4.19

(i) Let $B \subset D$. We shall denote the uniform completion of B in $(D, \mathcal{E}_D)$ by $\widetilde{B}$. Then, by Theorem A4.24, $\widetilde{B}$ is a closed subset of $\widetilde{D}$. It is therefore a closed subset of $\widetilde{M}$.

(ii) Let A be a closed subset of D. We define $\check{A} = \widetilde{A} \cap \check{D}$. That is, if A is closed in M, then $\check{A}$ is closed in $\check{M}$.

(iii) Let U be an open subset of D. We define $\check{U} = \operatorname{int}(\widetilde{U} \cap \check{D})$. That is, if U is open in D, then $\check{U}$ is open in $\check{D}$ (and therefore in $\check{M}$). Note that $\check{U}$ is also the order completion of the ordered space U.

When we have extended the order of M to $\check{M}$, we shall find that the above notations are consistent with our earlier ones.

We now have an ordered space M densely embedded in a space $\check{M}$ which is not yet ordered. Until that is done, we shall find it very convenient to distinguish between points of M and those of $\check{M} \setminus M$ – and to know when we should not make this distinction. We therefore adopt the following notations.

Still more notations 4.20

(i) Points in M will be denoted by lower-case Latin letters, *excepting* the letters x, y, z.

(ii) Lower case Greek letters ξ, η, ζ will denote points in $\check{M}$ which are *not* in M.

(iii) The letters x, y, z will represent points in $\check{M}$ which *may or may not be* in M.

Definition 4.21 (Light rays, local cones and D-intervals)

(a) **Light rays** The extension of order from M to $\check{M}$ will be carried out, as announced earlier, 'D-set by D-set'. There is one important exception to this, and it applies to light rays $l \in M$. As they are totally ordered sets that have the density property, any completion process on them is equivalent to Dedekind completion, and can therefore be carried out globally. We shall denote the completion of the light ray l in M by $\tilde{l}$, and by $\check{l}$ the intersection $\tilde{l} \cap \check{M}$. Taking this intersection excises the boundary points, if any, that were introduced by the completion process. $\check{l}$ will be called the *order completion* of l. The notations ${}^{l}\!>$, $<^{l}$ and $<^{ll}$ will apply to $\check{l}$ as well.

(b) **Local cones through points** $a \in M \subset \check{M}$ Let $a \in D$ and $C_a^{\pm}$ the forward and backward cones at a. We define

$$C^{\pm}_{a;D} = C^{\pm}_{a} \cap D, \tag{4.4}$$

and call them *local cones* at a in M. By Theorem 3.6 the relations

$$\beta C^{\pm}_{a;D} = \partial C^{\pm}_{a;D} \cap D \tag{4.5}$$

always hold for local cones. Define successively

$$C^{\pm}_{a;\check{D}} = C^{\pm}_{a;D} \text{ considered as a subset of } \check{D};$$

$$\widetilde{C}^{\pm}_{a;\check{D}} = \text{the uniform completion of } C^{\pm}_{a;\check{D}};$$

$$\check{C}^{\pm}_{a;\check{D}} = \widetilde{C}^{\pm}_{a;\check{D}} \cap \check{D}; \tag{4.6}$$

$$\tau\check{C}^{\pm}_{a;\check{D}} = \operatorname{int} \check{C}^{\pm}_{a;\check{D}}. \tag{4.7}$$

Equation (4.6) deserves a remark. The uniform completion $\widetilde{C}^{\pm}_{a;\check{D}}$ of the local cone $C^{\pm}_{a;\check{D}}$ not only completes the cone mantle; it also introduces a 'base', which is part of the topological boundary $\partial\widetilde{C}^{\pm}_{a;\check{D}}$. Intersection with $\check{D}$ excises this 'base' but keeps the completed mantle (which lies in $\check{D}$). This is precisely what we want.

(c) **Local cones through points** $\eta \in \check{M} \setminus M$ Let η be a new point introduced by the order completion. Remarkably, the same kind of limiting process that leads to the definition of η allows one to define local cones at η *before light rays through* η *are defined.* One needs the order structure of M, the topology of $\check{M}$ and the completeness of $\check{D}$. These definitions are given below.

Let $\eta \in \check{D} \setminus D$. Then there exists a filter base of closed order intervals $I[r_\alpha, s_\alpha]$ in D that converges to $\eta \in \check{M}$, and we define the forward local cone $\check{C}^+_{\eta;D}$ at η as

$$\check{C}^+_{\eta;\check{D}} = \left(\bigcap_{r_\alpha} \widetilde{C}^+_{r_\alpha;\check{D}}\right) \cap \check{D}. \tag{4.8}$$

The backward local cone at η will then be defined as

$$\check{C}^-_{\eta;\check{D}} = \left(\bigcap_{s_\alpha} \widetilde{C}^-_{s_\alpha;\check{D}}\right) \cap \check{D}. \tag{4.9}$$

(d) **D-intervals in $\check{M}$** If $x, y \in \check{C}^+_{a,\check{D}}$, we define

$$\begin{aligned} \check{I}[x,y] &= \check{C}^+_{x;\check{D}} \cap \check{C}^-_{y;\check{D}}, \\ \check{I}(x,y) &= \widetilde{C}^+_{x;\check{D}} \cap \widetilde{C}^-_{y;\check{D}}, \end{aligned} \tag{4.10}$$

and call them $\check{D}$-intervals. The terminology will be justified later. Recall that x and y may be any two points in $\check{M}$ (see Notations 4.20).

The definitions of local cones and $\check{D}$-intervals given above is purely topological. For the moment, the cone mantle is available to us only locally, and then too as part of its topological boundary. It is desirable to define a *mantle operator* which is the local analogue of the β-boundary operator in M. This is accomplished as follows:

Definition 4.22 (Mantle operator $\check{\partial}$)

$$\check{\partial}\check{C}^{\pm}_{x;\check{D}} = (\partial\check{C}^{\pm}_{x;\check{D}}) \cap \check{D}, \tag{4.11}$$

where ∂ is the standard topological boundary operator.

Our last definition in this section is that of spacelike hyperspheres in $\check{D}$, which is similar to the definition in $D \subset M$ (see (3.3)):

Definition 4.23 For $x, y \in \check{C}^+_{a;\check{D}}$, we define

$$\check{S}(x,y) = \check{\partial}\check{C}^+_{x;\check{D}} \cap \check{\partial}\check{C}^-_{y;\check{D}}. \tag{4.12}$$

4.5 Extension of order: topological results

In this section we shall state, mostly without proof, some basic relations among the objects defined in Section 4.4, using the completeness of $\widetilde{M}$ and the order structure and topology of M. These results have been established under the following restrictive assumption:[7]

Assumption 4.24 From now on, all ordered spaces M will be assumed to satisfy the first axiom of countability (page 274).

4.5.1 Symmetry properties

We begin by extending Definition 2.9 of the order $<$ (or $>$) to local cones in $\check{M}$:

Definition 4.25

$$\begin{aligned} x < y \quad &\text{iff} \quad y \in \check{C}^{+}_{x;\check{D}}; \\ x > y \quad &\text{iff} \quad y \in \check{C}^{-}_{x;\check{D}}. \end{aligned} \tag{4.13}$$

In M, the equivalence $y \in C_x^+ \Leftrightarrow x \in C_y^-$ followed from the fact that an ascending l-polygon from x to y is equally a descending l-polygon from y to x. l-polygons are not yet available to us in $\check{D}$. However, the result – which in M is order-theoretic but with profound topological consequences – can be established here by purely topological means. The proof is based on the following lemma:

Lemma 4.26 *Let $\xi \in \check{D}$. Then there exist convergent sequences $\{a_n\}, a_n \ll a_{n+1}$ and $\{b_n\}, b_n \gg b_{n+1}$ in D that converge to ξ from below and from above respectively.*

Using the above lemma, we prove the desired result in two stages:

Lemma 4.27 *Let $x, y \in \check{D}$. Then:*

(1) *If $y \in \tau\check{C}^{+}_{x;\check{D}}$, then $x \in \tau\check{C}^{-}_{y;\check{D}}$, and the same with order reversed.*
(2) *If $y \in \eth\check{C}^{+}_{x;\check{D}}$, then $x \in \eth\check{C}^{-}_{y;\check{D}}$, and the same with order reversed.*

It follows immediately from the above that

Corollary 4.28

$$y \notin \check{C}_{x;\check{D}} \Leftrightarrow x \notin \check{C}_{y;\check{D}}.$$

Notations and terminology 4.29 The above results extend the notations $x \ll y, x \not\ll y, y \gg x$ and $y \not\gg x$ to $\check{D}$-sets, and are equivalent. We shall also express Corollary 4.28 in words as 'x and y are spacelike to each other'. As before, the terminology will be fully justified only when the order has been extended to $\check{M}$.

[7] Mathematically, the assumption of first countability may not be necessary; all results established using Cauchy sequences appear to be provable using Cauchy filters. However, it is not clear, at least at the time of writing, what physical significance such a generalization may have.

4.5.2 Separation theorems

The sets $\check{D}$ enjoy some fairly strong separation properties which we shall express in forms that will be most useful to us: Theorems 4.32, 4.33 and 4.34. The proofs of these theorems proceed via the two inclusion lemmas that are stated below.

Lemma 4.30 *Let D be a D-set in M and r, s, t, u points in D such that $r \ll s \ll t \ll u$. Then*

$$\check{I}[s,t] \subset \check{I}(r,u).$$

The important point is that $\check{I}[s,t]$ is a closed set.

Let $S(a,b) \subset D$. We consider the space $S(a,b)$ in its relative uniformity and topology. Let O be an open set in $S(a,b)$, and denote its closure in $S(a,b)$ by $\overline{O}$. Denote by $\widetilde{O}$ the uniform completion of $O \subset S(a,b)$ (note that $\overline{O} \neq \widetilde{O}$). Then $\widetilde{O}$ is closed in $\widetilde{S}(a,b)$. Denote the interior of $\widetilde{O}$ by $\check{O}$. Clearly,

$$\check{O} \subset \widetilde{O} \subset \check{S}(a,b).$$

Lemma 4.31 *Let $S(a,b)$ be a spacelike hypersphere in $D \subset M$ and $\check{S}(a,b)$ its uniform completion. If V, W are open sets in $S(a,b)$ such that $\overline{W} \subset V$, then*

$$\widetilde{W} \subset \check{V}.$$

Theorem 4.32 *Let $U \subset M$ be a D-set, and $a, b, a', b' \in U$ such that $a \ll b$, $a' \ll b'$ and $I(a,b) \cap I(a',b') = \emptyset$. Then*

$$\check{I}(a,b) \cap \check{I}(a',b') = \emptyset.$$

Theorem 4.33 *Let U be a D-set in M and W, W' open sets in $S(a,b) \subset U$ such that $W \cap W' = \emptyset$. Then $\check{W} \cap \check{W}' = \emptyset$.*

Theorem 4.34 *Let $D \subset M$ be a D-set and p, q a pair of spacelike points in it, i.e., $C^+_{p;D} \cap C^-_{q;D} = \emptyset$. Then $\check{C}^+_{p;\check{D}} \cap \check{C}^-_{q;\check{D}} = \emptyset$.*

4.5.3 Density lemmas

Finally, we state two density lemmas:

Lemma 4.35 *Let $\xi, \zeta \in \check{D}$ such that $\xi \ll \zeta$. Then there exists $\eta \in \check{D}$ such that $\xi \ll \eta \ll \zeta$.*

Lemma 4.36 *Let $z \in \check{\partial}\check{C}^+_{x;\check{D}}$, $z \neq x$. Then there exists $y \in \check{\partial}\check{C}^+_{x;\check{D}}$, $x \neq y, y \neq z$ such that $z \in \check{\partial}\check{C}^+_{y;\check{D}}$.*

4.6 Extending the order to $\check{M}$

In order to extend the order from M to $\check{M}$, we have to define light rays in $\check{M}$ and verify that the space $\check{M}$, so equipped, satisfies the order, identification, cone and local structure axioms. Some light rays through points $a \in M \subset \check{M}$ have already been defined: the order completions $\check{l}$ of light rays l in M. However, in $\check{M}$ there are more light rays through a than the $\check{l}_a$. Let $a, b \in D$, $a \ll b$ and consider the spacelike hypersphere $\check{S}(a,b)$. Except in the trivial case $M = \check{M}$, $\check{S}(a,b)$ contains points that do not belong to $S(a,b)$. Let $\xi \in \check{S}(a,b)$ be such a point. It would be reasonable to expect that ξ is connected to a and b by light-ray segments, and this is indeed the case.

We shall use a different font to distinguish between the light rays $\check{l}$ and those that we are about to define:

Notations 4.37 ℓ will denote a new light ray in $\check{M}$, i.e., one which is *not* the order completion $\check{l}$ of any light ray l in M.

4.6.1 The segment $\ell[a, \eta]$

The segment $\ell[a,\eta]$ is defined in the following manner. Since $\check{S}(a,b)$ is the uniform completion of $S(a,b)$, there is a Cauchy sequence $\{t_n\}$, $n \in \mathbb{N}$ on $S(a,b)$ that converges to $\eta \in \check{S}(a,b) \setminus S(a,b)$. The light-ray segments $l[a,t_n]$, $n \in \mathbb{N}$ are well defined on $\beta I[a,b]$ (Fig. 4.1). The segment $\ell[a,\eta]$ is the 'limit' of the sequence

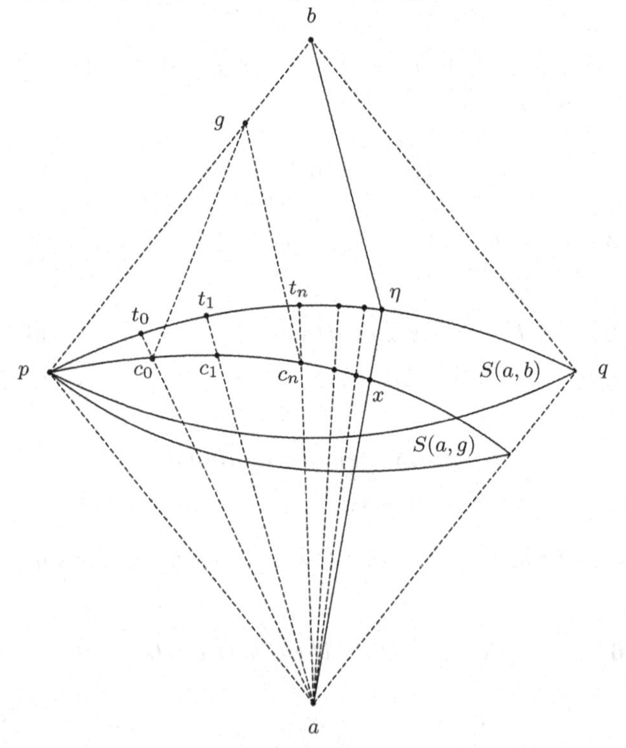

Fig. 4.1. Defining the segment $\ell[a,\eta]$

of segments $\{l[a, t_n]\}$ as $n \to \infty$. Of course, the word limit is used above only in an intuitive sense, and it has to be proved that the limit actually exists in a precise mathematical sense. This can be done, but the proof is long and involved. One first constructs the object $\ell[a, \eta]$, and then proves that it has the following properties, which justifies calling it a light-ray segment:

4.6.1.1 *Properties of the segment* $\ell[a, \eta]$

(i) $\ell[a, \eta]$ is totally ordered. (Upon examination of the construction, this turns out to be true by definition.)

(ii) $\ell[a, \eta]$ is homeomorphic to every $\check{l}[a, t_n]$, and therefore to a closed interval on the real line.

(iii) $\ell[a, \eta] \subset \breve{\partial}\check{I}[a, b]$. (This is true by construction.)

(iv) $\ell[a, \eta] \subset \breve{\partial}\check{C}^-_{\eta;\check{D}}$, where $\check{C}^-_{\eta;\check{D}}$ was defined by (4.9). (This requires proof.)

(v) Let $h \in \tau C^+_{a;D}$ such that $u_n \notin C^-_{h;D}$ for $n > N_0$ (see Fig. 4.2). Then $\breve{\partial}\check{C}^-_{h,\check{D}} \cap \ell[a, \eta]$ consists of a single point. (This requires proof.)

(vi) Let $\xi \in \tau C^+_{a;\check{D}}$ such that $u_n \notin \check{C}^-_{\xi;\check{D}}$ for some n. Then $\breve{\partial}\check{C}^-_{\xi;\check{D}} \cup \ell[a, \eta]$ consists of a single point. (This requires proof.)

Some further properties of the segment $\ell[a, \eta]$ are stated in the following.

Theorem 4.38 *Let* $\{w_n\}$, $w_n \in C^+_{a;D}, w_n \gg w_{n+1} \gg a$ *be a sequence that converges to* η *(see Fig. 4.2). Then*

$$\ell[a, \eta] = \bigcap_{n \in \mathbb{N}} \check{I}[a, w_n].$$

The following theorem shows that $\ell[a, \eta]$ does not belong to the completion of any light ray l in M, which justifies the term *new light ray*.

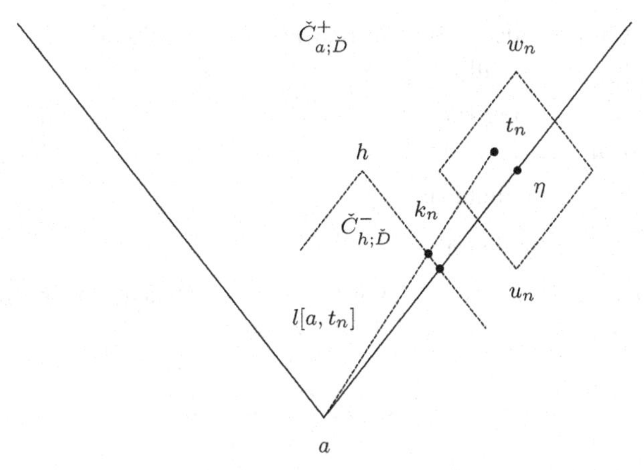

Fig. 4.2. Properties of the segment $\ell[a, \eta]$

Theorem 4.39 *Let $U \subset M$ be a D-set and $\ell[a,\eta] \subset \check{U}$. If $x \in \ell[a,\eta], x \neq a$, then $x \notin M$.*

Theorem 4.38 can also be used as an alternative definition of the segment $\ell[a,\eta]$. Subsegments of $\ell[a,\eta]$ can be defined similarly:

Proposition 4.40 *Let $\xi \in \ell(a,\eta)$ and let $\{v_n\}$, $n \in \mathbb{N}$, $v_{n+1} \ll v_n$ be a sequence of points in $C^+_{a;D}$ that converges to ξ. Then*

$$\ell[a,\xi] = \bigcap_{n\in\mathbb{N}} \check{I}[a,v_n].$$

We end this section with the following theorem:

Theorem 4.41 *Let $\eta \in \check{\partial}\check{C}^+_{a;\check{D}}$. Then*

$$\ell[a,\eta] = \check{\partial}\check{C}^+_{a;\check{D}} \cap \check{\partial}\check{C}^-_{\eta;\check{D}}.$$

4.6.2 The segment $\ell[\xi,\eta]$

The segments $\ell[\xi,\eta]$ are defined somewhat differently. Let $\xi \in \check{D}$ and $\eta \in \check{\partial}\check{C}^+_{\xi;\check{D}}$, $\xi \neq \eta$.

Definition 4.42

$$\ell[\xi,\eta] = \check{\partial}\check{C}^+_{\xi;\check{D}} \bigcap \check{\partial}\check{C}^-_{\eta;\check{D}}.$$

4.6.2.1 Properties of the segment $\ell[\xi,\eta]$

The segment $\ell[\xi,\eta]$ has the following properties:

(i) $\ell[\xi,\eta]$ is totally ordered by the relation $<$ (equivalently, by $>$). It is this fact which justifies calling $\ell[\xi,\eta]$ a light-ray segment.

(ii) $\ell[\xi,\eta]$ is closed, being the intersection of the sets $\check{\partial}\check{C}^+_{\xi;\check{D}}$ and $\check{\partial}\check{C}^-_{\eta;\check{D}}$, which are closed in $\check{D}$.

(iii) $\ell[\xi,\eta] \subset \check{\partial}\check{C}^+_{\xi;\check{D}}$, by definition.

(iv) $\ell[\xi,\eta] \subset \check{\partial}\check{C}^-_{\eta;\check{D}}$, by definition.

(v) Let $\{a_n\}, a_{n+1} \gg a_n$ and $\{b_n\}, b_{n+1} \ll b_n$ be Cauchy sequences in D that converge to ξ and η respectively in $\check{D}$. Then

$$\ell[\xi,\eta] = \bigcap_{n\in\mathbb{N}} \check{I}[a_n,b_n].$$

(vi) Let $p \in \tau\widetilde{C}^{p}_{\xi;\check{D}} \cap D$ such that $b_n \notin C^{-}_{p;D}$ for all $n \in \mathbb{N}$, where the b_n are as in (v) above. Then

$$\check{\partial}\check{C}^{+}_{p;\check{D}} \bigcap \ell[\xi,\eta]$$

is a single point. (This requires proof.)

(vii) Let $\zeta \in \tau\widetilde{C}^{+}_{\xi;\check{D}}$ such that $b_n \notin C^{-}_{\zeta;\check{D}}$ for $n \in \mathbb{N}$. Then

$$\check{\partial}\check{C}^{+}_{\zeta;\check{D}} \bigcap \ell[\xi,\eta]$$

is a single point. (This requires proof.)

4.6.3 Extending light-ray segments

Whereas the light rays $\check{l}$ are defined globally in $\check{M}$, a segment $\ell[x,y]$, where at least one of x and y do not belong to M, is defined only locally in a $\check{D}$-set. The definition of ℓ (any ℓ) has to be extended to all of $\check{M}$ by means of overlapping covers. The extension process is based on the following lemma:

Lemma 4.43 *Let $y \in \check{\partial}\check{C}^{+}_{x;\check{D}}$, $y \neq x$. Then the set $\check{\partial}\check{C}^{+}_{x;\check{D}} \cap \check{\partial}\check{C}^{+}_{y;\check{D}}$ is totally ordered by $<$.*

Proof Let $z \in \check{\partial}\check{C}^{+}_{x;\check{D}} \cap \check{\partial}\check{C}^{+}_{y;\check{D}}$. Then the light ray segments $\ell[x,z]$ and $\ell[y,z]$ are well defined, and ordered by $<$. The result follows. □

The procedure for step-by-step extension of new light-ray segments should now be fairly obvious. We shall denote the total order on ℓ by ${}^{l}>$ or $<^{l}$. The symbol $<^{ll}$ will have the same meaning on ℓ as it does on l and $\check{l}$.

This completes the definition of new light rays in $\check{M}$.

4.7 Verification of the axioms

Our last task is to verify that the space $\check{M}$ is indeed an ordered space in our sense of the term.

4.7.1 The order axiom

It is easy to see that the light rays $\check{l}$ and ℓ satisfy the order axiom 2.1. All rays $\check{l}$ and ℓ in $\check{M}$ are locally homeomorphic with $\mathbb{R}$. Parts (a) and (b) of Axiom 2.1 are basic properties of $\mathbb{R}$. Part (c) follows from the fact that order completion excises all boundary points that are introduced by uniform completion. From the definitions of the segments $\ell[a,\eta]$ and $\ell[\xi,\eta]$ it is clear that the order on these segments is consistent with the order on segments of $\check{l}$, which ensures that Axiom 2.1(d) is satisfied.

4.7.2 The identification axiom

Since the identification axiom 2.6 holds for any pair of rays l, l' in M, it holds trivially for $\check{l}, \check{l}'$ in $\check{M}$. From Theorem 4.39 it follows that two rays $\check{l}$ and ℓ cannot overlap in an open segment in any $\check{D}$-set. By construction, two distinct segments $\ell[\xi, \eta]$ and $\ell[\xi, \zeta]$ have only the point ξ in common in a $\check{D}$-set. The extension of these segments to overlapping $\check{D}$-sets uses the identification axiom as a tool. It follows that Axiom 2.6 is satisfied everywhere in $\check{M}$.

4.7.3 The cone axiom

Recall that the local cones were defined topologically in Definition 4.21, owing to which the polygon lemma was not available. Having defined the rays $\check{l}$ and ℓ, we can make use of the notion of l-polygons, and provide an order-theoretic characterization of local cones.

Lemma 4.44 *Let $y \in \check{C}^+_{x;\check{D}}$. Then*

(1) *If $y \in \eth\check{C}^+_{x;\check{D}}$, then there exists a light ray through x and y.*

(2) *If $y \in \tau\check{C}^+_{x;\check{D}}$, then there exist ascending l-polygons from x to y.*

The same assertions hold with order reversed.

We now define cones globally as follows:

$$\check{C}^+_x = \bigcup_{\text{all } P^\uparrow_x \subset \check{M}} \{y | y \in P^\uparrow_x\},$$

$$\check{C}^-_x = \bigcup_{\text{all } P^\downarrow_x \subset \check{M}} \{y | y \in P^\downarrow_x\}$$

and

$$\check{C}_x = \check{C}^+_x \bigcup \check{C}^-_x .$$

The example of the anti-de-Sitter space given earlier (Fig. 2.5), which was locally $\mathbb{M}^2$, shows that the cone axiom is independent of completeness or incompleteness. It depends, basically, on the global topology of the space.

It can be proved that if the cone axiom is satisfied in M, then it is satisfied in $\check{M}$. One assumes that the set

$$Q = \check{C}^+_x \cap \check{C}^-_x \setminus \{x\}$$

is nonempty, and derives a contradiction.

4.7.4 D-sets in $\check{M}$

We shall prove that a set $\check{D} = \operatorname{int}\widetilde{D} \subset \widetilde{M}$, where D is a D-set in M, is indeed a D-set in $\check{M}$. To do so, one has to verify that $\check{D}$ satisfies the six conditions in Definition 2.22 of D-sets. The verifications are simple, and are given below.

(i) **The l-convexity condition** $x, y \in \check{D}$, $x \ll y \Rightarrow \check{I}[x,y] \subset \check{D}$. Let U_x, U_y be open sets in $\check{M}$ such that $x \in U_x \subset \check{D}, y \in U_y \subset \check{D}$ and $U_x \cap U_y = \emptyset$. Then there exist Cauchy sequences $\{a_n\} \subset U_x, a_{n+1} \gg a_n$ and $\{b_n\} \subset U_y, b_{n+1} \ll b_n$ for all $n \in \mathbb{N}$ that converge to x and y respectively. Then $a_m \ll a_{m+1} \ll b_{n+1} \ll b_n$ for all $m, n \in \mathbb{N}$, and

$$\begin{aligned} I(x,y) &= \left(\bigcap \check{C}^{+}_{a_m;\check{D}}\right) \bigcap \left(\bigcap \check{C}^{-}_{b_n;\check{D}}\right) \\ &= \bigcap_{n\in\mathbb{N}} \left(\check{C}^{+}_{a_n;\check{D}} \bigcap \check{C}^{-}_{b_n;\check{D}}\right) \\ &= \bigcap_{n\in\mathbb{N}} \check{I}[a_n, b_n]. \end{aligned}$$

(ii) **The openness condition** Since $\check{D}$ is open, the intersection of a light ray with it cannot have end-points.

(iii) **The intersection condition** Let $x, y, z \in \check{D}$, $x \ll y \ll z$. Then $\check{l}^{+}_{y}$ (or ℓ^{+}_{y}, as the case may be) intersects $\partial\check{C}^{-}_{z;\check{D}}$ at a single point, and $\check{l}^{-}_{y}$ (or ℓ^{-}_{y}, as the case may be) intersects $\partial\check{C}^{+}_{x;\check{D}}$ at a single point. (See Subsections 4.6.1.1 and 4.6.2.1.)

(iv) **The convexity condition** The convexity condition clearly remains true for the rays $\check{l}$. New rays ℓ are defined segmentwise by intersections

$$\ell[x,y] = \breve{\partial}\check{C}^{+}_{x;\check{D}} \bigcap \breve{\partial}\check{C}^{-}_{y;\check{D}},$$

and therefore the ray ℓ will also satisfy the convexity axiom on $\breve{\partial}\check{C}^{\pm}_{z;\check{D}}$ for any $z \in \ell \cap \check{D}$.

(v) **The uniqueness condition** This is trivially true for the rays $\check{l}$. For the rays ℓ, it is true by construction: if $y \in \breve{\partial}\check{C}^{+}_{x;\check{D}}$ or $x \in \breve{\partial}\check{C}^{-}_{y;\check{D}}$, then there is only one ℓ through x and y.

(vi) **The dimension condition** The cardinality of the set of light rays through $a \in \check{M}$ cannot be smaller than that of the set of light rays through $a \in M$.

This concludes our verification that $\check{D}$ is indeed a D-set in $\check{M}$, and justifies the notation (and terminology, for those who still remember where the D of a D-set comes from).

4.7.5 The local structure axiom

We have just established that every $\check{D}$-interval is a D-set in $\check{M}$. Every point ξ introduced by order completion is contained in a nonempty $\check{D}$-interval $\check{I}(a,b)$. Observe that the family of $\check{D}$-intervals

$$\{\check{I}(a,b)|a,b \in D \subset M, a \ll b\}$$

is a base for the order topology of $\check{M}$.

This concludes our verification that $\check{M}$ satisfies the axioms that define ordered spaces. The verification also justifies the suggestive notations and terminology that we employed earlier.

5

Structures on order-complete spaces

In this chapter we shall study some mathematical structures on order-complete spaces. The mathematical developments that follow are concerned only with order-complete spaces, and therefore it will no longer be necessary to adhere to the special notations of Chapter 4. We shall revert to our original notations: M,l,$C_x^{\pm}$,D,I and S, and no special significance will be attached to any lower-case Greek or Latin letter. However, we shall continue to distinguish between ordered and order-complete spaces, and shall be explicit in stating whether M is merely ordered or order-complete.

Our first step will be to define *timelike curves*. We shall then use these to define a parametrization of order intervals. This will allow us to prove that all D-intervals are homeomorphic with each other, and with an open ball in some $\mathbb{R}^n$. It will follow immediately from the latter that D-intervals in finite-dimensional order-complete spaces are differentiable manifolds. This, in turn, will imply that ordered spaces are embedded in spaces that have locally the structure of differentiable manifolds. As experiments cannot distinguish between ordered and order-complete spaces, it becomes difficult to tell whether the differential calculus is a discovery or an invention. Of course, this presumes that the special theory of relativity is a discovery, and not an invention.

5.1 Timelike curves in M

A **curve** in a topological space X is a continuous map $\varphi : [0,1] \to X$ from the closed real interval $[0,1]$ into X. By analogy, we may define a curve in an ordered space M which is not necessarily order-complete as a continuous map $\varphi: l[a,b] \to M$, where $l[a,b]$ is any closed light-ray segment in M with $a <^{ll} b$. A timelike curve in M will then be a curve which is ordered by $\ll$. The definition of a timelike curve in an order-complete space follows from the above:

Definition 5.1 Let M be an order-complete space, $U \subset M$ a D-set and $a, b \in U$ such that $a \ll b$. A continuous map $\theta : [0,1] \to U$ such that $\theta(0) = a$ and $\theta(1) = b$ is called a *timelike curve* if $t_1 < t_2$ implies that $\theta(t_1) \ll \theta(t_2)$.

We shall denote the image of θ in M by Θ, and shall, by abuse of language, call $\Theta[a,b]$ a timelike curve joining a with b in M, where $a \ll b$.

Thus a timelike curve is, by definition, locally homeomorphic to a light ray. It turns out to be annoyingly difficult to establish the existence of timelike curves in arbitrary order-complete spaces. The reason can be understood by looking at two-dimensional Minkowski space.

Figure 5.1 shows an order interval $I[O,P]$ in two-dimensional Minkowski space; it is a rectangle. Y and X are the other two vertices of this rectangle. y, y' are two distinct points as shown on $l(O,Y)$, and the second forward rays from them (i.e., not $l_{O,Y}$) intersect $l_{X,P}$ at s and s' respectively. Similar statements apply to the points x, x', r, r' and the light-ray segments $l[x,r], l[x',r']$, as shown in the figure.

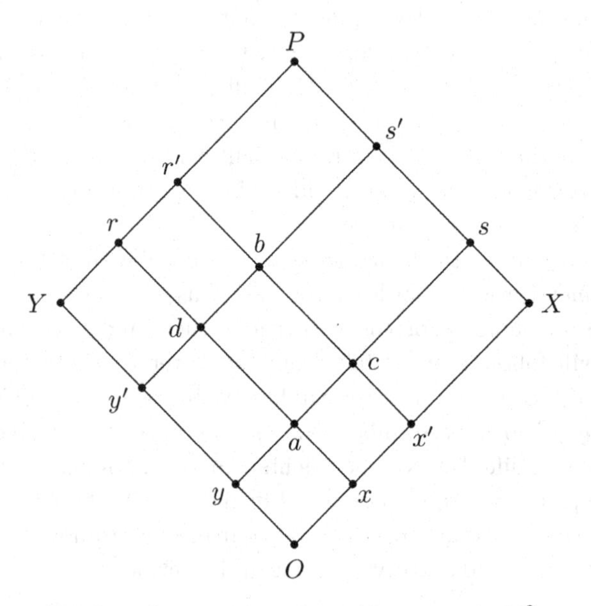

Fig. 5.1. Construction of timelike curves in $\mathbb{M}^2$

Now $x \ll P$, and therefore the backward ray $l^-_{s,y}$ from s intersects the forward ray $l^+_{x,r}$ at the point a on the figure. Then $O \ll a \ll P$. A similar consideration may be applied to the order interval $I[a,P]$. The backward rays $l^-_{s',y'}$ and $l^-_{r',x'}$ intersect each other at b, and $a \ll b \ll P$; $P^{\uparrow}(a,c,b)$ and $P^{\uparrow}(a,d,b)$ are two distinct ascending l-polygons from a to b. We have obtained two points a, b such that $O \ll a \ll b \ll P$. The whole process described above is completely trivial.

In fact, we can go further. Since light rays are locally homeomorphic with $\mathbb{R}$, we can consider O to be the origin of coordinates, take the ray $l_{O,Y}$ to be one axis and the ray $l_{O,X}$ the other. Then any point on $I[O,P]$ will be the intersection of a forward ray from $l[O,X]$ with a forward ray from $l[O,Y]$. If we identify the segments $l[O,X]$ and $l[O,Y]$ with the real interval $[0,1]$, the set of points $\{(p,p)|p \in [0,1]\}$ – the straight line joining O with P – will be a timelike curve.

The reason is that on $\mathbb{R}^2$ any forward ray $l^+_{x,r}$ is bound to intersect any forward ray $l^+_{y,s}$; this is a simple consequence of the intermediate value theorem of calculus. However, *this may no longer be true in an arbitrary order-complete space.*

Let M be an order-complete space, $D \subset M$ a D-interval, $a, b \in D$, $a \ll b$, $p, q \in S(a, b)$, $p \neq q$, $x \in l(a, q)$ and $y \in l(a, p)$. Figure 5.2 shows the forward rays $l^+_{x,r}$, $r \in l(p, b)$ and $l^+_{y,s}$, $s \in l(q, b)$ by dashed curves. The existence of these rays and of the points r, s is assured by the fact that $I[a, b]$ lies in a D-set. But *nothing assures us that the rays $l_{x,r}$ and $l_{y,s}$ intersect in D.* In fact, so far we have *not* been able to prove that they do, unless M is a Minkowski or a de Sitter space. This unsolved problem has been called the *cushion problem* in (Borchers and Sen, 2006).

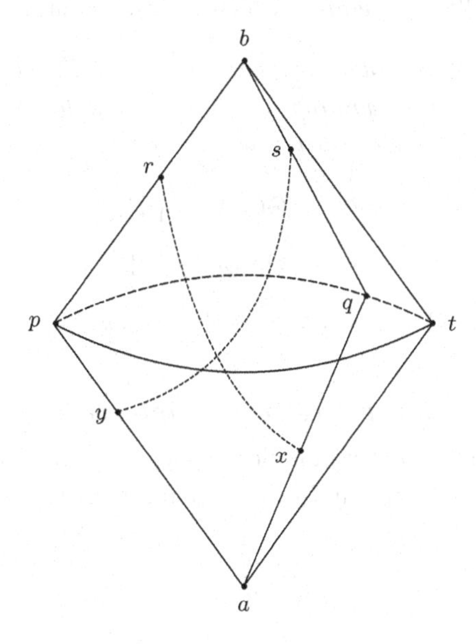

Fig. 5.2. The cushion problem

Despite the cushion problem, it is still possible to prove that timelike curves exist. The proof is by construction, and is long and involved. We shall content ourselves with quoting the final result, referring the interested reader to (Borchers and Sen, 2006) for details.

Theorem 5.2 *Let U be a D-set in an order-complete space M, $a, b \in U, a \ll b$. Then there exists a continuous timelike curve $\Theta(t)$ on M from a to b, with $a = \Theta(0)$ and $b = \Theta(1)$.*

The theorem quoted below follows from the details of the construction, which we have omitted:

Theorem 5.3 *The topology of the order $\ll$ on a timelike curve Θ is the same as the subspace topology that Θ inherits from M; i.e., Θ is locally homeomorphic with $\mathbb{R}$.*

It should be noted that the concatenation of two timelike curves is again a timelike curve. Therefore a timelike curve can be extended indefinitely, by overlapping D-covers, in both forward and backward directions. The cone axiom will ensure that the forward and backward continuations never meet; there are no closed timelike curves.

We shall use the obvious notation for open timelike curves: $\Theta(a, b)$ will denote the curve $\Theta[a, b]$ with its end-points deleted.

5.2 Parametrization of D-intervals

We begin with the following lemma, which is fundamental:

Lemma 5.4 *Let M be an order-complete space, U a D-set in it, $a, b \in U, a \ll b$ and $\Theta[a, b]$ a timelike curve joining a with b. Then, for any point $w \in I[a, b]$, there exist points $u, v \in \Theta[a, b]$ such that*

$$\begin{aligned} \beta C_w^- \cap \Theta(a, b) &= \{u\}, \\ \beta C_w^+ \cap \Theta(a, b) &= \{v\}. \end{aligned} \tag{5.1}$$

The situation is depicted in Fig. 5.3. Clearly, $u \ll v$.

5.2.1 2-cells in D-intervals

Let U be a D-set in the order-complete space M, and let $a, b \in U$, $a \ll b$. Figure 5.4 shows the D-interval $I[a, b]$, a point $p \in S(a, b)$, the ray segments

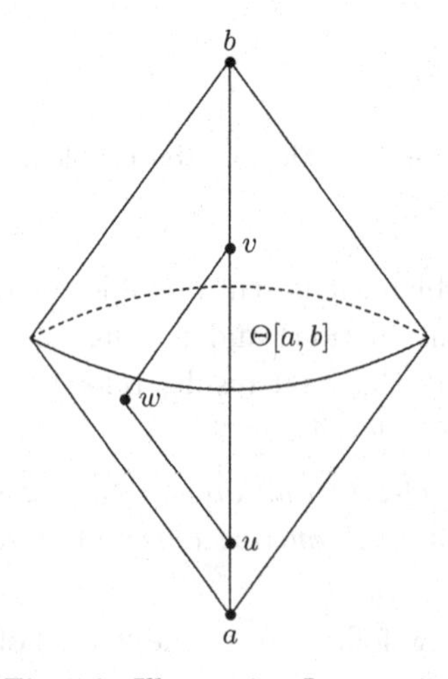

Fig. 5.3. Illustrating Lemma 5.4

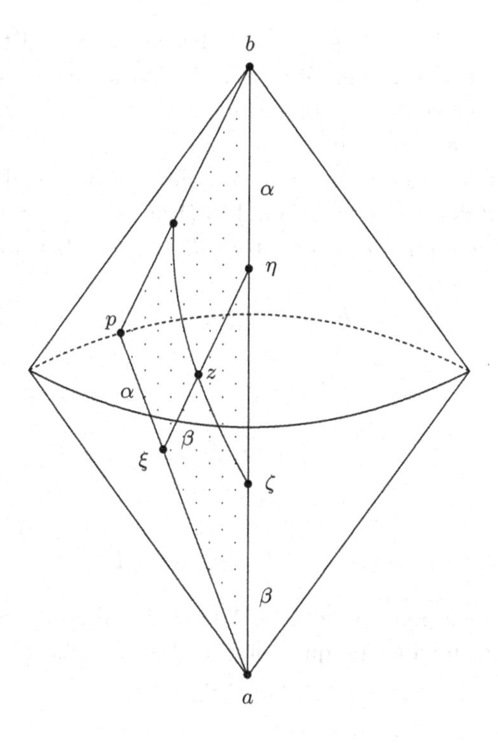

Fig. 5.4. 2-cells in D-intervals

$l[a,p]$ and $l[p,b]$ and a timelike curve $\Theta[a,b]$. Since both $l[a,p]$ and $\Theta[a,b]$ are homeomorphic with closed intervals on the real line, they are homeomorphic with each other, but there is a natural homeomorphism, defined as follows, which will be of special interest to us. Let $\xi \in l[a,p]$ (Fig. 5.4). By Lemma 5.4, the point

$$\vartheta(\xi) = \beta C_\xi^+ \cap \Theta[a,b] \tag{5.2}$$

exists and is unique. The map

$$\vartheta : l[a,p] \to \Theta[a,b] \tag{5.3}$$

so defined is bijective and order-preserving, and therefore a homeomorphism.

Let p and ξ be as above (Fig. 5.4), and set $\eta = \vartheta(\xi)$, where $\vartheta(\xi)$ is defined by (5.2). Define now

$$F[a,b;p] = \{x | x \in l[\xi,\eta], \xi \in l[a,p]\}. \tag{5.4}$$

In Fig. 5.4, the set $F[a, b; p]$ appears as a delicately shaded triangle. We shall give it the subspace topology inherited from U. With this topology, it is what is called a 2-**cell** in topology, i.e., a topological space that is homeomorphic with a nondegenerate triangle in the Euclidean plane.

The proof of this assertion is long and complicated. We shall mention only the key stages here; full details may be found in (Borchers and Sen, 2006).

We define a metric d on $\Theta[a, b]$ such that $d(a, b) = 1$. Let now

$$d(\eta, b) = \alpha, \quad d(a, \zeta) = \beta,$$

and assign coordinates to ξ as follows:

$$\xi = (\alpha, \beta). \tag{5.5}$$

Then

$$0 \leq \alpha, \beta \leq 1, \quad \alpha + \beta \leq 1, \tag{5.6}$$

so that $\alpha + \beta = 1$ for a point on $\Theta[a, b]$. A pair of real numbers (α, β) satisfying conditions (5.6) determines a unique point on $F[a, b; p]$. Let us denote by $\mathcal{C}_\alpha$ the subset of $F[a, b; p]$ that is defined as follows:

$$\mathcal{C}_\alpha = \{z = (\alpha, \beta) | \alpha \text{ fixed}, \ 0 \leq \beta \leq 1 - \alpha\}. \tag{5.7}$$

This set is clearly the ray segment $l[\xi, \eta]$ for which $d(\eta, b) = \alpha$. In the same way, we can define the subset of $F[a, b; p]$

$$\mathcal{C}_\beta = \{z = (\alpha, \beta) | \beta \text{ fixed}, \ 0 \leq \alpha \leq 1 - \beta\}. \tag{5.8}$$

Owing to the cushion problem, we cannot assume that this set is a light-ray segment. However, we can prove that:

Lemma 5.5 *The set $\mathcal{C}_\beta$ defined by* (5.8) *is a continuous curve.*

We may therefore assert that (5.5) defines a coordinate system on $F[a, b; p]$ in the usual sense. The curve $\mathcal{C}_\beta$ is the curve originating at ζ and passing through z in Fig. 5.4.

The pairs of real numbers (α, β) may also be regarded as points on a Euclidean plane. With the usual system of Cartesian coordinates, they are easily seen to define the triangle Δ with vertices $(0, 0), (0, 1)$ and $(0, 1)$, and there is a self-evident bijection Π between Δ and $F[a, b; p]$. With some further effort, one proves that

Theorem 5.6 *The bijection Π between Δ and $F[a, b; p]$ is a homeomorphism.*

Observe that this result holds for any $p \in S(a, b)$.

5.2.2 Cylindrical coordinates on D-intervals

In the notation of Lemma 5.4, if $w \in I(a,b)$, then $u, v \in \Theta(a,b)$. Clearly, $w \in S(u,v)$. It follows from the definitions of u, v and $F[a,b;p]$ that

$$S(u,v) \cap F[a,b;p] = \{w\}.$$

Conversely, given $w \in I[a,b]$, we may determine the 2-cell $F[a,b;p]$ on which it lies as follows. Let $\xi = l^-_{v,w} \cap \beta C^+_a$. Then $p = l^+_{a,\xi} \cap S(a,b)$. That is, w is uniquely determined by the triple $\{u, v, p\}$.

This triple may be cast into a form more suitable for our purposes by the introduction of cylindrical coordinates on $I[a,b]$ as follows. We first map $\Theta[a,b]$ homeomorphically onto the real interval $[-1,1]$ (instead of $[0,1]$, as earlier), so that

$$\begin{aligned} a &\mapsto -1, \\ b &\mapsto +1. \end{aligned} \tag{5.9}$$

Then every point of Θ is assigned a numerical coordinate lying between -1 and $+1$. Let now $w \in I[a,b]$ and u, v as before, i.e., as defined by (5.1). We shall use the same letters u, v to define their numerical coordinates on $\Theta[0,1]$. Then

$$-1 \le u \le v \le 1.$$

Clearly, if $w \in \Theta[a,b]$, then $u = v$. If $w = a$, then $u = v = -1$, and if $w = b$, then $u = v = 1$. We define

$$\begin{aligned} r &= \frac{v-u}{2}, \\ h &= \frac{v+u}{2}. \end{aligned} \tag{5.10}$$

Then

$$\begin{aligned} -1 \;&\le\; h \;\le\; 1, \\ 0 \;&\le\; r \;\le\; 1-|h|. \end{aligned} \tag{5.11}$$

The variables h and r will be called the *level* and *radius* respectively, as is common with cylindrical coordinates. Note that $r = 0$ means that $w \in \Theta[a,b]$. The upper bound for r, given h, follows from the fact that u is constrained by $u \le v$.

It is clear that for fixed h the sets of constant r are the corresponding hyperspheres $S(u,v)$. Therefore, to complete the parametrization of $I[a,b]$, we have to parametrize $S(u,v)$. However, we do not need to do this explicitly. From the second homogeneity property (Theorem 3.22), we know that all spacelike hyperspheres in D-sets are homeomorphic with each other. We may therefore choose a fiducial hypersphere S in a D-set and parametrize it in any way that is possible (usually this will require more than one coordinate chart). Let $\phi(x)$ be the coordinates of the point x on the fiducial hypersphere. We may parametrize the points $w \in I[a,b]$

$$w = \{h; r, \phi(w)\} \tag{5.12}$$

in such a way that $\phi(w)$ *is constant on the* 2*-cell* $F[a,b;p]$ *on which* w *lies.*

Since every closed D-interval $I[x,y]$ has the same parametrization, the theorem below follows immediately.

Theorem 5.7 (Third homogeneity property) *Let* $I[a,b]$ *and* $I[x,y]$ *be any two closed D-intervals in* M. *Then* $I[a,b]$ *and* $I[x,y]$ *are homeomorphic with each other.*

This is a very useful result, as we shall see below. Unfortunately, we have not been able to find a coordinate-free proof for it.

5.3 Final results

Having established the third homogeneity property, we may proceed to obtain our final results.

5.3.1 Contractibility of order intervals

We start with the trivial observation that the concatenation of two timelike curves is again a timelike curve. Therefore, if $I[a,b]$ is a D-interval and o is any point in $I(a,b)$, then there is a timelike curve $\Theta[a,b]$ that passes through o. We may now choose a homeomorphism of $\Theta[a,b]$ with the interval $[-1,1]$ such that $h(o) = r(o) = 0$. Let now $w \in I[a,b], w = \{h, r, \phi\}$, and consider the maps

$$[h, r, \phi] \mapsto [th, tr, \phi] \tag{5.13}$$

for $1 \leq t \leq 0$. Recalling that $\phi = \text{const}$ on 2-cells $F[a,b;p]$, we see that for any t such that $1 \leq t < 0$, (5.13) determines a homeomorphism of order intervals, whereas for $t = 0$ the entire ordered interval is mapped onto the point o. Since o is any point in $I(a,b)$, we have proved that:

Theorem 5.8 *The order interval* $I[a,b]$ *is contractible upon itself to any point* $o \in I(a,b)$.

5.3.2 The local differentiable structure

Since we assumed that the ordered spaces of Chapter 3 were locally precompact, our order-complete spaces are locally compact. Therefore a D-interval $I[a,b]$ with nonempty interior is compact, and is therefore a closed connected subset of $\mathbb{R}^n$ for some (finite) n. Since this set is homogeneous (i.e., every pair of points in its interior have neighbourhoods that are homeomorphic to each other) and contractible upon itself to every point in its interior, *it is homeomorphic to a closed ball in* $\mathbb{R}^n$. The interior of this ball is nonempty and has a differentiable structure, and this structure may be transferred automatically to $I(a,b)$, for example by choosing rectangular Cartesian coordinates on $\mathbb{R}^n$ such that the origin is at the centre of the ball. The radius is arbitrary. We conclude that:

Theorem 5.9 *Every nonempty open D-interval* $I(a,b)$ *is diffeomorphic to a nonempty open ball in* $\mathbb{R}^n$. *The* dimension n *is the same for all open D-intervals in* M.

We may therefore state our final result as follows:

Theorem 5.10 *Every locally precompact ordered space can be densely and uniformly embedded in an n-dimensional order-complete topological manifold which has the local structure of an n-dimensional differentiable manifold.*

We cannot say anything about the existence, or otherwise, of a *global* differentiable structure without making additional assumptions.

Part II

Geometrical points and measurement theory

Introduction to Part II

The argument of Part I of this book may be summed up as follows. The notion of causality can be defined as a partial order on an infinite set of geometrical points.[1] Defining causality is the same as assigning a light cone at each point (of a space). A light cone is determined by the set of light rays through its vertex. A light ray is totally ordered by the natural past–future order on it. The partial order on the whole space, called the causal order, may be reconstructed – i.e., axiomatized – from the properties of light rays and their intersections. A key property of a light ray is that between any two points of it lies a third, i.e., the cardinality of the set of points constituting a light ray is at least $\aleph_0$. It turns out that the entire causal structure can be defined on a set of cardinality $\aleph_0$. In common parlance (though not in the technical topological sense), such spaces would be called discrete, as opposed to continuua. Analysis of the causal spaces so defined shows that the locally compact ones among them can be densely embedded in continuua that have the local structure of finite-dimensional differentiable manifolds. Thus, although causality does not imply that space-time is a differentiable manifold, it comes as close as possible, mathematically, to implying it without actually doing so; a principle that most people would consider purely physical has far-reaching *mathematical* implications.

The whole structure is contingent upon the assumption that space-time is made up of geometrical points. This assumption has been questioned twice between 1930 and 1960. Between the wars, application of noncovariant perturbation theory to quantum electrodynamics (QED) produced divergent integrals in all orders except the lowest. The seemingly intractable nature of the problem led several physicists, Pauli and Heisenberg among them, to propose a radical overhaul of the notion of space-time at small distances.[2] After the second world war, Tomonaga, Schwinger, Feynman and Dyson created a perturbation theory that was both covariant and gauge-invariant, which allowed the divergences to be eliminated without modifying the structure of space-time at small distances; the latter idea had, if anything, been detrimental to progress.[3] Then, in 1952,

[1] A geometrical point is defined to be a point in the sense of Euclidean geometry.

[2] See (Schweber, 1994, Chapter 2). Relevant quotations from Pauli and Heisenberg will be found on pp. 84–85 and p. 101 respectively of that book.

[3] See (Schweber, 1994) for a riveting historical account. Some later theoretical developments, following the realization that the perturbation series diverges (Dyson 1952; Haag 1955), are described briefly in the Epilogue of this book.

Wigner asserted that an additive conservation law limits the precise measurement of an operator which does not commute with the conserved quantity.[4] This implied that the position of a point-particle could seldom be measured precisely, which in turn led Wigner to question the whole notion of geometrical points. As he put it to Haag, 'Some of us believe that there are no points'.[5]

The reader will recall that von Neumann's measurement theory contains a metaphysical part, namely an appeal to the observer's conscious ego, which was accepted – perhaps not wholeheartedly, but accepted nevertheless – by Wigner. The mathematical content of the theory was expressed by Wigner almost as an epigram: 'The state of the object is mirrored by the state of an apparatus'. More importantly, Wigner also captured what the theory *did not* achieve: '...neither apparatus nor object is in a state which has a classical description'.[6] To sum up, von Neumann's *mathematical* theory could neither determine the state of the apparatus nor account for collapse of the state vector; these required, at the end of a finite or infinite 'von Neumann chain', the intervention of the observer's conscious ego.

The result which led Wigner to assert that 'Some of us believe that there are no points' is known as the Wigner–Araki–Yanase theorem, which was proven by Araki and Yanase *within* von Neumann's measurement theory.[7] The object was assumed to be in an eigenstate φ_μ of the observable A to be measured.[8] The apparatus was prepared in the initial state ξ independently of the object. Araki and Yanase showed that the equation (page 198)

$$U(t)[\varphi_\mu \otimes \xi] = \varphi_\mu \otimes \zeta_\mu,$$

where ζ_μ is the unique final state of the apparatus that corresponds to the state φ_μ of the object, cannot hold unless A commutes with all additively conserved quantities. In the above, t is greater than the time it takes for the apparatus to reach its final state.

Having established the above-mentioned limitations on precise measurements, Araki and Yanase proceeded to define a notion of approximate measurements. In classical physics, measurement error is defined to be the difference between the true and the measured value. Since the state of the apparatus has no classical description, this concept of measurement error was not available in von Neumann's theory. Araki and Yanase circumvented this difficulty by

[4] See (Wigner, 1952).

[5] Quoted by Haag at the end of his talk at the 1995 Wigner conference in Goslar; Haag seemed to suggest that this could be the reason why Wigner received the algebraic formulation of relativistic quantum field theory less than enthusiastically. It was Professor R J Eden, who knew Wigner well, who first told me that Wigner's doubts arose from the inexactness of *single* measurements in quantum mechanics.

[6] Both quotations may be found in (Wigner, 1970); they are on pp. 158 and 187 respectively.

[7] See (Araki and Yanase, 1960).

[8] Araki and Yanase considered the eigenvalue μ to be degenerate. For the sake of clarity of exposition, we are ignoring the degeneracy; it changes nothing of significance to us.

quantifying a concept of *malfunctioning of the apparatus.*[9] Let the combined object–apparatus system evolve as follows:[10]

$$U(t)[\varphi_\mu \otimes \xi] = \varphi_\mu \otimes \zeta_\mu + \psi \otimes \vartheta_\mu,$$

where $(\zeta_\mu, \vartheta_\nu) = 0$ and $||\psi \otimes \zeta_\mu||^2 < \varepsilon$. This means that the state of the object is mirrored by the state of the apparatus not with probability 1, as in (11.7) quoted above, but with probability $1 - \varepsilon$. They then gave an example in which ε could be made as small as one wished. The nonvanishing quantity ε represented the probability of malfunctioning of the apparatus, and not an error of measurement in the conventional sense. Their final conclusion was that, for any given ε, an approximate measurement was always possible. This result was controverted by Shimony and Busch,[11] who considered measurement theories that differed from von Neumann's in that they allowed the initial state of the apparatus to be a *mixed* state, and came to the opposite conclusion: even approximate measurements were not possible in their modified versions of von Neumann's theory!

What the work of Shimony and Busch had in common with that of Araki and Yanase was the following: *in none of them was the state of the apparatus required to have a classical description.*

There is, however, a physical conception of measurement – not quite amounting to a theory – known as the *Heisenberg cut* or the Bohr–Heisenberg cut[12] in which the state of the apparatus does have a classical description and an error of measurement its classical meaning. Heisenberg asserted that one had to divide the world into an observed and an observing system, and it was this division which prevented a causal description of the act of measurement. This much was accepted by von Neumann.[13] Indeed, one may be forgiven for thinking that von Neumann's **2**nd and **1**st interventions (page 138) constituted an interpretation of the Heisenberg cut in mathematical terms; on one side of the cut, the state vector evolved under the Schrödinger equation; on the other side of the cut, it collapsed. Where von Neumann and Wigner parted company with Bohr and Heisenberg was in their differing conceptions of the measuring apparatus. Bohr insisted that the states of a measuring apparatus must have a classical description; Wigner was equally adamant that 'neither apparatus nor object is in a state which has a classical description'.[14]

[9] See (Yanase, 1961, Sec. 2) also (Wigner 1983, Eqn (63), p. 302).

[10] This is discussed in greater detail on page 200.

[11] See (Shimony, 1974; Busch and Shimony, 1996). These results are discussed in greater detail in Section 11.4.

[12] This is discussed in greater detail in Remark 8.3, pages 140-141.

[13] See (von Neumann, 1955, p. 420).

[14] Von Neumann *may* have been a bit less categorical; see the last quotation in Subsection 8.3.3.2, page 157.

What remains to be explored[15] is the possibility of incorporating a classical description of the states of the measuring apparatus as an integral part of a mathematical theory of measurement – i.e., of *building, mathematically, a bridge over the Heisenberg cut.* The mathematical theories that would have suggested that this task is worth attempting were not available to Bohr and Heisenberg, or to von Neumann. Bohr and Heisenberg would have looked to the correspondence principle to guide their physical intuition, but the correspondence principle loses all meaning if the limit of large quantum numbers is not uniquely defined. Von Neumann may even have been sceptical of the task itself; in the section on 'Radiation Theory' of his book, he wrote:[16]

> Now it is inconvenient formally *and of doubtful validity* [emphasis added] to admit systems with infinitely many degrees of freedom, or wave functions with infinitely many arguments. Our original discussions were always based on a finite number of coordinates.

But it is precisely when one admits systems with infinitely many degrees of freedom that the possibility of reconciling the divergent views of von Neumann and Wigner, and of Bohr and Heisenberg, begins to appear on the horizon. Ironically, the basic tools required for this enterprise were forged by von Neumann himself.

We shall call an N-particle quantum-mechanical system *large* if it is not possible to tell, experimentally, whether the system contains N or $N+1$ particles. This will mostly be the case when N is within a few orders of magnitude of Avogadro's number, and in such situations qualitative features of the theory are often thrown into sharper relief in the limit of infinitely many particles (e.g., the thermodynamic limit). The canonical commutation and anticommutation relations for infinitely many degrees of freedom have so many representations that they cannot be handled effectively by Hilbert space techniques. A result obtained by Haag in 1955 showed that in relativistic quantum field theories, which were necessarily theories with infinitely many degrees of freedom (since particle numbers were not conserved), it was essential to use non-Fock representations.[17] Following the lead of Segal,[18] these representations were tamed by using the

[15] Considering the very many avenues that *have* been explored in the literature, this statement requires clarification. The clarification is that we are searching for a conservative mathematical theory that *causes* the wave packet to collapse owing to its interaction with an observing system in the sense of Bohr and Heisenberg, and not a radical departure. We include nonlinear modifications of the Schrödinger equation and interaction with the rest of the universe among radical departures.

[16] The quotation may be found in (von Neumann, 1955, p. 265).

[17] Haag's theorem established that the interaction picture, which was developed in the Fock representation, existed only if there was no interaction. For QED this reproduced, from a different perspective, Dyson's result that the renormalized perturbation series diverged (Dyson, 1952). A very readable account of Haag's theorem is given in (Barton, 1963).

[18] An account of Segal's work from his own perspective may be found in (Segal, 1963).

theory of operator algebras, which was initiated by von Neumann himself in 1936.[19] While the initial impetus may have come from quantum field theory, the theory of operator algebras quickly chalked up impressive successes in applications to quantum statistical mechanics. Three of the early landmarks were the papers by Araki and Woods in 1963, by Haag, Hugenholtz and Winnink in 1967 and the book by Ruelle in 1969.[20] In 1932, when von Neumann wrote his book, or in 1963, when Wigner wrote his article, there was little hint of the richness of this theory or of the breadth of its applications. It led, in the nonrelativistic domain, to what Sewell has described as a 'generalized quantum-mechanical framework', commonly (but misleadingly) known as *algebraic quantum theory*. The extension of Newtonian particle mechanics to continuous media, initiated by Daniel Bernoulli, may be viewed as a similar generalization in the classical domain.

Within the generalized quantum-mechanical framework, the existence of observables 'with a classical description' was first established by Lanford and Ruelle in 1969.[21] These authors called them *observables at infinity*, because they could make their presence felt outside any bounded region of space. They were 'weak limits', which are defined on representations and not on the algebra itself, i.e., their existence or nonexistence depended entirely on the representation – and this is where one parts company with quantum mechanics on a single Hilbert space.[22]

When these weak limits did exist they belonged to the centre of the representation, justifying the adjective 'classical'. They could, in some sense, be likened to intensive variables of thermodynamics. These observables were used by Hepp to give a rigorous demonstration of the reduction of the wave packet in the limit $t \to \infty$. To the best of the present author's knowledge, this work of Hepp was the first to demonstrate, mathematically, that the wave packet could collapse without the intervention of the observer's conscious ego.[23] Such observables of the apparatus are often called *pointer observables*.

If one is secure in the knowledge that pointer observables exist, one can try to approximate them, or their thermodynamic conjugates (which would again be pointer observables) in large systems; indeed, such approximations are a *sine qua non* for establishing contact with experimental physics. With hindsight, one sees that they had been devised, and used to considerable effect, long before algebraic

[19] The first of the papers on *Rings of operators* was (Murray and von Neumann, 1936). In the introduction to this paper, the authors wrote: '... various aspects of the quantum mechanical formalism suggest strongly the elucidation of this subject.'

[20] The relevant references are (Araki and Woods, 1963; Haag, Hugenholtz and Winnink, 1967; Ruelle, 1969).

[21] The work of Lanford and Ruelle is discussed briefly in Chapter 12.

[22] One may have to part company with the single Hilbert space formalism even in the quantum mechanics of one spinless particle. See Reeh's example, given in Section 7.5.2, and the discussion of superseparability in the Epilogue.

[23] See (Hepp, 1972).

quantum theory was formulated: by Pauli in 1928, in formulating the master equation[24] and by van Kampen in 1954, in his theory of irreversible processes. The observables used by van Kampen were coarse-grained averages of extensive variables of parts of the system. Ten years later, Emch gave a rigorous discussion of quantum-mechanical systems on a *single* Hilbert space that contained, by assumption, a set of commuting macroscopic observables (without going into their genesis) and used it to derive generalized master equations.[25]

It is sometimes said that passage to the thermodynamic limit smooths out the fluctuations that are always present in a finite system. However, these fluctuations can also be eliminated by a different, but equally unrealizable device: by cooling down to absolute zero. But, even if one disregards human errors, the atomic structure of matter and thermodynamic fluctuations, one cannot run away from the real number system.[26] It is this fact that results in the gap between theoretical and experimental physics, as discussed in the Prologue: almost every value of a continuous real variable is irrational, and therefore a continuous classical variable cannot be measured precisely in a finite time with finite physical resources. Experimental physics requires one to replace the notion of precise measurements by that of *measurements with arbitrary precision*, i.e., measurements in which the error, while nonzero, can be made as small as one wishes. The example of the thermodynamic limit suggests that the error ε and the size of the system N should vary in opposite directions. In quantum theory, N should be large enough for the apparatus to have pointer observables.

A measurement is an interaction between object and apparatus which drives the pointer of the apparatus. This effect is qualitatively different from the usual interactions between two quantum-mechanical or two classical systems. The lack of characterization of such interactions was described by Wigner as 'the principal conceptual weakness of the orthodox view'.[27] A dynamical theory of measurement has to specify at least a minimal set of conditions which a measurement interaction has to satisfy.

Sewell has proposed a measurement theory in which the apparatus does have pointer states, and in which interaction of the quantum object with the (macroscopic) pointer observables of the apparatus causes the state vector of the object to collapse. This statistical assertion of the theory is contained in a conditional expectation functional, the time evolution of which is determined by

[24] The master equation made its first appearance in (Pauli, 1928). The term is now used for a number of similar equations in the theory of stochastic processes. We shall use the term only as it was used by van Kampen and Emch.

[25] The references to van Kampen and Emch are (van Kampen, 1954; Emch, 1964).

[26] As far as I know, the first physicist to take this seriously was Max Born. It is mentioned explicitly in his Nobel lecture (1954), and some of its consequences developed in several subsequent publications. A summary of Born's conclusions is presented in Section 6.5.

[27] See (Wigner, 1970, p. 167, 1983, p. 338).

the Schrödinger–von Neumann equations.[28] The notion of measurement with arbitrary precision has the same meaning in this theory as it does in classical physics. Sewell's theory was originally formulated for the measurement of observables with discrete spectra, but it was pointed out by the present author that it applied equally to the measurement of observables with continuous spectra – even those that did not commute with additively conserved quantities – in the approximate sense defined by von Neumann.[29] In Sewell's theory, the apparatus is an N-particle quantum-mechanical system, where N is of the order of Avogadro's number, but the theory also assumes that *the Hilbert spaces of object and apparatus are finite-dimensional.* These assumptions are more restrictive than the general principles of quantum mechanics. They obviously imply that the canonical commutation relations cannot be precisely implemented on these spaces, which is tolerable only if measurement errors are much larger than the uncertainty principle bounds. The result suggests that, as far as geometrical points are concerned, the problem of measurement does not make the situation significantly worse in quantum mechanics than it is in classical mechanics, but falls short of a complete proof. It falls short precisely because of the finite-dimensionality assumptions. A complete proof would seem to require lifting the finite-dimensionality assumptions of Sewell's theory, which would take us into waters as yet uncharted. Some reflections on this matter will be found in the Epilogue.

This part is organized as follows. Chapter 6 begins with a backward glance at quantum mechanics from our present vantage point. It then discusses the measurement of continuous variables in classical physics, for purposes of comparison with the measurement of observables with continuous spectra in quantum mechanics. The last section of this chapter is devoted to Born's observation that it makes no sense to talk about measuring an irrational number, and of the far-reaching conclusion he drew from it. As far as the present author knows, this work has not entered into the consciousness of the community of physicists. Chapter 7 is a collection of topics in nonrelativistic quantum mechanics that are not always discussed in introductory texts; it includes the example by Reeh that shows that, contrary to a widespread belief, the Stone–von Neumann uniqueness theorem does not hold at the Lie algebra level, and that this fact has observable physical consequences.[30] It also includes a section on the probability interpretation of quantum mechanics that serves to define our terms. (Later we shall find that Sewell's theory of measurement dictates a subtle change in the standard concept of what is measured.) Chapter 8 provides a detailed account of

[28] See (Sewell, 2005, 2006).

[29] See (Sen, 2008).

[30] The reference to Reeh's work is (Reeh, 1988). The Stone–von Neumann theorem applies to the exponentiated 'Weyl form' of the canonical commutation relations for a finite number of degrees of freedom, i.e., for the group, and not for its Lie algebra.

von Neumann's measurement theory, including the historical background of the collapse postulate. The latter is intended as a reminder that a theory of measurement that does not include wave function collapse – either as an assumption or as a theorem – may have to abandon the conservation of energy and momentum in individual collisions, and this point is most tellingly made by a flashback to the Bohr–Kramers–Slater proposal and the experiment of Compton and Simon. Chapter 9 deals with macroscopic observables in quantum mechanics, based primarily on von Neumann's work. The approach we have adopted is rather formal, and should be complemented by a study of the relevant parts of van Kampen's works (which has not been carried out in this book).[31] Chapter 10 is devoted to Sewell's measurement theory, which is the focus of this part. Computations in this chapter, of a kind that may be unfamiliar to many readers, are spelled out in considerable detail. It is followed by one which deals (albeit sketchily) with impossibility theorems and approximate measurements mentioned earlier, and relates the results of Part II to the fundamental assumption of Part I. The last chapter, Chapter 12, provides an introduction to the mathematical treatment of the symmetries and dynamics of large quantum-mechanical systems; its purpose is to illustrate how the qualitative differences between microscopic and macroscopic systems in nonrelativistic quantum mechanics may be brought under mathematical control, without which it would scarcely be possible to lift the finite-dimensionality assumptions of Sewell's theory.

[31] The works in question are (van Kampen, 1954, 1962).

6

Real numbers and classical measurements

6.1 The impact of quantum theory

Till quantum mechanics came along, it would not have occurred to many that the structure of a physical theory itself may be constrained by limitations on the precision of measurements. In classical mechanics, dynamical variables were implicitly assumed to be precisely measurable at all times, from which it followed that any mathematically well-defined function of the dynamical variables was precisely measurable at any time. Quantum mechanics severed the 'natural' link between the two concepts. Two observables, well-defined at all times,[1] would be simultaneously measurable only if they commuted with each other. However, a *single* observable could always be measured precisely. Although the details are well known, it may be worth recalling that while it was the uncertainty principle that unsettled the theorists,[2] what the experimentalists were unearthing were *effects of the superposition principle*, with measurement errors that far exceeded the uncertainty principle constraints.

What may be less well known (except to quantum field theorists) is that relativistic quantum field theory demands a severance of the link between being well defined and being measurable at yet another level. The classical electromagnetic field is well defined – and therefore measurable – at every point of space-time. However, no component of the quantized electric or magnetic field is precisely measurable at any point.[3] The last fact was discovered by Landau and Peierls, but it remained for Bohr to point the way to the correct conclusion to be drawn.

An electric or magnetic field strength is measured by determining the force it exerts on a charge or on a current (or permanent magnet). To measure the electric field strength at a point, one would need a charged point-particle of finite mass and measure the rate of change of its momentum caused by the field at that point; not the momentum itself, which of course cannot be measured with any precision at all, but its derivative. In classical electrodynamics, an accelerated charge radiates, and this radiation causes a 'radiation reaction' upon the charge

[1] Recall that in the Schrödinger picture observables are assumed to be time-independent.

[2] See, for instance, 'The Bohr–Einstein dialogues', in (Wheeler and Zurek, 1983, pp. 1–49).

[3] They are well defined mathematically as operator-valued distributions on space-time. We shall not go into details; the interested reader may consult the splendid elementary text (Barton, 1963), the more advanced standard text (Streater and Wightman, 1964) or the more recent book (Haag, 1993).

itself. When the velocity of light cannot be regarded as infinite, Landau and Peierls showed that, owing to this radiation reaction, the force exerted by the field on a test particle (charge or magnetic dipole) could not be measured precisely. It followed that electric and magnetic field strengths at a point could not be measured precisely. The conclusion that they drew from this was expressed as follows in the abstract of their paper (Landau and Peierls, 1931), which bears the title 'Extension of the uncertainty principle to relativistic quantum theory':

> It is shown by considering possible methods of measurement that all the physical quantities occurring in wave mechanics can in general no longer be defined in the relativistic range.

Landau and Peierls attempted to link the above to the 'well-known failures' of relativistic quantum mechanics, which were: (i) the appearance of negative-energy states in Dirac's electron theory, (ii) the self-energy problem for a charged particle, and (iii) the infinite zero-point energy of the radiation field.

Landau and Peierls went to Copenhagen with the manuscript of their paper in late 1930 or early 1931. However, they could not get Bohr to agree. A humorous, if slightly partisan account of the whole story is given by Rosenfeld in *Niels Bohr and the Development of Physics* (Rosenfeld, 1955).

Bohr refused to admit that quantum mechanics could forbid the precise measurement of any single quantity. Rosenfeld writes:

> My first task was to lecture Bohr on the fundamentals of field quantization; the mathematical structure of the commutation relations and the underlying physical assumptions of the theory were subject to unrelenting scrutiny. After a very short time, needless to say, the roles were reversed and he was pointing out to me essential features to which nobody as yet had paid sufficient attention.
>
> His first remark, which threw decisive light on the problem, was that field components taken at definite space-time points are used in the formalism as idealizations without immediate physical meaning; the only meaningful statements of the theory concern averages of such field components over finite space-time regions. This meant that in studying the measurability of field components we must use as test bodies finite distributions of charge and current, and not point charges as had been loosely done so far. The consideration of finite test-bodies immediately disposed of Landau and Peierls' argument concerning the perturbation of the momentum measurement by the radiation reaction: it is easily seen that this reaction is so much reduced for finite test bodies, as to be always negligible.

The scheme of canonical field quantization (devised by Dirac, Jordan, Heisenberg and Pauli) upon which Rosenfeld must have lectured to Bohr was based upon the Lagrangian formalism. The electromagnetic field was described by the vector potential $A_\mu = A_\mu(\mathbf{x}, x_0)$. Its canonical conjugate, according to the Lagrangian formalism, was (see, e.g., (Schweber, 1961))

$$\pi^\mu = \frac{\partial \mathcal{L}}{\partial A_{\mu,0}} = -A^{\mu,0},$$

where an index σ (zero in the above) after the comma in a superscript or subscript indicates the partial derivative with respect to x_σ. The equal-time commutation relations (ETCR) then become (in their original, noncovariant form)

$$[A^{\mu,0}(x), A_\nu(x')]_{x_0=x_0'} = -\mathrm{i}\hbar\delta^\mu_\nu\delta(\boldsymbol{x}-\boldsymbol{x}'). \tag{6.1}$$

The electric and magnetic field strengths are components of the antisymmetric tensor $F_{\mu\nu} = A_{\mu,\nu} - A_{\nu,\mu}$. Their commutators therefore involve the δ-function and its derivatives. Bohr and Rosenfeld interpreted this as follows (Bohr and Rosenfeld, 1933):

> The occurrence of the δ-function in the commutation relations [between the field components] brings to the fore the fact that the quantum theoretical field quantities are not to be considered as true point functions but that unambiguous meaning can be attached only to space-time integrals of the field components. With a view to the simplest possibility of testing the formalism we shall consider only averages of field components over simply connected space-time regions whose spatial extension remains constant during a given time interval.

We note that the nonmeasurability of fields at points did not cause Bohr to question the notion of geometrical points. Wigner, as we shall see later, was inclined to be more pessimistic. We end this section by pointing out that in 1950 Bohr and Rosenfeld published an extended version of their paper which took advantage of the covariant formulation of QED in the interaction picture that had been developed in the intervening years (Bohr and Rosenfeld, 1950).

6.2 Precise measurements in classical mechanics

We saw in the previous section that a central assumption of classical mechanics, namely that a dynamical variable is well defined only if it is precisely measurable,[4] is no longer tenable in quantum mechanics. But what exactly does a 'precise measurement' mean? We begin by analysing this problem, which does

[4] The statement 'A is true if B is true' means that the condition B is *sufficient* for A to be true. The statement 'A is true only if B is true' means that the condition B is *necessary* for A to be true. The statement 'A is true if and only if B is true' means that the condition B is *necessary and sufficient* for A to be true.

not seem to have been discussed much by physicists.[5] To avoid misunderstanding, we should add that we shall use the term classical mechanics in the sense of nonrelativistic, or Newtonian mechanics, unless the contrary is stated explicitly.

6.2.1 Space, time and measurement in classical mechanics

We begin by defining our terms. The words *point* and *line* will be used in the sense of Euclidean geometry. The Euclidean line will be related to the number system we shall use through the following assumption:

Assumption 6.1 The line of Euclidean geometry is the same as the linear continuum $\mathbb{R}$ of mathematical analysis.

This assumption, which may explain why $\mathbb{R}$ is called the *real line*, is always made but seldom made explicit. The mathematical structure of $\mathbb{R}$ was elaborated only in 1872, by Cantor and Dedekind, 2000 years after Euclid, and almost 200 years after Newton and Leibniz. Modern treatments of Euclidean geometry – which serve mainly to study departures from it – are an exception to this rule (see, for instance, the text (Martin, 1975)). The assumption is discussed, from the physicist's point of view, in (Sen, 1999). The phrases *real line* and *real interval* will be used in the mathematical sense.

It will be assumed that the notion of *empty space* makes sense. Space will be assumed to be three-dimensional, and time, one-dimensional. Measuring instruments interact with physical objects and not with empty space, and therefore to measure the distance between two points in space we need the notion of *test-particles*, i.e., objects of the size of a geometrical point but with physical attributes like charge or mass that may interact with measuring devices.

Since a measuring device interacts with test-particles, and not with points in space, why should the motion of a test-particle provide information about space? That it does is an assumption, which may be regarded as the physical equivalent of the intermediate-value theorem of calculus:

Assumption 6.2 If a test-particle travels from a point A to a point B in space, its trajectory is a continuous curve joining the points A and B.

This assumption allows us to use the mathematical structures on Euclidean space $\mathbb{R}^3$ to describe the behaviour of systems of test-particles.[6] A *rigid configuration* of test-particles will be defined as one that is invariant under the three-dimensional Euclidean group and the group of time translations.

[5] A notable exception was Max Born. His work on the subject was published mainly after his retirement from Edinburgh. It is discussed briefly in Section 6.5. The author was not aware of this work when he wrote the initial version of this chapter, and would like to thank Professor H Roos for bringing it to his attention and for providing him with the references.

[6] The notation $\mathbb{R}^3$ will be used both for the Cartesian product $\mathbb{R} \times \mathbb{R} \times \mathbb{R}$ and this Cartesian product endowed with the Euclidean metric (6.2). This will not cause any confusion.

6.2.2 Perfect rulers

The notion of a *perfect ruler*, i.e., an instrument to measure lengths, will be taken as understood. If the points P, P' are assigned rectangular Cartesian coordinates (x, y, z) and (x', y', z') respectively using a perfect ruler, then the distance $d(P, P')$ measured by the same ruler will be given by the formula

$$d(P, P') = \{(x - x')^2 + (y - y')^2 + (z - z')^2\}^{\frac{1}{2}}. \tag{6.2}$$

A perfect ruler will be regarded as a black box which accepts pairs of points (P, P') as input and produces the real number $d(P, P')$ of (6.2) as output. We may consider this process to be mediated by *internal states* ξ of the device (the notation $\mapsto$ is defined on page 241),

$$(P, P') \longmapsto \xi \stackrel{\mu}{\longmapsto} d(P, P'),$$

where μ, the *calibration map*, is bijective. The pairs $(\xi, \mu(\xi))$ have to be stored in the permanent memory of the device.

Since d is a nonnegative real number, the assumption that the Euclidean line is the same as the real line implies that the cardinality of the set of internal states of $\mathcal{R}$ cannot be less than $\aleph$.[7] Now almost every real number is irrational,[8] and an irrational number has a unique strictly infinite binary representation. It will require countably many bits of memory to store a *single* irrational number. It follows that:

Condition 6.3 The device $\mathcal{R}$ described above will need to have $\aleph$ cells of permanent memory, each cell containing $\aleph_0$ bits.

It will not have escaped the reader's attention that this condition is rather strong.

6.3 Discussion

Fabrication and calibration of the ruler are independent processes. It is also possible to calibrate the device $\mathcal{R}$ so that it implements the so-called *discrete metric*

$$d(P, P') = \begin{cases} 0, & \text{for } P = P', \\ 1, & \text{for } P \neq P' \end{cases}$$

[7] The notion of cardinality is defined in Subsection A1.5.1, pages 245–246.

[8] In the technical sense that the rational numbers form a countable subset of $\mathbb{R}$ (Subsection A2.1), and a countable subset of $\mathbb{R}$ has Lebesgue measure zero (Subsection A5.6.2).

on $\mathbb{R}^3$. In this case the device will have only two distinct internal states, and will require only a finite amount of internal memory.

There has never been any doubt – at least, to the best of this author's knowledge – that infinitely precise measurements cannot be made in the laboratory. The reason, it was generally believed, was the impossibility of fabricating test-particles and perfect rulers. The analysis carried out above shows that, even in a hypothetical laboratory furnished with test-particles and perfect rulers, infinitely precise measurements *cannot* be carried out unless the following additional conditions are fulfilled:

(i) The ruler is calibrated with infinite precision.[9]
(ii) Irrational numbers are recorded precisely.

The impossibility of meeting these conditions does not spring either from quantum-mechanical roots, or from the atomistic structure of inanimate nature. It springs from the nature of the real number system. Quantum mechanics cannot improve upon this situation; it can only make it worse. Whether or not it does make matters worse will be investigated in the next few chapters.

6.4 The role of the experimentalist

Let us now try to look at the implications of these limitations from the experimentalists' point of view. The numbers they read on the dials of their instruments will necessarily be rational. The decimal expression for a rational number is either recurrent, e.g., $1/7 = 0.\overline{142857}$ (the bar over the group 142857 meaning that the group is repeated indefinitely) or finite, in which case the infinite tail of zeroes is often omitted in daily life; one writes $2/5 = 0.4$ rather than the more precise $0.4\overline{0}$. However, the experimentalist can never be sure whether the number on the dial represents a true rational or a rational approximation to an irrational, and therefore the standard practice in physics is to report numbers 'correct to d significant figures', *according to a fixed, predetermined system of units.* Increasing the accuracy of a particular measurement then translates into increasing the number of significant figures in the reading.

The effort to increase the accuracy of measurements is a driver of, and is also driven by, technological changes. The latter can seldom be modelled theoretically. However, in a large class of experiments the period of 'design of the experiment' is short by comparison with the pace of technological change, which effectively eliminates technological change from the equation. Under these circumstances,

[9] This would require an algorithm for labelling the internal states of the device by real numbers, which *may* be equivalent to well-ordering the real numbers (see page 248). As no well-ordering of the real numbers is known (Subsection A2.2.4), it is not possible to assume that a ruler can be calibrated with infinite precision without further analysis. We shall not attempt to resolve this problem.

the experimentalist's concern is fundamentally the same as that of the economist: *efficient allocation of scarce resources.*

Had available memory been infinite, this discussion would not have been necessary. The problem arises from the finiteness of available memory. If the number of internal states of the apparatus is n, and each internal state requires r units of memory to calibrate it, then the total memory requirement for the instrument is m units, where

$$m = nr. \tag{6.3}$$

The quantity m is to be considered as a *fixed resource*, and the allocation problem is that of dividing it between r and n. That is, the resolving power of the instrument (accuracy of observation) may be increased only by decreasing the number of internal states. If m is held constant, n has to be divided by 10 to increase the resolving power by a single decimal digit.

We may therefore conclude that, for fixed memory resources, the resolving power of an instrument is inversely proportional to the *range of observation*, i.e., the number of data points that may be recorded by the instrument. This is a fundamental constraint in the design of an experiment at any given 'state of the art'. We shall find that this fact plays an important role in the theory of measurement in quantum mechanics.

6.5 Born's probability interpretation of classical mechanics

Max Born retired from his Edinburgh chair in 1953, and was awarded the Nobel Prize in 1954. Subsequent to his retirement, he published several articles in which he asserted that classical mechanics too was essentially a probabilistic theory. His argument was based on two pillars:

(i) The concept of infinitely precise measurements was devoid of meaning, owing to the nature of the real number system. The best one could claim about (say) the position of a point particle was that it was a probability distribution with a peak at a certain point and a certain spread (variance).

(ii) Nonlinear classical systems exhibited a *sensitive dependence on initial conditions.*[10] He gave a very simple example (which he attributed to Einstein) to illustrate this phenomenon.

From these, he concluded that a second measurement of the position of the particle effected a *reduction of the probability*, in the sense that the second measurement caused the probability distribution, as a function of time, to change dramatically (the assumption being that the second measurement *had* detected the particle).

[10] Born did not use this term. He first mentioned the fact (as far as I know) in his Nobel prize lecture of 1954; Lorenz's paper on 'Deterministic nonperiodic flow', which set off the chaos revolution, appeared in 1963 (Lorenz, 1963).

I have not been able to find any clear indications, in Born's own writings, of when he began to develop this viewpoint.[11] It does not seem to be there either in his 1948 Waynflete lectures (Born, 1949), or in the letters he wrote to Einstein (Einstein died on 18 April 1955). Born said the following in his Nobel lecture (Born, 1954, p. 264):

> ...Thus determinism lapses completely into indeterminism as soon as the slightest inaccuracy in the data on velocity is permitted...Is one justified in saying that the coordinate $x = \pi$ cm where $\pi = 3.1415\ldots$ is the familiar transcendental number...? As a mathematical tool the concept of a real number represented by a nonterminating decimal fraction is exceptionally important and fruitful. As the measure of a physical quantity it is nonsense...

However, in the comments Born added to *The Born–Einstein Letters* as it was being prepared for publication, he wrote (Einstein, 1971, item 116, pp. 227–228):

> My own manuscript seemed to me to contain certain thoughts which I had not yet come across elsewhere. I rewrote it completely...
>
> At this time I received an invitation from the Danish Academy to contribute to a volume...to be published on the occasion of Niels Bohr's seventieth birthday. I therefore sent my paper...to the Danish Academy in Copenhagen...In a letter from Zürich of 11th December 1955,...Pauli writes...'Your paper in the Danish presentation volume to Bohr makes very pleasant reading now; its epistomological content has now become very clear, and I agree with all of it. I had used the mathematics of the example of the mass point between two walls, and of the wave packets which belong to it, in my lectures in such a way that the transformation formula of the theta-function comes into play. But that is a mere detail.' It is more than a detail. It shows that Pauli had long been familiar with all that I had to say...ever since the time he had been my assistant in Göttingen, I had been aware that he was a genius, comparable only with Einstein himself...

The above quotation may not reflect accurately the role Pauli may have played in the development of Born's ideas; the interested reader will have to read all of items 111–116, pp. 216–228, of (Einstein, 1971), especially the part from paragraph 3 on p. 224 to the end of Born's remarks on p. 228.

Since I have not been able to delve deeper into this fascinating bit of the history of physics, I am adding a few references for the interested reader. A brief account of the Born–Einstein correspondence is given in the reminiscences of Max Born and his wife Hedwig (Born and Born, 1969, pp. 134–136). The phrase 'reduction of probabilities' does not appear in Born's Nobel prize lecture (Born,

[11] The article (Born, 1955) refers to an article by Einstein in (Einstein, 1953). The example used by Born appears to be taken from this source.

1954, pp. 264–265). It appeared a year later, in the Danish Niels Bohr Festschrift (Born, 1955) and again in an article in the Heisenberg Festschrift (Born, 1961, §§ 3–4, pp. 107–111).[12] More improbably, Born translated a book of Wilhelm Busch's poems, *Klecksel the Painter*, into English (currently out of print). An absorbing account of Born's family history, life and work, including the work he did after retirement, has been provided by his son (Born, G., 2002, esp. p. 257).

Note added in proof

I was able to obtain a copy of Einstein's paper (Einstein, 1953) after this book was sent to press. The first sentence of the last paragraph of this paper reads as follows:[13]

> To my mind it is not satisfactory to build on such a [quantum-mechanical] view of physics because one cannot dispense with an objective description of individual *macroscopic* systems [emphasis in the original] (description of the "real" state) without letting the physical view of the world dissolve into fog.

Einstein was distinguishing between individual systems and ensembles of systems. His conclusion was summarized by Born as follows (Born, 1955, p. 8): '...Einstein then discusses the question whether quantum mechanics leads to a description of the behaviour of macro-bodies which corresponds to... [his] notion of reality, and his answer is no.'

Notice the singling-out of *macroscopic* systems by Einstein, and recall that the title of the (Einstein, Podolsky and Rosen, 1935) paper was "Can quantum-mechanical description of physical reality be considered complete?' Much of the controversy caused by this paper has devolved around the word 'reality', and has been more metaphysical than physical in content. By contrast, the description of macroscopic systems within quantum mechanics is – or so the present author contends – a task for the physicist.[14] Einstein's 1953 essay was quite possibly the last words he wrote on the subject; even at that late date, a mathematically rigorous and physically meaningful description of macroscopic systems had not been extracted from the formalism of quantum mechanics, and, in that sense, quantum mechanics was manifestly 'incomplete'.[15] A concise state-of-the-art account of how far the task of 'completing' quantum mechanics in this sense has progressed, and the results that the formalism has led to, may be found in Sewell's book, appropriately titled *Quantum Mechanics and its Emergent Macrophysics* (Sewell, 2002).

[12] Due to illness, Born was not able to proofread this article. In the footnote on p. 106, Born quotes a verse from Wilhelm Busch. Professor Roos, who provided me with most of these references, has pointed out that there is an 'annoying misprint' in this quotation. A correct rendition, as well as Born's own translation of it, will be found in (Born, G., 2002, p. 260).

[13] I am grateful to Professor H Roos for this translation.

[14] This is the subject of Chapter 9 of the present work.

[15] In this context, see also the quotation from Einstein on page 234.

7

Special topics in quantum mechanics

The chief aim of this chapter is to give brief accounts of topics in nonrelativistic quantum mechanics that are not always treated in elementary texts. We begin with the Hilbert space formulation of quantum mechanics as set down by von Neumann in his book *Mathematical Foundations of Quantum Mechanics* (von Neumann, 1955),[1] which will henceforth be referred to as *von Neumann's book.* Much of our concern will be with continuous spectra, which cannot be discussed adequately in the Dirac formalism. The density matrix, which will play a key role in Chapters 8 and 10, will be treated in some detail.

The section on formalism is followed by one on the probability interpretation; the latter is included because Sewell's theory of measurement suggests a subtle reformulation of a part of it. These are followed by sections on superselection rules and the Galilei group, which is the relativity group of quantum mechanics. They are based on the pioneering works of Wigner and Bargmann. The last section is devoted to the fundamental theorems of von Neumann and Stone, and to Reeh's observation on the physical significance of the failure of the Stone–von Neumann uniqueness theorem at the Lie algebra level.

7.1 The formalism of quantum mechanics

By *quantum mechanics* we shall mean the nonrelativistic quantum theory of a system with a finite number, N, of particles. The number N is assumed fixed. It will be convenient, for later recall, to divide the material into subsections.

7.1.1 Pure states

The *pure states* of the system will be unit rays in a separable, infinite-dimensional Hilbert space $\mathfrak{H}$ over the complex numbers (see Appendix A6, pages 316–318).[2] We are emphasizing the separability condition (which is equivalent to the existence of a countable orthonormal base) because in current mathematical

[1] This is the title of the English translation. In the translator's preface (dated December 1949), the translator stated that 'the translated manuscript has been carefully revised by the author... and any deviations from the original text are... due to the author'. The German original, *Mathematische Grundlagen der Quantenmechanik,* was published in 1932.

[2] If errors due to the uncertainty principle can be ignored, e.g., when only the spins are of interest, the system can often be modelled conveniently on a finite-dimensional Hilbert space.

literature Hilbert spaces are no longer assumed to be separable,[3] contrary to von Neumann's original definition. The problem of understanding '*why the imaginary unit enters quantum theory*' has been addressed by Stueckelberg and collaborators (Stueckelberg, 1960; Stueckelberg *et al.*, 1961; Stueckelberg and Guenin, 1961, 1962) by constructing quantum theories on real Hilbert spaces. The following quotation is from von Neumann's book, pp. 196–198:

> In this method of description, it is evident that everything which can be said about the state of a system must be derived from its wave function $\phi(q_1, \ldots, q_k)$... Furthermore, it should be pointed out regarding this that while ϕ is dependent on the time t, as well as the coordinates $q_1, \ldots, q_k$ of the configuration space of our system, nevertheless the Hilbert space involves only the $q_1, \ldots, q_k$ (because the normalization is related to these alone). *Hence the dependence on t is not to be considered in forming the Hilbert space* [emphasis added]. Instead of this, it is rather to be regarded as a parameter... Because of this, we shall occasionally indicate the parameter t in ϕ... by writing ϕ_t.

We shall have occasion to refer to this quotation.

7.1.2 Mixed states

Let $\{u_i | i \in \mathbb{N}\}$ be an orthonormal basis for $\mathfrak{H}$ and $\psi = \sum_i c_i u_i$ a normalized vector, $\sum_i |c_i|^2 = 1$. Let A be a bounded self-adjoint operator on $\mathfrak{H}$ (page 326). The mean or expectation value $E(\psi; A)$ of A in the state ψ is defined to be $E(\psi; A) = (\psi, A\psi)$, and may be written as

$$E(\psi; A) = \sum_{i=0}^{\infty} |c_i|^2 (u_i, Au_i) + \sum_{i \neq j} \bar{c}_i c_j (u_i, Au_j). \tag{7.1}$$

The second sum on the right consists of the quantum interference terms, which are trivially absent if the basis $\{u_i\}$ consists of eigenvectors of A.

The quantum interference terms will also be absent if, instead of the linear superposition ψ of the states u_i, we consider a statistical ensemble formed from the states $\{u_i\}$, the probability of occurrence of the state u_i being p_i, where $0 \leq p_i$ and $\sum_i p_i = 1$. In this case the ensemble average of the operator A will be given by

$$E(p; A) = \sum_{i=0}^{\infty} p_i (u_i, Au_i). \tag{7.2}$$

[3] However, the Hilbert spaces used in the theory of infinite-dimensional group representations, as developed by Mackey (Mackey, 1968, 1976, 1978), are separable. We shall use an elementary form of Mackey's theory in Section 7.4. Professor Mackey once told the author that 'I always assume separability, whether I need it or not'.

If we choose $p_i = |c_i|^2$, then (7.2) becomes (7.1), minus the quantum interference terms. This may be interpreted as follows: every pair $(\psi, \{u_i\})$ on $\mathfrak{H}$, where ψ is a normalized pure state and $\{u_i\}$ an orthonormal basis, gives rise to a statistical ensemble of states on $\mathfrak{H}$ (van Kampen, 1962, p. 185). These statistical ensembles are known as *mixed states*, and the u_i their *components*.[4] They can be described, equivalently, by a class of *operators* on the Hilbert space $\mathfrak{H}$, called *density matrices*.[5] Define an operator ρ on $\mathfrak{H}$ via its matrix elements in the basis $\{u_i\}$:

$$(u_i, \rho u_j) = \rho_{ij} = \begin{cases} p_i, & i = j, \\ 0, & i \neq j. \end{cases} \tag{7.3}$$

Then $\operatorname{Tr} \rho = \sum_i p_i = 1$. By the invariance of the trace, this equality is independent of the basis in $\mathfrak{H}$. The operator ρ is clearly self-adjoint. Using ρ, (7.2) can be written compactly as

$$E(p; A) \equiv E(\rho, A) = \operatorname{Tr}(\rho A). \tag{7.4}$$

The right-hand side of (7.4) is independent of the basis in $\mathfrak{H}$. Recall also that $\operatorname{Tr}(\rho A) = \operatorname{Tr}(A\rho)$. The quantity $E(\psi; A)$ is manifestly independent of the basis.

When there is no risk of confusion (and often even when there is), the quantities $E(\psi; A)$ and $E(p; A)$ are written simply as $E(A)$.

Let now u_a be a fixed vector in $\{u_i\}$, and consider a second probability distribution p' on $\{u_i\}$, defined by

$$p'(u_i) = \delta_{ia}. \tag{7.5}$$

The matrix ρ now has only one nonzero element, which is $[\rho]_{aa} = 1$. It satisfies the condition $\rho^2 = \rho$, so that it is a projection operator. For visual impact, we write ψ $(= u_a)$ for u_a. Formula (7.2) now shows that in this case

$$E(p'; A) = (\psi, A\psi) = E(\psi; A), \tag{7.6}$$

i.e., the ensemble average of A is the same as its quantum-mechanical expectation value in the state ψ. In this case the operator ρ represents a pure state. In an arbitrary orthonormal basis $\{\varphi_n\}$, the density matrix of the pure normalized state $\varphi = \sum_n a_n \varphi_n$ has the matrix elements

$$\rho_{mn}[\varphi] = \bar{a}_m a_n.$$

[4] The components $\{\psi_i\}$ of a mixed state need not be orthogonal to each other, but nonorthogonal components will not be eigenvectors of the density matrix.

[5] The density matrix was introduced independently in 1927 by von Neumann, Landau and Weyl. See (Landau and Lifshitz, 1959, §12, pp. 35–38) and (Weyl, 1931, pp. 77–79). Von Neumann calls it the *statistical operator*.

After this discussion, we proceed to the formal definitions and main results; proofs may be found in most textbooks on the subject, and also in (London and Bauer, 1939); English translation in (Wheeler and Zurek, 1983):

Definition 7.1 (Density matrix) A *density matrix* (more correctly, a density operator) on $\mathfrak{H}$ is a positive self-adjoint operator of unit trace.

Let the density matrix ρ have the spectral decomposition

$$\rho = \sum_{n=0}^{\infty} w_n E_n, \tag{7.7}$$

where E_n are projection operators on $\mathfrak{H}$. Then $w_n \geq 0$, and

$$\begin{aligned} k_n &= \dim E_n < \infty \text{ if } w_n \neq 0, \\ \sum_{n=0}^{\infty} w_n k_n &= 1. \end{aligned} \tag{7.8}$$

In words, *zero* is the only eigenvalue of a density matrix that can be infinitely degenerate.

Theorem 7.2 *The set of density matrices on $\mathfrak{H}$ is* convex, *i.e., if ρ_0, ρ_1 are density matrices, then so is*

$$\rho_\lambda = (1-\lambda)\rho_0 + \lambda\rho_1$$

for any $\lambda \in [0,1]$.

The expression on the right-hand side of the above equation is called a **convex combination** of ρ_0 and ρ_1. An element in a convex set which cannot be expressed as a convex combination of two distinct elements is called an **extremal element** of the set.

The notion of convexity in linear spaces is a generalization of the notion of convexity in n-dimensional Euclidean spaces. A body in Euclidean n-space is called **convex** if the line segment joining any two points in the body lies wholly within the body. For details, see, e.g., (Kitchen, 1968), or almost any book on advanced calculus.

Definition 7.3 (Pure and mixed states) A density matrix ρ is said to represent a *pure state* iff it is also a projection operator, i.e., iff $\rho^2 = \rho$. Otherwise, it is said to represent a *mixed state.*

Theorem 7.4 *A pure state is an extremal state in the set of density matrices.*

This theorem means that a pure state cannot be represented as a convex combination of mixed states.

We shall denote the set of density matrices on $\mathfrak{H}$ by $\mathcal{D}_{\mathfrak{H}}$.

7.1.3 Partial traces

Let $\mathfrak{H}^{\mathrm{I}}$, $\mathfrak{H}^{\mathrm{II}}$ be Hilbert spaces and $\mathfrak{H} = \mathfrak{H}^{\mathrm{I}} \otimes \mathfrak{H}^{\mathrm{II}}$ be their tensor product (page 321). Let $\{\varphi_m^{\mathrm{I}}\}$, $\{\varphi_n^{\mathrm{II}}\}$ be orthonormal bases in $\mathfrak{H}^{\mathrm{I}}$ and $\mathfrak{H}^{\mathrm{II}}$ respectively. Then $\{\varphi_{m,n} = \varphi_m^{\mathrm{I}} \otimes \varphi_n^{\mathrm{II}}\}$ is an orthonormal basis for $\mathfrak{H}$. Let A be a bounded operator on $\mathfrak{H}$. We shall write its matrix elements in the basis $\{\varphi_{m,n}\}$ as

$$A_{mn,m'n'} = (\varphi_m^{\mathrm{I}} \otimes \varphi_n^{\mathrm{II}}, A\,\varphi_{m'}^{\mathrm{I}} \otimes \varphi_{n'}^{\mathrm{II}}). \tag{7.9}$$

Suppose now that A is a positive trace class operator on $\mathfrak{H}$ (page 327). Then

$$\sum_{m,n=1}^{\infty} A_{mn,mn} < \infty,$$

so that

$$\sum_{n=1}^{\infty} A_{mn,m'n} < \infty \quad \text{and} \quad \sum_{m=1}^{\infty} A_{mn,mn'} < \infty.$$

We may therefore define an operator $A^{\mathrm{I}} = \mathrm{Tr}^{\mathrm{II}} A$ on $\mathfrak{H}^{\mathrm{I}}$ as follows:

$$A_{mm'}^{\mathrm{I}} = (\varphi_m^{\mathrm{I}}, \mathrm{Tr}^{\mathrm{II}} A\,\varphi_{m'}^{\mathrm{I}}) = \sum_{n=1}^{\infty} \left(\varphi_m^{\mathrm{I}} \otimes \varphi_n^{\mathrm{II}}, A\,(\varphi_{m'}^{\mathrm{I}} \otimes \varphi_n^{\mathrm{II}})\right). \tag{7.10}$$

The definition of this operator does not depend on the chosen basis $\{\varphi_n^{\mathrm{II}}\}$ of $\mathfrak{H}^{\mathrm{II}}$. The proof is the same as that for the ordinary trace.

The reader is invited to verify that A^{I} is a trace class operator on $\mathfrak{H}^{\mathrm{I}}$. It is called the **partial trace** of A over $\mathfrak{H}^{\mathrm{II}}$. The partial trace $A^{\mathrm{II}} = \mathrm{Tr}^{\mathrm{I}} A$ of A on $\mathfrak{H}^{\mathrm{II}}$ is defined analogously.

Partial traces are operators. To avoid confusion, we shall temporarily denote the ordinary trace by a subscript; thus, if $\mathfrak{K}$ is any Hilbert space and B a trace class operator on it, $\mathrm{Tr}_{\mathfrak{K}} B$ will denote the trace of B on $\mathfrak{K}$. It follows immediately from the definitions that

$$\mathrm{Tr}_{\mathfrak{H}^{\mathrm{I}}}(\mathrm{Tr}^{\mathrm{II}} A) = \mathrm{Tr}_{\mathfrak{H}^{\mathrm{II}}}(\mathrm{Tr}^{\mathrm{I}} A) = \mathrm{Tr}_{\mathfrak{H}}\, A \tag{7.11}$$

and

$$\mathrm{Tr}^{\mathrm{II}}(A^{\mathrm{I}} \otimes A^{\mathrm{II}}) = (\mathrm{Tr}_{\mathfrak{H}^{\mathrm{II}}} A^{\mathrm{II}}) A^{\mathrm{I}}. \tag{7.12}$$

The above formula also holds with I and II interchanged.

The following lemma is often useful:

Lemma 7.5 *Let ρ be a trace class operator on $\mathfrak{H} = \mathfrak{H}^{\mathrm{I}} \otimes \mathfrak{H}^{\mathrm{II}}$. Then*

$$E(\rho, A^{\mathrm{I}} \otimes I^{\mathrm{II}}) \equiv \mathrm{Tr}\,(\rho(A^{\mathrm{I}} \otimes I^{\mathrm{II}})) = E(\rho^{\mathrm{I}}, A^{\mathrm{I}}),$$

where ρ^{I} is the partial trace of ρ over $\mathfrak{H}^{\mathrm{II}}$ and I^{II} is the identity operator on $\mathfrak{H}^{\mathrm{II}}$.

Proof By direct computation. Using the orthonormal bases $\{\varphi^{\mathrm{I}}_m\}$ and $\{\varphi^{\mathrm{II}}_n\}$ in $\mathfrak{H}^{\mathrm{I}}$ and $\mathfrak{H}^{\mathrm{II}}$ respectively, we have

$$\begin{aligned}
\mathrm{Tr}(\rho(A^{\mathrm{I}} \otimes I^{\mathrm{II}})) &= \sum_{m,n=1}^{\infty} (\varphi^{\mathrm{I}}_m \otimes \varphi^{\mathrm{II}}_n, \rho(A^{\mathrm{I}} \otimes I^{\mathrm{II}})(\varphi^{\mathrm{I}}_m \otimes \varphi^{\mathrm{II}}_n)) \\
&= \sum_{m=1}^{\infty} \left(\sum_{n=1}^{\infty} (\varphi^{\mathrm{I}}_m \otimes \varphi^{\mathrm{II}}_n, \rho(A^{\mathrm{I}}\varphi^{\mathrm{I}}_m \otimes \varphi^{\mathrm{II}}_n)) \right) \\
&= \sum_{m=1}^{\infty} (\varphi^{\mathrm{I}}_m, (\mathrm{Tr}^{\mathrm{II}}\rho)A^{\mathrm{I}}\varphi^{\mathrm{I}}_m) \\
&= \mathrm{Tr}_{\mathfrak{H}^{\mathrm{I}}}((\mathrm{Tr}^{\mathrm{II}}\rho)A^{\mathrm{I}}) = E(\rho^{\mathrm{I}}, A^{\mathrm{I}}),
\end{aligned}$$

where the definition (7.10) of the partial trace was used to transform the second equation into the third. □

7.1.4 Dynamics

As noted on page 111, time does not enter in the definition of the Hilbert space $\mathfrak{H}$ in von Neumann's axiomatization of quantum mechanics. For example, for a single spinless particle, $\mathfrak{H} = L^2(\mathbb{R}^3, \mathrm{d}^3x)$. Dynamics is introduced via an additional assumption which, for a conservative system, is as follows: the Hilbert space $\mathfrak{H}$ carries a continuous unitary representation of a one-parameter group;[6] this parameter is time. This group of unitary transformations defines the Hamiltonian of the system, and conversely; the relation is discussed in Section 7.5.3.

In the Schrödinger picture, state vectors are time-dependent, whereas observables are not. Since the vectors $\{u_i\}$ of any orthonormal basis for $\mathfrak{H}$ are time-independent, the time dependence of a Schrödinger picture state vector $\psi(t) = \sum_i c_i(t)u_i$ must be carried entirely by the expansion coefficients $c_i(t)$. The time evolution of the state vector is governed by the Schrödinger equation. The density matrix is an operator, and not a state vector. Its time evolution is

[6] The term *one-parameter group* means a Lie group which is either the real line $\mathbb{R}$ or the circle S_1.

governed by the *von Neumann equation*

$$\frac{\mathrm{d}\rho}{\mathrm{d}t} = -\mathrm{i}[H, \rho]. \qquad (7.13)$$

(See von Neumann's book, pp. 350–351.) This is to be contrasted with Heisenberg's equation of motion (in the Heisenberg picture) for the operator A:

$$\frac{\mathrm{d}A}{\mathrm{d}t} = \mathrm{i}[H, A]. \qquad (7.14)$$

If H is not time-dependent, then the initial-value problem for (7.13) can be solved explicitly. The solution is

$$\rho(t) = U(t)\rho(0)U^\star(t), \qquad (7.15)$$

where $U(t) = \exp(-\mathrm{i}Ht)$, and $U^\star(t) = [U(t)]^\star$. In this case:

Theorem 7.6 *If time evolution is described by a time-independent Hamiltonian, then the eigenvalues of the density matrix are time-independent.*

This theorem plays an important role in von Neumann's measurement theory, which is the subject of the next chapter.

In the literature, one encounters the statement that an operator A is time-independent. The statement as it stands is slightly ambiguous, and we would like to remove this ambiguity. In this work, this statement will be taken to mean only that the operator A does not depend *explicitly* on the time parameter t; it will *not* imply that A commutes with the Hamiltonian.

7.1.5 Observables

In classical physics, any quantity that admits of an equation of motion may be called a dynamical variable. The notion of an observable is, in some sense, its analogue in quantum mechanics, but there are important differences between the two. The term *observable* seems to have been introduced by Dirac. In the first edition of his *Principles of Quantum Mechanics*, he wrote (Dirac, 1930, p. 25):

> In quantum mechanics it is more convenient to deal with something that refers to one particular time instead of all times, analogous to the value of a classical variable at a particular instant of time. We shall call such a quantity an *observable*.

In von Neumann's formulation, observables become operators on the Hilbert space $\mathfrak{H}$. Since $\mathfrak{H}$ bears no reference to time, observables are, at least at the first stage of definition, time-independent operators. Furthermore, since the result of any measurement is a real number,[7] observables have to be self-adjoint operators. We leave aside, for the time being, the following questions:

(i) Can every self-adjoint operator be called an observable?
(ii) Can every observable be measured in the laboratory?

Observables do not have to be bounded. This introduces severe mathematical complications, some of which are discussed in Appendix A6, and will not be repeated here.

It should be stated that the terms self-adjoint operator and observable are used interchangeably in many works on quantum mechanics. This may (in rare instances) lead to confusion. For example, the Schrödinger picture is often defined as one in which the states carry the time dependence, whereas the operators do not. There is clearly a terminological conflict, for a density matrix is by definition a self-adjoint operator, but it is also time-dependent in most nontrivial cases. Formally, the difficulties can be avoided if one agrees to specify the observables *before* introducing the dynamics, and this is the procedure that we shall follow.[8] The reason for labouring upon this point will become clear in Section 11.1.

7.2 The probability interpretation of quantum mechanics

In this section we shall recapitulate the standard probability interpretation of quantum mechanics. The latter, we recall, consists of a set of assumptions that do not follow from the principles of quantum mechanics. The goal of quantum-mechanical measurement theory may then be stated as follows: *determine the weakest possible set of assumptions*[9] *that have to be adjoined to the principles (axioms) of quantum mechanics to obtain the assumptions of the probability interpretation as theorems.*

Let $H\psi(\boldsymbol{x}) = \mathrm{i}\partial\psi(\boldsymbol{x})/\partial t$ be the Schrödinger equation for a single spinless particle. In this case, the probability interpretation asserts that

$$\int_V |\psi(\boldsymbol{x})|^2 \mathrm{d}\boldsymbol{x}$$

is the probability for finding the particle in the volume V. Implicit in this statement are the assumptions that (a) it is possible to measure (precisely) the position of the particle, and (b) it is possible to prepare the system in the same initial state as often as one wants. In quantum mechanics, the term *measurement*

[7] Technically, this statement can sometimes be generalized to include complex numbers. However, this generalization will not add an iota of physical insight, and will therefore be ignored.

[8] This specification requires a bit of care, and is given in Sections 12.5–12.7, pages 218–225. Until then, we shall not make any use of the technicalities of the definition, and the reader will not be misled by understanding the terms *algebra of observables* and *set of observables* in an intuitive sense.

[9] The term *weakest* is being used here in the mathematical sense.

(of an observable) generally means a set of identical measurements carried out upon every member of an *ensemble* of identically prepared systems, as opposed to a *single* measurement; the adjective 'single' cannot be omitted.

Further development of the probability interpretation[10] proceeds as follows. It is assumed that every observable can be measured.[11] (Von Neumann was apparently the first to make this assumption explicit.) It can then be shown that two or more observables can be measured simultaneously if and only if they commute with each other (von Neumann 1955, p. 201). Single measurements are supposed to be *instantaneous*, at least in the first instance; an operator which does not commute with the Hamiltonian (which generates the time evolution) will change during any finite interval, no matter how small. Let A be a self-adjoint operator with a purely discrete and nondegenerate spectrum, λ_k its eigenvalues and ψ_k the eigenvector corresponding to the eigenvalue λ_k. It is assumed that the system can be prepared repeatedly in a state

$$\psi = \sum_k c_k \psi_k \tag{7.16}$$

in which the coefficients c_k are fixed, but the observer has no a priori knowledge of them (except the normalization condition $\sum_k |c_k|^2 = 1$). The probability interpretation now makes the following assertions:

(i) A single measurement of the observable A yields one of the eigenvalues, say λ_k, as its result. A second single measurement made immediately after the first will yield the same value.[12]
(ii) Repeated measurements of A (on identically prepared systems) yield the value λ_k with probability $|c_k|^2$.

Before the measurement, the ensemble of identically prepared systems is described by the pure state (7.16); after the measurement, it is described by the mixed state

$$\sum_k |c_k|^2 P_{[\psi_k]}, \tag{7.17}$$

where $P_{[\psi]}$ is the projection operator on $\mathfrak{H}$ onto the one-dimensional subspace spanned by the vector ψ:

$$P_{[\psi]} f = (\psi, f) \frac{\psi}{||\psi||} \tag{7.18}$$

[10] Called the *statistical interpretation* by von Neumann.

[11] That is, it is not required that a measuring device corresponding to a given observable be realizable in the laboratory. We shall return to this question in Chapter 10.

[12] We remind the reader that we are dealing with nonrelativistic quantum mechanics; if two successive measurements are performed upon the same system, the time difference between them will not depend on the frame of reference.

for all $f \in \mathfrak{H}$. The transformation

$$\sum_k c_k \psi_k \longrightarrow \sum_k |c_k|^2 P_{[\psi_k]} \tag{7.19}$$

is known as the *reduction* or *collapse of the wave packet.*

The probability interpretation of quantum mechanics was initiated by Max Born in 1926 (Born 1926). We quote from the English translation of this paper in (Wheeler and Zurek, 1983, p. 54):

> If one translates this result into terms of particles, only one interpretation is possible: $\Phi_{n,m}(\alpha, \beta, \gamma)$ gives the probability* for the electron...

The star on the word 'probability' above refers to a footnote, which reads:

> Addition in proof: More careful consideration shows that the probability is proportional to the square of the quantity $\Phi_{n,m}$.

In the history of the printed word, there could have been few footnotes as consequential as this.

7.2.1 Quantum theory of measurement

The term *measurement* is used in a very restricted sense in what is known as the quantum theory of measurement. This theory has only one aim, and that is to account for the process (7.19). It does not mean that determinations of e/m or of scattering cross-sections are not measurements; it does mean that such measurements are well outside the scope of the theories of measurement that we shall discuss in subsequent chapters. The problem that has exercised physicists for the better part of a century is whether or not the process (7.19) can be accounted for by the standard assumptions of quantum mechanics.

7.2.2 Remarks on notations

In the rest of this book, the letters E and P will sometimes be used simultaneously in two different senses each: E for expectation values and projection operators, and P for projection and momentum operators. When E is used to denote expectation values, its argument will be enclosed in brackets, as in Section 7.1.2; when used to denote projection operators, its argument will be denoted by a subscript, which may be a single letter, as in Section 7.3, or a more complicated expression (see Appendix A6). The letter P, with or without subscripts, will stand for momentum operators; but when the subscript is a vector in some Hilbert space *enclosed within square brackets*, as defined in (7.18) above, it will denote the projection operator onto the one-dimensional subspace spanned by that vector. In addition, the letter E will be used to denote the energy, a spectral value of the generator of time translations, in Section 7.4.

7.3 Superselection rules

Suppose that the Hilbert space of a quantum-mechanical system can be decomposed into the direct sum of two nonzero subspaces, $\mathfrak{H} = \mathfrak{H}_1 \oplus \mathfrak{H}_2$, such that the relative phase of any two vectors $\varphi_1 \in \mathfrak{H}_1$ and $\varphi_2 \in \mathfrak{H}_2$ is nonmeasurable, i.e., has no observable consequences. Put differently, for any two fixed vectors φ_1 and φ_2, any observable O and any $\alpha \in \mathbb{R}$, it is not possible to distinguish, experimentally, between $(\varphi_1, O\varphi_2)$ and $(\varphi_1, O\mathrm{e}^{\mathrm{i}\alpha}\varphi_2)$. In particular (choosing $\alpha = \pi$), it is not possible to distinguish, experimentally, between $(\varphi_1, O\varphi_2)$ and $-(\varphi_1, O\varphi_2)$. This will surely be the case *if no observable has a nonvanishing matrix element between* $\varphi_1 \in \mathfrak{H}_1$ *and* $\varphi_2 \in \mathfrak{H}_2$.

Such a situation will arise if there exists a self-adjoint operator on $\mathfrak{H}$ which is not a multiple of the identity and commutes with every observable of the system. Let M be such an operator, and assume that the spectrum of M is discrete. Let the eigenvalues of M be μ_j, $j \in \mathbf{N}$. We may assume, without loss of generality, that $\mu_i \neq \mu_j$ for $i \neq j$. We may then write

$$\mathfrak{H} = \bigoplus_j \mathfrak{H}_j \text{ and } M = \sum_j \mu_j E_j,$$

where $\mathfrak{H}_j$ is the subspace of $\mathfrak{H}$ belonging to the eigenvalue μ_j of M, and E_j the projection operator onto $\mathfrak{H}_j$. Let O be any observable, $i \neq j$, $\psi_i \in \mathfrak{H}_i$, $\psi_j \in \mathfrak{H}_j$. From $[O, M] = 0$ we find that

$$(\mu_i - \mu_j)(\psi_i, O\psi_j) = 0. \tag{7.20}$$

But $\mu_i - \mu_j \neq 0$, and therefore

$$(\psi_i, O\psi_j) = 0. \tag{7.21}$$

Equation (7.21) describes what is known as a *superselection rule*,[13] and the subspaces $\mathfrak{H}_j$ are called *superselection sectors* of the Hilbert space $\mathfrak{H}$. If φ and ψ are vectors belonging to different superselection sectors, then their relative phase cannot be determined.

Suppose that a given Hilbert space can be decomposed into the direct sum of two subspaces, one that has only states of half-integral spin and the other that has only states of integral spin. A rotation by 2π (about any axis) will change the sign of a half-integral spin state, but will leave the sign of an integral spin state unchanged. But this rotation is the identity operation, which should change nothing. It follows that no observable can have a nonvanishing matrix element between a half-integral spin state and an integral spin state. In plain words, it is physically meaningless to superpose states with integral spin and

[13] In the example given, the superselection rule is generated by the operator M.

those with half-odd integral spin. This result is known as the superselection rule for *univalence*.[14]

If an observable is additively conserved and commutes with every observable, then there can be no operator that effects a transition between states with different eigenvalues of this observable; the result will be a superselection rule. It is believed that electric charge and baryon number generate superselection rules.

The concept of superselection rules was introduced in (Wick, Wightman and Wigner, 1952). Superselection sectors have become very significant in relativistic quantum field theory. For an introduction, see (Haag 1993).

7.4 The Galilei group

The invariance group of nonrelativistic quantum mechanics is the universal covering group of the inhomogeneous Galilei group G, together with the discrete operations of space inversion and time reversal. The covering group of G is obtained by replacing the subgroup of rotations in 3-space by its covering group $SU(2)$, and has the effect of introducing half-integral spins. Space inversion leads to the concept of parity (see (Wigner, 1959)), but we shall not be concerned with it. Time inversions are special: according to the theorem of Wigner which lies at the foundation of the theory of symmetry in quantum mechanics,[15] a symmetry operation is represented by a unitary operator on Hilbert space – *with the exception of time reversal*, which has to be represented by an *anti*unitary operator. Formally, it lies beyond the theory of unitary group representations, but is easily accommodated. See (Wigner, 1959, Chapter 26).

The inhomogeneous Galilei group G is a ten-parameter group of linear transformations on space and time, which may be displayed explicitly as follows:

$$\begin{aligned} t' &= t + b, \\ \boldsymbol{x}' &= R\boldsymbol{x} + \boldsymbol{v}t + \boldsymbol{a}. \end{aligned} \tag{7.22}$$

In the above, b is a time translation, $\boldsymbol{a}$ a space translation, $\boldsymbol{v}$ a velocity boost or simply boost[16] and R a rotation in three-dimensional space.

If we write an element $g \in G$ as

$$g = (b, \boldsymbol{a}, \boldsymbol{v}, R), \tag{7.23}$$

[14] The superselection rule for univalence can also be derived from invariance under time reversal (Wick, Wightman and Wigner, 1952). The argument is the one used for deriving (7.21); the (antiunitary) operator of time reversal is not a multiple of the identity. We have chosen the argument using the 2π rotation because of its similarity to the one leading to Bargmann's superselection rule (Subsection 7.4.2), which is probably the most important superselection rule in nonrelativistic quantum mechanics. A full treatment of time reversal may be found in Chapter 26 of (Wigner, 1959).

[15] See (Wigner, 1959); for clearer proofs see (Emch and Piron, 1963) or (Bargmann, 1964).

[16] The term *boost* is attributed to Wightman. Inönü and Wigner used the terms *pure Galilei transformation* or *acceleration* (Inönü and Wigner, 1952).

then the multiplication law may be written as

$$(b', \boldsymbol{a}', \boldsymbol{v}', R')\,(b, \boldsymbol{a}, \boldsymbol{v}, R) = (b' + b,\ \boldsymbol{a}' + R'\boldsymbol{a} + \boldsymbol{v}'b,\ \boldsymbol{v}' + R'\boldsymbol{v},\ R'R). \tag{7.24}$$

The identity element is[17]

$$e = (0, 0, 0, 1)$$

and the inverse of g is

$$g^{-1} = (-b, -R^{-1}(\boldsymbol{a} - b\boldsymbol{v}), -R^{-1}\boldsymbol{v},\ R^{-1}).$$

It is easily seen that (i) time translations, (ii) space translations, (iii) boosts and (iv) rotations constitute subgroups of G. The group of rotations in n-dimensional Euclidean space is identical with the group of orthogonal $n \times n$ matrices of determinant 1, which is denoted by $O(n, \mathbb{R})$. The Euclidean group in three dimensions is the subgroup of G consisting of rotations and space translations: $\{(0, \boldsymbol{a}, 0, R)\}$. It is evidently isomorphic with the subgroup $\{(0, 0, \boldsymbol{v}, R)\}$ of the boosts and rotations.

Since G is a real Lie group, its Lie algebra is also an algebra over the reals. In physics one is interested in unitary representations in which the generators are represented by Hermitian operators.[18] Since the commutator of two Hermitian operators is necessarily anti-Hermitian, it may be written as i times a Hermitian operator, where $\mathrm{i} = \sqrt{-1}$. It is customary in physics to write Lie algebras in this manner. We shall follow this practice for the Lie algebra $\mathfrak{g}$ of the group G.

We shall denote by J_i, K_i and P_i, $i = 1, 2, 3$, the generators of the rotations, boosts and space translations respectively, and by H the generator of time translations. They satisfy the commutation relations

$$\begin{aligned}
[J_i, J_j] &= \mathrm{i}\epsilon_{ijk} J_k, \\
[J_i, K_j] &= \mathrm{i}\epsilon_{ijk} K_k, \\
[J_i, P_j] &= \mathrm{i}\epsilon_{ijk} P_k, \\
[K_i, H] &= \mathrm{i} P_i,
\end{aligned} \tag{7.25}$$

where ϵ_{ijk} is the Levi-Civita symbol (the completely antisymmetric tensor in three dimensions with $\epsilon_{123} = 1$). All commutators that do not appear in (7.25)

[17] We are using the ordinary numerals 0 and 1 to denote the zero vector and the identity rotation respectively.

[18] Unless the group is compact, some of these generators will be unbounded operators which are not defined everywhere on the Hilbert space (see Appendix A6). It is fairly easy to avoid the pitfalls. We shall omit the cautionary remarks to achieve a more fluent exposition.

vanish. It may be verified from (7.25) that the quantities

$$\boldsymbol{P}^2 \text{ and } (\boldsymbol{K} \times \boldsymbol{P})^2 \tag{7.26}$$

commute with all elements of $\mathfrak{g}$; the symbols $\boldsymbol{K}$ and $\boldsymbol{P}$ denote 3-vectors.

It follows that $\boldsymbol{P}^2$ must be a multiple of the identity in any irreducible representation of G. The irreducible unitary representations[19] of G were analysed by Inönü and Wigner, who showed that they did not contain states that were localizable either in space or in velocity space, and concluded that none of them could admit of a particle interpretation (Inönü and Wigner, 1952).[20]

7.4.1 Projective representations of the Galilei group

It was shown by Bargmann (Bargmann, 1954) that a nonrelativistic particle of nonzero mass is described, not by a true or vector representation of Galilei group G but rather by a projective or ray representation of it. These will be defined below. We shall confine ourselves to connected Lie groups; however, much of what follows applies to disconnected Lie groups and to finite groups as well.

7.4.1.1 Factor systems and group exponents

Let F be a connected Lie group, $\mathfrak{H}$ a finite- or infinite-dimensional Hilbert space over the complex numbers,[21] and $\mathcal{D}(\mathfrak{H})$ the set of invertible linear transformations of $\mathfrak{H}$ onto itself. An **up-to-a-factor representation** (D, ω) of F upon $\mathfrak{H}$ is a map $D : F \to \mathcal{D}$ that satisfies the condition

$$D(x_1)D(x_2) = \omega(x_1, x_2)D(x_1x_2), \tag{7.27}$$

where $D(x)$ is the image of $x \in F$ in $\mathcal{D}$, and $\omega(x_1, x_2)$ is a complex-valued function of modulus unity which is jointly continuous in x_1 and x_2. The function ω, called a **factor system** on F, has to satisfy the condition

$$\omega(x_1, x_2x_3)\,\omega(x_2, x_3) = \omega(x_1, x_2)\,\omega(x_1x_2, x_3), \tag{7.28}$$

which follows from the associativity of multiplication in F.

Let now $D'(x) = \varphi(x)D(x)$, where $\varphi(x)$ is a continuous complex-valued function on F of unit modulus. Straightforward calculations show that the pair

[19] Henceforth the term *representation* of a group will always denote a *unitary* representation, unless the contrary is stated explicitly.

[20] A nonrigorous but basically correct argument is as follows. Since $\boldsymbol{P}$ is the generator of space translations, its spectral values $\boldsymbol{p}$ are momenta, and if $\boldsymbol{p}^2 = \text{const}$ there are simply not enough momenta available to form a δ-function in three-dimensional space.

[21] Here we are using the term *Hilbert space* as it was originally defined by von Neumann (Appendix A6, pages 316–317. Nowadays many physicists use this term *only* for infinite-dimensional separable Hilbert spaces, and mathematicians no longer assume separability.

(D', ω') is an up-to-a-factor representation of F, with the factor system

$$\omega'(x_1, x_2) = \frac{\varphi(x_1)\varphi(x_2)}{\varphi(x_1 x_2)}\,\omega(x_1, x_2). \tag{7.29}$$

We now come to the essential definition.

Definition 7.7 (Equivalence of up-to-a-factor representations) The up-to-a-factor representations (D, ω) and (D', ω') are said to be **equivalent**, written $D \sim D'$, if there exists a continuous function φ on F of modulus unity such that the conditions (7.29) are satisfied.

It is easily seen that the relation $\sim$ is a true equivalence relation in the mathematical sense (see page 263). It therefore divides the set of up-to-a-factor representations of F into pairwise-disjoint equivalence classes. If an equivalence class contains an up-to-a-factor representation with the factor system $\omega(x_1, x_2) = 1$ for all $x_1, x_2 \in F$, then its members are called **true** or **vector** representations. If it does not, then its members are called **ray** or **projective** representations.[22]

Henceforth the term *representation* will mean a true representation. Projective (or ray) representations will be explicitly specified as such.

Factor systems can be simplified as follows. Setting $x_1 = x_2 = e$ in (7.27), we find that $D(e) = \exp(\mathrm{i}\gamma)$ for some $\gamma \in \mathbb{R}$. Using this fact, we find from (7.27) that $\omega(e, x) = \omega(x, e) = \exp(\mathrm{i}\gamma)$ for all $x \in F$. Now, choosing $\varphi(x) = \exp(-\mathrm{i}\gamma)$ for all $x \in F$ in (7.29), we find that

$$\omega'(e, x) = \mathrm{e}^{-\mathrm{i}\gamma}\omega(e, x) = 1 = \omega'(x, e) \quad \text{for all} \quad x \in F.$$

That is, any up-to-a-factor representation is equivalent to a simplified one with $\omega(e, x) = \omega(x, e) = 1$ for all $x \in F$.

A continuous real-valued function $\xi : F \times F \to \mathbb{R}$ is called a **group exponent** if

$$\xi(e, e) = 0, \tag{7.30}$$

and, for any $x_1, x_2, x_3 \in F$,

$$\xi(x_1, x_2) + \xi(x_1 x_2, x_3) = \xi(x_2, x_3) + \xi(x_1, x_2 x_3), \tag{7.31}$$

which is the counterpart of (7.28). It follows from (7.31) and (7.30) that

$$\xi(x, e) = \xi(e, x) = 0 \tag{7.32}$$

[22] In the literature, true representations are often regarded as a subclass of ray representations. We wish to keep the two strictly separate, which we do by defining them to be distinct subclasses of up-to-a-factor representations. This term was introduced by Wigner, and is seldom used today.

and

$$\xi(x, x^{-1}) = \xi(x^{-1}, x) \tag{7.33}$$

for all $x \in F$. Factor systems and group exponents (in their simplified forms) are related as follows:

$$\omega(x, y) = e^{i\xi(x,y)}. \tag{7.34}$$

7.4.1.2 The extended Galilei group

A ray representation of the Galilei group G is a true representation of a larger group $\widetilde{G}$, which may be defined as follows. Form the Cartesian product $\mathbb{R} \times G$, and, given a group exponent ξ on G, define a multiplication on $\mathbb{R} \times G$ by

$$(\theta', g') \cdot (\theta, g) = (\theta' + \theta + \xi(g', g), g'g), \tag{7.35}$$

where $\theta, \theta' \in \mathbb{R}$ and $g, g' \in G$. Owing to (7.31), this multiplication is associative. It is also invertible; using (7.32) and (7.33) one finds that it has the identity $(0, e)$, and that the inverse of $\tilde{g} = (\theta, g)$ is

$$\tilde{g}^{-1} = (\theta, g)^{-1} = (-\theta - \xi(g^{-1}, g), g). \tag{7.36}$$

The **extended Galilei group** $\widetilde{G}$ is the set $\mathbb{R} \times G$ furnished with the multiplication law (7.35). One sees by straightforward computations that $\Theta = \{(\theta, e)\}$ is a normal subgroup of $\widetilde{G}$, and that $G \sim \widetilde{G}/\Theta$. The group $\widetilde{G}$ is called a **central extension**[23] of G by Θ. Moreover,

$$(\theta, g) = (\theta, e)(0, g). \tag{7.37}$$

Finally, we have to show that a ray representation of G is a true representation of $\widetilde{G}$. Let $D(g)$ be a ray representation of G on $\mathfrak{H}$ with factor system

$$\omega(g', g) = e^{i\eta(g',g)} \tag{7.38}$$

such that $\omega(g, e) = \omega(e, g) = 1$ for all $g \in G$. Set

$$\xi(g', g) = \frac{1}{m}\eta(g', g), \tag{7.39}$$

[23] Let F, G, H be groups such that H is a normal subgroup of F and G is isomorphic with F/H. Then F is called an **extension** of G by H. If H belongs to the centre of F, the extension is called **central**. A fairly detailed account, addressed to physicists, may be found in (Michel, 1964).

where $m \neq 0$, and define the operator $\widetilde{D}(\theta, g)$ on $\mathfrak{H}$ by

$$\widetilde{D}(\theta, g) = \mathrm{e}^{\mathrm{i}m\theta} D(g). \tag{7.40}$$

Then, using (7.37), (7.38) and (7.40) we find that

$$\begin{aligned}\widetilde{D}(\theta', g') \cdot \widetilde{D}(\theta, g) &= \mathrm{e}^{\mathrm{i}m(\theta'+\theta)} D(g') \cdot D(g) \\ &= \mathrm{e}^{\mathrm{i}m[\theta'+\theta+\xi(g',g)]} D(g'g) \\ &= \widetilde{D}(\theta' + \theta + \xi(g', g),\, g'g).\end{aligned} \tag{7.41}$$

Equation (7.41) shows that the set of operators $\{\widetilde{D}(\theta, g)|\theta \in \mathbb{R}, g \in G\}$ forms a true representation of $\widetilde{G}$ on $\mathfrak{H}$. This representation depends on a real parameter $m \neq 0$. Conversely, given a true representation $\widetilde{D}$ of $\widetilde{G}$ on $\mathfrak{H}$, the set of operators $\{D(g)|g \in G\}$ defined by

$$D(g) = \widetilde{D}(0, g)$$

constitute a ray representation of G, with the factor system $\omega(g', g)$ being defined by (7.38).

7.4.1.3 Mass in nonrelativistic quantum mechanics

The parameter m in (7.39) is identifiable with the nonrelativistic mass of a particle. To see this most clearly, consider the Lie algebra $\tilde{\mathfrak{g}}$ of $\widetilde{G}$. This differs from $\mathfrak{g}$ only in the following. In $\mathfrak{g}$, one set of vanishing Lie brackets is

$$[P_i, K_j] = 0.$$

In $\tilde{\mathfrak{g}}$, this is replaced by the nonvanishing bracket

$$[P_i, K_j] = \mathrm{i}\delta_{ij} m \cdot I, \tag{7.42}$$

where I is the eleventh element of $\tilde{\mathfrak{g}}$, the other ten being the same as those of $\mathfrak{g}$. The element I *commutes with every element of* $\mathfrak{g}$. Owing to the commutator (7.42), the Casimir operators of $\tilde{\mathfrak{g}}$ differ drastically from those of $\mathfrak{g}$. It is easily verified, by direct computation, that the quantity

$$H - \frac{\boldsymbol{P}^2}{2m} \tag{7.43}$$

commutes with every element of $\tilde{\mathfrak{g}}$. It is called the *internal energy*, and is a constant in every irreducible representation of $\widetilde{G}$. As there is no loss of generality in doing so, it is equated to zero in every such representation. The relation between mass, energy and momentum of a free particle in nonrelativistic quantum mechanics follows from this stipulation.

7.4.2 Bargmann's superselection rule; conservation of mass

The group exponents $\xi(g', g)$ in (7.35) are defined only up to equivalence. One possible choice of $\xi(g', g)$ is (we omit the calculations; for details, see (Lévy-Leblond, 1972))

$$\xi(g', g) = (\tfrac{1}{2}v'^2 b + \boldsymbol{v}' \cdot R'\boldsymbol{a}). \tag{7.44}$$

Let now $\mathfrak{H}_1$ and $\mathfrak{H}_2$ be Hilbert spaces that carry the irreducible projective spin-zero representations of G with masses m_1 and m_2 respectively, where $m_1 \neq m_2$, and form their direct sum $\mathfrak{H} = \mathfrak{H}_1 \oplus \mathfrak{H}_2$. Let $\varphi \in \mathfrak{H}_1$ and $\psi \in \mathfrak{H}_2$. Then $\varphi + \psi$ is a vector in $\mathfrak{H}$. Assume that we are using the group exponent (7.44). Apply, successively, the transformations

$$(0, \boldsymbol{a}, 0, 1),\ (0, 0, \boldsymbol{v}, 1),\ (0, -\boldsymbol{a}, 0, 1) \text{ and } (0, 0, -\boldsymbol{v}, 1)$$

(which together constitute the identity transformation) to the state $\varphi + \psi \in \mathfrak{H}$. An easy calculation shows the result to be

$$\varphi + \psi \longrightarrow \mathrm{e}^{-\mathrm{i}m_1 \boldsymbol{a}\cdot\boldsymbol{v}}\varphi + \mathrm{e}^{-\mathrm{i}m_2 \boldsymbol{a}\cdot\boldsymbol{v}}\psi. \tag{7.45}$$

Since $m_1 \neq m_2$, the relative phases of φ and ψ differ on the two sides of (7.45). Since this difference has resulted from an identity transformation, it follows that the relative phases of φ and ψ are undefined, so that no meaning can be attached to the superposition $\varphi + \psi$. In other words, *no observable can have a nonvanishing matrix element between states of different mass in nonrelativistic quantum mechanics.*

This result is known as *Bargmann's superselection rule.*

The pioneering work on ray representations of Lie groups is that of Bargmann (Bargmann, 1954). The book by Hamermesh (Hamermesh, 1962) contains an elementary account of the general theory (both for finite and Lie groups), as well as several applications. The review article by Lévy-Leblond (Lévy-Leblond, 1972) is devoted exclusively to the Galilei group. The subject of ray representations has been absorbed into the subject of group extensions, which in turn has been absorbed into a branch of mathematics called the cohomology of groups. An introductory account, aimed at physicists, may be found in Michel's Istanbul lectures (Michel, 1964).

It may be worth remarking that the notion of particle mass in nonrelativistic *classical* mechanics follows from the cohomology of the Galilei group (which does not depend on Hilbert space representations).

7.4.3 The Galilei group in two dimensions

Central extensions $\widetilde{G}_2$ of the Galilei group in two spatial dimensions, G_2, have been studied by Bose, motivated in part by the question of excitations with fractional (not half-integral) angular momenta. Although such angular momenta did not turn up, Bose found a significant difference between the groups in two and three dimensions. Central extensions of the Lie algebra $\mathfrak{g}_2$ of G_2 form a three-parameter family $\tilde{\mathfrak{g}}_2$, whereas those of G_2 itself form only a two-parameter family. It is the universal covering group G_2^{cov} of G_2 that has a three-parameter family of central extensions, and the Lie algebras of these groups are precisely the $\tilde{\mathfrak{g}}_2$. The third element of the extension of $\mathfrak{g}_2$ is *not a complete vector field on a* $\widetilde{G}_2$. For details, we refer the reader to his papers (Bose, 1995a,b) which are very readable if the reader is willing to accept a little bit of homological algebra on faith.[24]

7.5 Theorems of von Neumann and Stone; Reeh's example

In this section we shall state two fundamental theorems upon which the quantum mechanics of N particles has been based. They are known as the Stone–von Neumann uniqueness theorem and Stone's theorem.[25] The first of these has often been misinterpreted.

To avoid the difficulties associated with unbounded operators, Weyl proposed replacing the canonical commutation relations (CCR), for a finite number of degrees of freedom, by a *group* which has since become known as the *Weyl group.* This is a Lie group, and the CCR are its Lie algebra. The Stone–von Neumann uniqueness theorem asserted that the Weyl group had only one irreducible unitary representation. This was taken by some physicists to mean that the CCR had only one irreducible unitary representation, the familiar Schrödinger representation. Quantum field theories were seen to differ from quantum mechanics in that, for infinitely many degrees of freedom, the CCR (and the anticommutation relations as well) had infinitely many inequivalent representations.

Mathematicians have known, at least since 1967, that the CCR, as opposed to the Weyl group, possessed inequivalent irreducible representations even for one degree of freedom (Fuglede, 1967). In 1972, Reed and Simon gave an example of one that was not equivalent to the Schrödinger representation (Reed and Simon, 1972, pp. 274–276). In 1983 Schmüdgen constructed infinitely many inequivalent irreducible representations (Schmüdgen, 1983). However, no physical relevance seems to have been attached to these examples. In 1988 Helmut Reeh, using the magnetic Aharonov–Bohm effect, constructed a representation that was not

[24] A topological–geometrical interpretation of these phenomena in terms of vector fields on manifolds may be found in Appendix A8, particularly Section A8.7 and Subsection A8.7.3.

[25] Many authors refer to the Stone–von Neumann uniqueness theorem as von Neumann's uniqueness theorem. It was announced independently by Stone and von Neumann in 1930, but von Neumann gave a complete proof.

unitarily equivalent to the Schrödinger representation for two degrees of freedom (Reeh, 1988), showing that these representations were not mere mathematical curiosities that could be disregarded by the physicist. In 2001, Summers wrote that 'seventy years ago, this example would have been a bombshell' (Summers, 2001). We shall discuss Reeh's example in Section 7.5.2, and return to it in the Epilogue.

7.5.1 The Stone–von Neumann uniqueness theorem

Let the system consist of a finite number of spinless particles in one or more spatial dimensions. The subscripts j, k on the canonical coordinates and momenta Q_j, P_k will run from 1 to N, where N is a positive integer. The CCR for the system will then be

$$\begin{aligned} Q_j P_k - P_k Q_j &= \mathrm{i}\delta_{jk} I, \\ Q_j Q_k - Q_k Q_j &= 0, \\ P_j P_k - P_k Q_j &= 0, \end{aligned} \tag{7.46}$$

where we are using units in which $\hbar = 1$. The quantity I commutes with every Q_j and P_j, and physics dictates that it should be the identity. The last requirement is essential; the 3×3 matrices

$$Q = \begin{pmatrix} 0 & 0 & 0 \\ 0 & 0 & 1 \\ 0 & 0 & 0 \end{pmatrix}, \quad P = \begin{pmatrix} 0 & 1 & 0 \\ 0 & 0 & 0 \\ 0 & 0 & 0 \end{pmatrix}, \quad R = \begin{pmatrix} 0 & 0 & \mathrm{i} \\ 0 & 0 & 0 \\ 0 & 0 & 0 \end{pmatrix} \tag{7.47}$$

satisfy $[Q, P] = \mathrm{i}R$, but R is not the identity matrix.[26]

The problem is to determine the inequivalent irreducible unitary representations of the CCR (7.46) on a Hilbert space on which I is represented by the identity operator. This space cannot be finite-dimensional; for, if the Q_j, P_k are $n \times n$ matrices, then, taking the traces of both sides of the first equation in (7.46), we obtain $0 = \mathrm{i}n$. But if the Hilbert space is infinite-dimensional, then at least one of each canonically conjugate pair (P_i, Q_i) is unbounded, and therefore not defined everywhere on it (see Section A6.2, particularly Theorem A6.8, page 324). Weyl therefore chose to work instead with the exponentiated, or *Weyl forms* of the CCR, and his lead has been followed ever since. The Weyl forms are defined by (Weyl, 1931, p. 274)

$$A_j(a_j) = \exp(\mathrm{i}a_j Q_j) \quad \text{and} \quad B_k(b_k) = \exp(\mathrm{i}b_k P_k), \tag{7.48}$$

[26] The matrices Q, P and R form the Lie algebra of a group known as the *Heisenberg group*, which is a 3-parameter Lie group that generalizes to $2n + 1$ dimensions, and is of some mathematical interest. See (Tilgner, 1970) for an introduction.

where $j, k = 1, \ldots, N$, $a_j, b_k \in \mathbb{R}$, and *there is no summation over repeated indices*. The multiplication rules for the $A_j(a_j), B_k(b_k)$ are

$$\begin{aligned} A_j(a_j)A_j(a'_j) &= A_j(a_j + a'_j), \\ B_j(b_k)B_j(b'_j) &= B_j(b_j + b'_j), \\ A_j(a_j)B_k(b_k) &= B_k(b_k)A_j(a_j)\exp\left[\mathrm{i}a_j b_k \cdot \delta_{jk}\right]. \end{aligned} \tag{7.49}$$

The $A_j(a_j)$ form an N-parameter Abelian group; the group space is $\mathbb{R}^n$. The same is true of the $B_k(b_k)$. The $A_j(a_j)$ and $B_k(b_k)$ together form a $2N$-parameter group, with the identity element

$$I = A_i(0) = B_j(0), \ \ i, j = 1, \ldots, N,$$

where the I is the same I that appears in the first of the commutation relations (7.46). The last equation of (7.49) shows that this group is no longer Abelian. We shall denote it by W, or more explicitly by $W_N(Q, P)$ (W for Weyl). The *commutator* or *derived subgroup* of W is clearly $W' = \{zI; z \in \mathbb{C}, |z| = 1\}$, and $W'' = I$; the group W is *solvable* (see (Jacobson, 1974, pp. 238–239)). Recall that a group G is *simple* if the only normal subgroups it has are G and $\{e\}$, where e is the identity of G. The Weyl group is neither simple nor semisimple.

The Stone–von Neumann uniqueness theorem may be stated succinctly as follows; for details, the reader is referred to the historical review by Summers (Summers, 2001).

Theorem 7.8 (Stone–von Neumann uniqueness theorem) *The group $W_N(Q, P)$ has, up to unitary equivalence,* only one *continuous irreducible representation, and the operators Q_k, P_j of the canonical commutation relations are its infinitesimal generators.*

Recall that while the group action is analytic in a Lie group, it is only required to be continuous in a unitary representation (which is necessarily infinite-dimensional) if the group is noncompact.

The full proof of this theorem, given by von Neumann (von Neumann, 1930), is beyond the scope of this book.[27] However, there is a different proof of the first part, due to Mackey, which is based on ideas more familiar to physicists. The key idea of this proof is sketched below.

We begin by recalling the inducing construction, adapted by Wigner (from the works of Frobenius on finite groups) for his study of the unitary representations of the Poincaré group (Wigner, 1939), and developed into a complete

[27] The mathematically prepared reader will find an absorbing account of this theorem and its impact upon physics in (Summers, 2001).

mathematical theory of infinite-dimensional group representations by Mackey.[28] The representations of the whole group are induced from those of a certain subgroup, called the *little group* by Wigner (and *isotropy group* or *stabilizer* in the mathematical literature). Every irreducible representation of the little group gives rise to a representation of the whole group, and representations that arise from inequivalent representations of the little group are themselves inequivalent. If the little group consists of the identity alone, then the whole group will have only one irreducible representation. Mackey used this method to determine the irreducible representations of the group $W_N(Q,P)$, and found that the little group consisted of the identity alone. The Stone–von Neumann uniqueness theorem followed immediately. Mackey's proof may be found in (Mackey, 1968, pp. 53–57) or (Mackey, 1978, pp. 180–181). It is based on his imprimitivity theorem, which should be easily accessible to the reader who is familiar with projection-valued measures (defined on page 333).

7.5.2 Reeh's example

The discussion of Subsection 7.5.1 may be summarized in two sentences. The canonical commutation relations (7.46) constitute the Lie algebra $\mathfrak{w}$ of the Lie group $W = W_N(Q,P)$. The latter has only one irreducible unitary representation. The problem is that the correspondence between Lie groups and Lie algebras is not one-to-one; while every Lie group has a uniquely defined Lie algebra, the converse is not true. For example, if F is a simply connected Lie group and Δ any discrete normal subgroup of F, then F and F/Δ have the same Lie algebra. The fact that nonisomorphic Lie groups may be locally isomorphic (i.e., have the same Lie algebra) was shown by Michel to have implications for elementary particle physics (Michel, 1964). Reeh's example, outlined below, can be viewed as demonstrating that its implications for physics may be far wider.

As mentioned earlier, the example is based on the magnetic Aharonov–Bohm effect. We refer the reader to the book by Peshkin and Tonomura for early disputes and experimental confirmation of the effect (Peshkin and Tonomura, 1989), and consider an idealized model of a charged (spinless) particle moving in the xy-plane which is free of any magnetic field, except for a flux line of intensity α' at $z = 0$. The quantity α' cannot be changed by a gauge transformation. If $\alpha' \neq 0$, then the configuration space for the particle is the punctured plane $\mathbb{R}^2 \smallsetminus O$; if $\alpha' = 0$, i.e., there is no flux line, then the configuration space is the entire plane $\mathbb{R}^2$. In the first case ($\alpha' \neq 0$), the vector potential $\boldsymbol{A}(x,y)$ at the point (x,y) on the circle $C_r(O)$ with centre $O = (0,0)$ and radius r will be gauge-equivalent to $\alpha'\boldsymbol{a} = \alpha'\boldsymbol{e}/r$, $\boldsymbol{e}$ being the unit tangent vector[29] to $C_r(O)$

[28] In the following, all representations will be assumed to be unitary unless the contrary is explicitly stated; representations such as (7.47) are thereby excluded.

[29] In this section we shall denote two-vectors by boldface italic letters.

at (x, y). The Schrödinger operators for this case will be

$$\begin{aligned} \boldsymbol{P}' &= \mathrm{i}\frac{\partial}{\partial \boldsymbol{x}} + \alpha \boldsymbol{a}, \\ \boldsymbol{Q}' &= \text{multiplication by } \boldsymbol{x}, \end{aligned} \tag{7.50}$$

where we have set $\alpha = e\alpha'$, e being the electronic charge.[30] When the configuration space is the entire plane (when the vector potential is gauge-equivalent to zero), the Schrödinger operators will assume their standard form

$$\begin{aligned} \boldsymbol{P} &= \mathrm{i}\frac{\partial}{\partial \boldsymbol{x}}, \\ \boldsymbol{Q} &= \text{multiplication by } \boldsymbol{x}. \end{aligned} \tag{7.51}$$

The $\boldsymbol{P}'$ and $\boldsymbol{Q}'$ obey the same commutation relations as the $\boldsymbol{P}$ and $\boldsymbol{Q}$; $\boldsymbol{a}(\boldsymbol{x})$ obviously commutes with $\boldsymbol{x}$. But since the the magnetic flux α is independent of the choice of gauge, it follows that:

(i) When α is not an integer, the vector potential in the punctured plane cannot be gauged away (owing to the 'topological obstruction' at the origin), so that the representation of $\boldsymbol{P}', \boldsymbol{Q}'$ cannot be transformed unitarily into that of $\boldsymbol{P}, \boldsymbol{Q}$; the representation of $\boldsymbol{P}', \boldsymbol{Q}'$ is 'non-Fock'.

(ii) The representations of $\boldsymbol{P}', \boldsymbol{Q}'$ for different noninteger values of α cannot be transformed unitarily into each other.

Reeh proved the first of the above results by a close examination of the domains of unbounded self-adjoint operators involved. The details are beyond the scope of this book. The second result is not mentioned explicitly by Reeh, but follows from his analysis.

We should like to make a final comment before concluding this subsection. The examples constructed by Schmüdgen are based on the operators

$$Q = x - \mathrm{i}\frac{\partial}{\partial y}, \quad P = -\mathrm{i}\frac{\partial}{\partial x} \tag{7.52}$$

on $L^2(\mathbb{R}^2)$, on which they are well-defined unbounded operators. However, the operator Q of (7.52) *is not a derivation* of any algebra of functions on $\mathbb{R}^2$; it does not annihilate the constants, and therefore cannot be identified with a vector field on a two-dimensional manifold.[31]

[30] The spin of the electron is generally disregarded in the theory of the Aharonov–Bohm effect. The validity of this assumption appears to be confirmed by experiment. See (Peshkin and Tonomura, 1989).

[31] A brief discussion of differentiable manifolds, Lie groups and Lie algebras is given in Appendix A8. The operator of multiplication by x is never a derivation on the algebra of functions of x.

7.5.3 Stone's theorem

Let $\{U(t)\}$, $t \in \mathbb{R}$ be a one-parameter group of unitary operators ('unitaries', for short) on a Hilbert space $\mathfrak{H}$. This group is said to be *strongly continuous* in t (see Appendix A6) if

$$\lim_{t \to t_0} ||[U(t) - U(t_0)]\varphi|| = 0 \text{ for every } \varphi \in \mathfrak{H},\ t_0 \in \mathbb{R}.$$

Stone's theorem asserts that if $\{U(t)\}$ is a strongly continuous group of unitaries on $\mathfrak{H}$, then there exists a self-adjoint operator H on $\mathfrak{H}$ such that

$$U(t) = \mathrm{e}^{-\mathrm{i}Ht}.$$

The proof of this theorem requires considerably more machinery than we are able to develop here. (It requires care to define unbounded self-adjoint operators; the definition is given in Section A6.5.) The interested reader is referred to (Reed and Simon, 1972). We shall content ourself with a few comments.

(i) If A is a bounded operator, the formal power series for $\exp \mathrm{i}A$ converges in the norm, and may therefore be taken to define the exponential. This method fails if A is unbounded. In quantum mechanics, Hamiltonians are generally unbounded self-adjoint operators, and therefore one has, first of all, to devise a method for defining their exponentials (and other functions). We shall not enter into this problem, except to state that if $\int \lambda \mathrm{d}E_\lambda$ is the spectral resolution of the unbounded self-adjoint operator A, then the function $f(A)$, if it is definable at all, is given by $\int f(\lambda)\mathrm{d}E_\lambda$. For a full treatment, we refer the reader to the text by Reed and Simon cited above.

(ii) The operator H is called the infinitesimal generator of the group[32] $\{U(t)\}$.

(iii) In quantum mechanics, it is the converse of Stone's theorem that is most often used: the solution of the Schrödinger equation $\mathrm{i}\partial\psi/\partial t = H\psi$, subject to the initial condition $\psi(0) = \psi_0$, is given by

$$\psi(t) = U(t)\psi_0,$$

where $U(t) = \exp(-\mathrm{i}Ht)$.

(iv) The time t enters as a parameter in the explicit solution in the following manner. Let $\{\psi_n | n \in \mathbf{N}\}$ be an orthonormal base for $\mathfrak{H}$, and let

$$\psi(0) = \sum_{n=1}^{\infty} c_n \psi_n,$$

[32] The group $\{U(t)\}$ may also be regarded as a one-parameter real Lie group; it is possible to 'forget' the background of unitary operators and Hilbert spaces over the complex numbers. In this case, the group element would be written as $\exp tX$, without making any reference to complex numbers.

where $\sum |c_n|^2 = 1$. Then we may write $\psi(t)$ as

$$\psi(t) = \sum_{n=1}^{\infty} c_n(t)\psi_n,$$

where $c_n(0) = c_n$ and $\sum |c_n(t)|^2 = 1$ for all t. The burden of time dependence is carried entirely by the coefficients $c_n(t)$, and does not involve the vectors ψ_n, in conformity with the quotation from von Neumann on page 111.

Since $\mathfrak{H}$ is separable, it admits of complete orthonormal bases (see page 318), and any two orthonormal bases are related by a unitary or an antiunitary transformation. The Schrödinger and Heisenberg pictures correspond to bases in $\mathfrak{H}$ that are unitary transforms of each other, the transformations themselves being dependent on time. This fact was first observed by Dirac (Dirac, 1930).

8

Von Neumann's theory of measurement

Von Neumann's theory of measurement in Quantum Mechanics was spelled out in the last chapter of his book, which was published in 1932. This book was highly mathematical for its time, and in 1939 London and Bauer provided a simplified account of the measurement theory part of it (London and Bauer, 1939). Von Neumann died in 1957. Thirty years after the publication of von Neumann's book, Wigner published a review containing his views on the shortcomings of von Neumann's theory, but omitting any discussion of its mathematical core, namely von Neumann's analysis of composite systems (Wigner, 1963). He also published a set of lecture notes entitled *Interpretation of Quantum Mechanics* (Wigner, 1983) in which some of his concerns were spelled out in greater detail, and an article addressed 'to an audience of non-physicists' (Wigner, 1964). Wigner's own contributions to measurement theory were discussed by Shimony in a talk at the Wigner centennial conference (Shimony, 2002). The English translation of von Neumann's book was published in 1955.[1] The account that follows is based on these sources.

We shall assume that the reader is acquainted with notions such as wave function collapse and the Heisenberg cut, but we shall not assume familiarity with the technicalities of von Neumann's theory. This chapter is organized accordingly. Section 8.1 explains what we mean by the term *von Neumann's measurement theory* and gives an overview of the subject.[2] It is followed by Sections 8.2 and 8.3, in which the theory is spelled out in detail. Section 8.4 recounts Wigner's reservations. In Section 8.5, the last, von Neumann's main results are reformulated in the language of entanglement, and an apparent difference of perception between von Neumann and Wigner is pointed out; it concerns a crucial but often ignored point.

8.1 Overview

The subject we shall call *von Neumann's measurement theory* can be divided into four parts, the first two of which are seldom stated explicitly. The main mathematical part is contained in Chapter VI of his book, but we shall have

[1] See the footnote on page 110.

[2] In the literature, one sometimes encounters the term 'von Neumann's measurement theory' but is unsure of what the author means; we wish to avoid this pitfall.

occasion to refer to other chapters as well. For the convenience of the reader, we shall cite all references by chapter, section or page numbers.

(i) *Inference from observation.* Having analysed the experimental evidence, von Neumann concluded that the act of measurement changes the state of the object from a pure to a mixed one. (See also Remark 8.3 on page 140.)
(ii) *Measurement of operators with continuous spectra.* Since there are no eigenvectors that belong to a point in the continuous spectrum of a self-adjoint operator, von Neumann analysed what it means to measure such an operator. London and Bauer describe this analysis as '... sophistications, which do not concern questions of principle... ' (Wheeler and Zurek, 1983, footnote on p. 223). The present author contends that these 'sophistications' are at the heart of the matter; the question will be discussed in Chapter 9.
(iii) *Mathematical structure of composite systems.* Von Neumann described the 'combining of two systems' by the mathematical operation of forming the tensor product of Hilbert spaces of the individual systems. Analysing the structure of pure states of the tensor product, he found that the substates of the two parts were perfectly correlated with each other. As Wigner was to describe it later, 'the state of the apparatus mirrors the state of the object'.
(iv) *Intervention of the observer.* In his deductive development of the theory, von Neumann posited the human observer (who 'remains outside the calculation') as an integral part of the measurement process (Section 8.3.1). Let I be the object of measurement, II the measuring apparatus and III the human observer. There are two possibilities:

   (a) I is the 'observed system' and II + III the 'observer'. Then it is the wave function of I that collapses.
   (b) I + II is the 'observed system' and III the 'observer'. Then it is the wave function of I + II that collapses. The analysis of a composite system applies to I + II, and I and II mirror the states of each other.

   Von Neumann established that, under his assumptions – some of which are metaphysical – (a) and (b) lead to the same result.

Von Neumann's writing is clear; it is possible to separate the mathematical from the metaphysical parts, and we shall exploit this fact. The crux of his measurement theory is that the state vector of a system changes with time in two different ways: (i) smoothly, when it evolves under the Schrödinger equation; and (ii) abruptly, when one performs a measurement upon it. This abrupt change has become known as the collapse of the wave packet. If the Schrödinger equation cannot account for it – and von Neumann concluded that it could not – then one has to devise an alternative mechanism for the collapse. The one that von Neumann devised was the intervention of the observer's conscious ego. Faced with this unsettling suggestion, some physicists chose just to 'shut up and calculate',[3]

[3] A phrase attributed to Dirac.

while others looked towards philosophy. Prominent among the latter was Wigner, who explained the situation as follows in his 1964 article (reprinted in (Wigner, 1970); see pp. 186–187):

> Even though it is not strictly relevant, it may be useful to give the reason for the increased interest of the contemporary physicist in problems of epistemology and ontology. The reason is, in a nutshell, that physicists have found it impossible to give a satisfactory description of atomic phenomena without reference to the consciousness. This...refers...to the process called the 'reduction of the wave packet'. This [reduction of the wave packet] takes place whenever the result of an observation enters the consciousness of the observer...Alternatively, one could say that quantum mechanics provides only probability connections between the results of my observations as I perceive them. Whichever formulation one adopts, the consciousness evidently plays an indispensable role.[4]
>
> In outline, the situation is as follows. The interaction between the measuring apparatus and the...*object* of the measurement...results in a state in which there is a strong statistical correlation between the state of the apparatus and the state of the object. In general, *neither apparatus nor object is in a state which has a classical description* [emphasis added]. However, the state of the...apparatus plus object is, after the interaction, such that only one state of the object is compatible with any given state of the apparatus. Hence, the state of the object can be ascertained by determining the state of the apparatus after the interaction has taken place. It follows that the measurement of the state of the object has been reduced to the measurement of the state of the apparatus. However, *since the state of the apparatus has no classical description* [emphasis added], the measurement of the state of the apparatus is, from the conceptual point of view, no different from the measurement on the original object. In a similar way, the problem can be transferred from one link of a chain to the next, and so on. However, *the measurement is not completed until its result enters our consciousness* [emphasis added]. This last step is, at the present state of our knowledge, shrouded in mystery and no explanation has been given so far in terms of quantum mechanics, or in terms of any other theory.[5]

[4] *Footnote in the original*: 'This is not the proper place to give a detailed proof of this assertion...It should suffice, therefore, to mention that the fact was pointed out with full clarity first by von Neumann (see Chapter VI of his *Mathematical Foundations of Quantum Mechanics*)...' We shall dispute this assertion of Wigner's in Section 8.5.2.

[5] Many physicists who were not content to 'shut up and calculate' were ill at ease with the 'last step' described above by Wigner. Roland Omnès expressed himself forthrightly (Omnès, 1999, p. 69) in a book which set forth his counterproposal. Other counterproposals, or references to them, may be found in (Wheeler and Zurek, 1983; Joos *et al.*, 2003; Namiki *et al.*, 1999; Sinha, 1994; and Sinha and Goswami, 2007). The present author, who has no claim to expertise in this vast field, apologizes in advance to those whose works have not come to his notice.

In Chapter 9 we shall contest Wigner's assertion that 'the state of the apparatus has no classical description', and in Chapter 10 we shall investigate the implications of this contest for measurement theory.

8.2 Von Neumann's initial considerations

In quantum-mechanical measurement theory, one uses the term *measurement* to denote the exact or approximate determination of an eigenvalue or a spectral value of a self-adjoint operator. The theory encompasses self-adjoint operators for which no measurement procedure may be realizable in the laboratory. We shall return to this last point in Chapter 11.

8.2.1 The inference from observation

In von Neumann's book, equivalence of the Heisenberg and the Schrödinger pictures is established in Chapters I and II. The problem of measurement is discussed in Chapter VI; it is analysed exclusively in the Schrödinger picture.

As stated earlier, von Neumann postulated that the state vector of a quantum-mechanical system can change with time in two different ways:

(i) Reversibly, under the Schrödinger equation. In this evolution probability amplitudes evolve into probability amplitudes. Von Neumann called it the '**2**nd intervention'.
(ii) Irreversibly, by an act of measurement. Von Neumann called it the '**1**st intervention'. This change was described by (7.19) and discussed on page 119.

The Schrödinger equation is a cornerstone of quantum mechanics; item (ii) above is the famous (or infamous) *reduction of the state vector.*[6] Von Neumann was led to it by the Bohr–Kramers–Slater (hereafter BKS) paper (January 1924) and the experiments of Compton and Simon (June 1925).[7] Both of these papers *preceded* the article by Heisenberg (July 1925) which heralded the advent of quantum mechanics.[8] The following quotation is from the Introduction of the BKS paper (van der Waerden, 1967, p. 159):

> On the one hand, the phenomena of interference, on which the action of all optical instruments essentially depends, claim an aspect of continuity of the same character as that involved in the wave theory of light... On the other hand, the exchange of energy and momentum between matter and radiation, on which the observation of optical phenomena ultimately

[6] Also known as *collapse of the wave packet.* The terms state vector and wave packet are used interchangeably.

[7] The references are to (Bohr, Kramers and Slater, 1924) and (Compton and Simon, 1925).

[8] The reference is to (Heisenberg, 1925); English translation in (van der Waerden, 1967).

> depends, claims essentially discontinuous features. These have *even led to the introduction of the theory of light-quanta* [emphasis added], which in its most extreme form denies the wave constitution of light. At the present state of science it does not seem possible *to avoid the formal character of the quantum theory* [emphasis added] which is shown by the fact that the interpretation of atomic phenomena does not involve a description of the mechanism of discontinuous processes...

The 'aspect of continuity... in the wave theory of light' is best captured in the Huygens principle, according to which every point on a wavefront is the source of outgoing spherical waves (circular waves, in two dimensions). The quantum theory of radiation devised by BKS tried to preserve as much of the classical wave theory as possible, but at the cost of abandoning the conservation of energy and momentum. What BKS did manage to achieve was the following: while energy and momentum were not conserved in individual events, they were conserved statistically, i.e., when summed, or averaged, over many events.[9] The Compton–Simon experiment refuted this picture decisively. The closing paragraph of their article reads:

> These results do not appear to be reconcilable with the view of the statistical production of recoil and photo-electrons proposed by Bohr, Kramers and Slater. They are, on the other hand, in direct support of the view that *energy and momentum are conserved during the interaction between radiation and electrons* [emphasis in the original].

Basing himself on the BKS paper and the Compton–Simon experiment, von Neumann (pp. 213–214 of his book) argued that:

> ...three degrees of causality or non-causality may be distinguished [in nature]. First, the... [measured] value could be entirely statistical, i.e., the result of a measurement could be predicted only statistically; and if a second measurement were taken immediately after the first one, this would also have a dispersion, without regard to the value found initially – for example, the dispersion might be equal to the original one.[10] Second, it is conceivable that the value of... [the measured quantity] may have a dispersion in the first measurement, but that [an] immediately subsequent measurement is constrained to give a result which agrees with that of the

[9] The Bohr–Kramers–Slater paper is reprinted in the source book edited by van der Waerden (van der Waerden, 1967). Papers which originally appeared in German were translated into English for this volume. The volume contains a very helpful historical introduction by the editor.

[10] In this case the state of the system will not be altered by the measurement. If one measures the energy, and the initial state is a superposition of energy eigenstates with different eigenvalues, then a measurement which returns a unique value *and* leaves the state unchanged will obviously violate the law of conservation of energy.

> first. Third,... [the measured quantity] could be determined causally at the outset.
>
> The Compton–Simon experiment now shows that only the second case is possible in a statistical theory.

We shall now make a formal statement of the hypothesis of reduction of the state vector, which we shall refer to, for brevity, as the collapse postulate.

Hypothesis 8.1 (Collapse postulate) Let A be an observable with a discrete, nondegenerate spectrum[11] on the Hilbert space $\mathfrak{H}$, and u_k its eigenvectors. A measurement of A sends a pure state $\sum_k c_k u_k$ into the mixed state $\sum_k |c_k|^2 P_{[u_k]}$, where $P_{[u_k]}$ is the projection operator onto the vector $u_k \in \mathfrak{H}$.

Having formulated this hypothesis, we may summarize the key point of the discussion that preceded it as follows:

Conclusion 8.2 (Consistency with conservation laws) If an additively conserved quantity (like momentum or a component of the angular momentum) is being measured, then the negation of the collapse postulate may lead to a contradiction with the conservation law.

If the measured eigenvalue λ of A is nondegenerate, then the collapse postulate implies that the state of the system after (a single) measurement is a unique ray in the Hilbert space. If λ is a degenerate eigenvalue, then it only implies that the state after a single measurement lies in the eigenspace of λ (von Neumann, 1955, p. 218).

Remark 8.3 (The Heisenberg cut) The collapse postulate may be regarded as the precise mathematical formulation of the *Heisenberg cut.* In the published version of his Chicago lectures of 1929, Heisenberg wrote (Heisenberg, 1930, p. 58):

> The partition of the world into observing and observed system prevents a sharp formulation of the law of cause and effect. (The observing system need not always be a human being; it may also be an inanimate apparatus, such as a photographic plate.)

This partition is the Heisenberg cut (sometimes called the Bohr–Heisenberg cut). The idea of the cut was first expressed by Heisenberg at the Congresso Internazionale di Fisica, held in Como in September 1927, during the discussion following Bohr's talk (Heisenberg, 1928). In a letter dated 18 January 1933 (replying to a letter by Heisenberg which has been lost), Pauli stated that the result of a measurement does not depend on the precise location of this cut, and

[11] There is no difficulty in admitting degenerate eigenvalues; the statement becomes a little more complicated. Details are left to the reader.

referred to von Neumann's book in support (Pauli, 1985).[12] The mathematical results (see the quotation on pages 144–145) elaborated on pp. 439–445 of von Neumann's book offer no solution except the 'infinite von Neumann chain'. The Heisenberg cut avoids this infinite chain, but does not provide a bridge between the observing and the observed system. On the other hand, the observing system, 'such as a photographic plate', does have a classical description.[13] Shimony has emphasized that

> The insistence upon a classical description of the measuring apparatus, not as a convenient approximation but as a matter of principle, clearly differentiates Bohr's interpretation of quantum mechanics from that of von Neumann and of London and Bauer.

(Shimony, 1963; reprinted in Shimony, 1993, p. 24). We shall interpret 'a classical description of the measuring apparatus' to mean that the measuring apparatus possesses observables the values of which can be registered on displays which are incontrovertibly classical. Such observables will be called *classical observables.* Recall that the term *observable* itself means a self-adjoint operator on a Hilbert space, and therefore a classical observable is, by definition, a quantum-mechanical concept. The question of these observables will occupy us in Chapter 9.

8.2.2 Measurement of operators with continuous spectra

There is no loss of generality in assuming that the operator has no discrete spectrum. Consider the measurement of an operator H which has the spectral decomposition ((A6.23), page 339)

$$H = \sum_{n=1}^{\infty} \lambda_n E_n + \int_0^{\infty} \lambda \mathrm{d}E_\lambda,$$

e.g., the energy of an electron in the field of a positive point charge (hydrogen atom). Here the λ_n, which are negative, are the energies of the bound states. The Hilbert space $\mathfrak{H}$ of states of the electron decomposes into the direct sum $\mathfrak{H} = \mathfrak{H}_{\mathsf{bd}} \oplus \mathfrak{H}_{\mathsf{scatt}}$, where $\mathfrak{H}_{\mathsf{bd}}$ contains only the bound states and $\mathfrak{H}_{\mathsf{scatt}}$ (which is necessarily infinite-dimensional) contains only the scattering states. The restriction of H to $\mathfrak{H}_{\mathsf{scatt}}$ is an operator that has only a continuous spectrum.

[12] In the summer of 1935 Heisenberg wrote an article, 'Ist eine deterministische Ergänzung der Quantenmechanik möglich?', which, however, was not published. Heisenberg sent a copy of this paper to Bohr, and it was eventually published in Pauli's scientific correspondence as an appendix to Heisenberg's letter to Pauli of 2 July 1935 (Pauli, 1985, pp. 409–418.)

[13] I would like to thank Professors H Goenner, H Reeh and H Rechenberg (who is presently writing a scientific biography of Werner Heisenberg) for the references cited above.

In $\mathfrak{H}_{\text{scatt}}$, the operator H has no eigenvectors, only approximate eigenvectors. That is, for any λ in the continuous spectrum of H and any $\varepsilon > 0$, there exist vectors $\psi_\lambda \in \mathfrak{H}_{\text{scatt}}$ such that $||(H - \lambda I)\psi_\lambda|| < \varepsilon$. But what kind of measurement will cause the state vector of a system to collapse to an approximate eigenvector – and to which of the many approximate eigenvectors? Von Neumann addressed this problem as follows (p. 220 of his book):

> We have seen that a quantity A [a self-adjoint operator] can always (i.e., for each state ψ) be measured exactly if and only if it possesses a pure discrete spectrum. If it possesses none, *then it can be measured only with limited accuracy*... [emphasis added].

To prove the last assertion, von Neumann divides $\mathbb{R}$ into a countable set of intervals $[\lambda^{(n)}, \lambda^{(n+1)}]$, $n \in \mathbb{Z}$, with $\lambda^{(n)} < \lambda^{(n+1)}$. He then chooses, for each $n \in \mathbb{Z}$, a number λ_n such that $\lambda^{(n)} \leq \lambda_n \leq \lambda^{(n+1)}$, and defines a piecewise-constant function G as follows:

$$G(\lambda) = \lambda_k, \text{ for } \lambda \in (\lambda^{(k)}, \lambda^{(k+1)}) \ k \in \mathbb{Z}.$$

The values of $G(\lambda)$ at the points $\lambda^{(n)}$ are arbitrary. He then proves that the operator $G(A)$ has a purely discrete spectrum, and that its eigenvalues are the λ_n. (Operator functions $G(A)$ are defined via the spectral theorem; see Appendix A6.) It can therefore be measured precisely, and a precise measurement of $G(A)$ is equivalent to an approximate measurement of A, the measurement being accurate to ε, where

$$\varepsilon = \text{Max}\,(\lambda^{(n+1)} - \lambda^{(n)})$$

is the maximum spacing between two adjacent division points.

There are two points to be noted in this scheme:

(i) There is no canonical choice for the function G. It is entirely up to a human agency.

(ii) The numbers $\lambda^{(n)}$ and λ_n are arbitrary. They are not required to be either rational or irrational.

The experimentalist has to design, not only the apparatus to measure the operator, but also the operator itself (with a little help from the mathematician; see 'the converse of the spectral theorem', page 331).

8.2.3 The quantum measurement problem

The discussion of Subsections 8.2.1 and 8.2.2 may be summarized as follows:

(i) The collapse postulate, by allowing for the conservation of energy and momentum in radiative processes, distinguishes between quantum mechanics and the radiation theory of Bohr, Kramers and Slater.
(ii) If a quantum measurement is understood to be the assignment of a real number to a *vector* in Hilbert space, an operator with a continuous spectrum cannot be measured; one can only measure an approximation to it by an operator with a purely discrete spectrum (which we shall call a *von Neumann approximant* to the operator with a continuous spectrum).

We shall now give a verbal formulation of the measurement problem.

Problem 8.4 (Quantum measurement problem) Describe how probability amplitudes evolve into probabilities under the Schrödinger equation for the composite object–apparatus system.

Von Neumann could not resolve this problem mathematically. At the 'last step' mentioned by Wigner (last two sentences of the quotation on page 137), he had to invoke the human observer. His analysis of the structure of composite systems was mathematically unimpeachable; the seeds of his failure lay in the assumptions he made in translating the verbal formulation given above into a mathematical one.

We now turn to his analysis, which is given in Chapter VI of his book.

8.3 Von Neumann's Chapter VI

Chapter VI of von Neumann's book, 'The measuring process', is divided into three sections. In the following, we shall provide a section-by-section summary of this chapter. In Subsections 8.3.1–8.3.3, we shall adhere to von Neumann's headings and his labels I, II and III, but shall depart from his mathematical notations. The summary of the last section (our Subsection 8.3.3) will be incomplete. We shall omit von Neumann's proof of the proposition that his theory of measurement does not violate the principle of psycho-physical parallelism (see below).

Von Neumann's book appeared in 1932, three years before the term *entanglement* was coined by Schrödinger (Schrödinger, 1935). As we shall see, what von Neumann called a *combined system* would be called an entangled system today. We shall exploit this fact, but shall restrict use of the concept to Remarks 8.8 and 8.9, and to Section 8.5.

Definitions of the term entanglement that one finds are sometimes ambiguous, even when set down by workers in the field. We shall use the term in the following sense, which seems to conform to to its actual use in the literature:

Definition 8.5 (Entangled states) A state of a system which consists of n distinguishable subsystems is called *separable* if it is a tensor product of the states of the subsystems; otherwise it is called *entangled.*[14]

The above definition – which is the one we shall adopt – makes no statement about systems consisting of n subsystems that are indistinguishable from each other, such as n identical particles obeying Bose or Fermi statistics. The notion of entanglement for systems consisting of two identical particles is under investigation; see, for example, (Li *et al.*, 2001) and (Schliemann *et al.*, 2001).[15]

8.3.1 Formulation of the problem

After remarking that time evolution of a state vector under the Schrödinger equation is both causal and reversible, whereas the changes induced by a measurement are neither, von Neumann writes (p. 418 of his book):

> Let us now compare these circumstances with those which actually exist in nature or in its observation. First of all... the measurement or... [its] subjective perception is a new entity relative to the physical environment *and is not reducible to the latter* [emphasis added]... Nevertheless, *it is a fundamental requirement of the scientific viewpoint* [emphasis added] – the so-called principle of the psycho-physical parallelism – that it must be possible... to describe the extra-physical process of the subjective perception *as if it were in reality in the physical world* [emphasis added]...

As an example, von Neumann discusses the measurement, by a human observer, of the temperature of a body by a mercury thermometer, pointing out that the process can be thought of as consisting of a succession of steps,[16] the steps themselves not being uniquely defined. He finally concludes that (pp. 420–421 of his book):

> That is, we must always divide the world into two parts, the one being the observed system, the other the observer... The boundary between the two is arbitrary to a very large extent...[17]

[14] To avoid misunderstandings, we should like to repeat that we are concerned exclusively with nonrelativistic quantum mechanics, in which there is no upper limit to signal velocity. Under this hypothesis, it is difficult to distinguish between action at a distance and field action unless the field itself happens to be measurable.

[15] I would like to thank Dr C F Roos for this clarification, as well as the references cited above.

[16] Such as expansion of the mercury column, light reflected by it striking the observer's eye, a signal from eye to brain through the optic nerve, etc.

[17] If the observer were to have a classical description, this division would be precisely the Heisenberg cut.

Now quantum mechanics describes the events which occur in the observed portion of the world, so long as they do not interact with the observing portion, with the aid of the [Schrödinger equation], but *as soon as such an interaction occurs,* [emphasis added] i.e., a measurement, it requires the application of [the collapse] process. The dual form is therefore justified.[18] However, the danger lies in the fact that *the principle of psycho-physical parallelism is violated* [emphasis added], so long as it is not shown that the boundary between the observed system and the observer can be displaced arbitrarily in the sense given above.

In order to discuss this, let us divide the world into three parts: I, II and III. Let I be the system actually observed, II the measuring instrument, and III the actual observer.[19]... [The phrase *abstract ego*, often rendered as the *conscious ego*, appears for the first time in the sentence omitted.] It is to be shown that the boundary can be drawn just as well between I and II + III as between I + II and III... That is, in one case [the Schrödinger equation] is to be applied to I, and [collapse] to the interaction between I and II + III; and in the other case [the Schrödinger equation] is to be applied to I + II, and [collapse] to the interaction between I + II and III. (*In each case,* III *itself remains outside of the calculation* [emphasis added].) The proof of this assertion, that both procedures give the same results regarding I (this and only this belongs to the observed part of the world in both cases), is then our problem. But in order to accomplish this successfully, we must first investigate more closely the process of forming the union of two physical systems (which leads from I and II to I + II).

Next, we shall give an account of von Neumann's analysis of 'forming the union of two physical systems'. This analysis is purely mathematical, and is independent of the proposition that the principle of psycho-physical parallelism is 'a fundamental requirement of the scientific viewpoint'.

8.3.2 Composite systems

Von Neumann considers two physical systems I and II, 'not necessarily the ones of Section 8.3.1', with k and l degrees of freedom respectively and with Hilbert

[18] At this point there is a footnote (p. 420, footnote 207), which I was not able to understand. I therefore enlisted the help of Professor H Roos of the University of Göttingen, who translated this footnote from the German original into English as follows:

> N Bohr, Naturwiss. Vol. 17 (1929) was the first to point out that the dual description of nature – necessitated by the formalism of quantum mechanics – is also justified by the physical nature of things, and he pointed to the connection with the psycho-physical parallelism.

[19] *Footnote 208 in the original*: 'The discussion carried out in the following, as well as that in [Section 8.3.3], contains essential elements which the author owes to conversations with L Szilard... '

spaces $\mathfrak{H}^{\mathrm{I}}$ and $\mathfrak{H}^{\mathrm{II}}$. The Hilbert space of the composite system is the tensor product $\mathfrak{H} = \mathfrak{H}^{\mathrm{I}} \otimes \mathfrak{H}^{\mathrm{II}}$ (see Appendix A6, pages 321–322). All three are Hilbert spaces of complex-valued square-integrable functions, i.e., L^2-spaces, but it will often be economical to treat them as abstract Hilbert spaces over the complex numbers. When regarded as complex-valued functions, vectors in $\mathfrak{H}^{\mathrm{I}}$ and $\mathfrak{H}^{\mathrm{II}}$ will be denoted by lower-case Greek letters with one argument,

$$\begin{aligned} \varphi^{\mathrm{I}}(q) &= \varphi^{\mathrm{I}}(q_1, \dots, q_k), \\ \varphi^{\mathrm{II}}(r) &= \varphi^{\mathrm{II}}(r_1, \dots, r_l), \end{aligned} \tag{8.1}$$

and vectors in $\mathfrak{H}$ by capital Greek letters with two arguments, e.g.,

$$\Phi(q, r) \in \mathfrak{H}. \tag{8.2}$$

In this case, the product of $\varphi^{\mathrm{I}}(q)$ and $\varphi^{\mathrm{II}}(r)$, which will be a vector in $\mathfrak{H}$, will be denoted multiplicatively as $\varphi^{\mathrm{I}}(q)\,\varphi^{\mathrm{II}}(r)$. When $\varphi^{\mathrm{I}}(q)$ and $\varphi^{\mathrm{II}}(r)$ are regarded as vectors in abstract Hilbert spaces, the variables q and r will be omitted, and the product of φ^{I} and φ^{II} will be written as $\varphi^{\mathrm{I}} \otimes \varphi^{\mathrm{II}}$. The inner product on $\mathfrak{H}^{\mathrm{I}}$ will be written as $(\varphi^{\mathrm{I}}, \psi^{\mathrm{I}})$ in the abstract case, and as the integral

$$\int \overline{\varphi^{\mathrm{I}}(q)}\psi^{\mathrm{I}}(q)\mathrm{d}q$$

in the concrete case, and similarly for $\mathfrak{H}^{\mathrm{II}}$ and $\mathfrak{H}$.

1. Physical quantities; rules of correspondence

Next, von Neumann discusses *physical quantities.* We shall present the beginning in his own words (pp. 422–423 of his book).

> ...The physical quantities of I, II [and] I + II are correspondingly the [self-adjoint][20] operators A^{I}, A^{II} and A in $\mathfrak{H}^{\mathrm{I}}$, $\mathfrak{H}^{\mathrm{II}}$ and $\mathfrak{H}$ respectively.
>
> Each physical quantity in I is naturally also one in I+II, and in fact its A is obtained from its A^{I} in this way: to obtain $A\Phi(q, r)$ consider r as constant and apply A^{I} to the function $\Phi(q, r)$.[21] This rule of transformation is correct in any case for the coordinate and momentum operators $Q_1, \dots, Q_k$ and $P_1, \dots, P_k$, i.e.,
>
> $$q_1, \dots, q_k, \quad -\mathrm{i}\frac{\partial}{\partial q_1}, \dots, -\mathrm{i}\frac{\partial}{\partial q_k}$$
>
> (cf. I.2) [The section referred to is entitled *The original formulations of quantum mechanics*, pp. 7–17 of von Neumann's book] and it conforms

[20] Hypermaximal Hermitian in the original; see Definition A6.31.

[21] *Footnote* 209 *in the original*: 'It can easily be shown that if A^{I} is hermitian or hypermaximal, A is also.'

with principles **I**, **II**[22] in IV.2.[23] We therefore postulate this generally. (This is the customary procedure in quantum mechanics.)

In the same way, each physical quantity in II is also one in I + II, and its A^{II} gives rise to its A by the same rule: $A\Phi(q,r)$ equals $A^{\mathrm{II}}\Phi(q,r)$ if in the latter expression, q is taken as constant, and $\Phi(q,r)$ is considered as a function of r.

Following this, von Neumann exhibits his correspondences $A^{\mathrm{I}} \to A$ and $A^{\mathrm{II}} \to A$ explicitly in matrix form. The formulae make it clear that he is dealing with the operators $A^{\mathrm{I}} \otimes I^{\mathrm{II}}$ and $I^{\mathrm{I}} \otimes A^{\mathrm{II}}$ on $\mathfrak{H}$ (where $I^{\mathrm{I,II}}$ are the identity operators on $\mathfrak{H}^{\mathrm{I,II}}$ respectively), and the natural correspondences

$$\begin{aligned} A^{\mathrm{I}} &\longrightarrow A^{\mathrm{I}} \otimes I^{\mathrm{II}}, \\ A^{\mathrm{II}} &\longrightarrow I^{\mathrm{I}} \otimes A^{\mathrm{II}}. \end{aligned} \tag{8.3}$$

Note that von Neumann does *not* assert that a physical quantity in I is also one in II, or vice versa, leaving open the possibility that one of them is microscopic and the other macroscopic. This is the case that he considers (briefly) in the second quotation in Subsection 8.3.3.2, page 157.

2. Density matrices; rules of correspondence

Let $\{\varphi^{\mathrm{I}}_m\}$ and $\{\varphi^{\mathrm{II}}_n\}$ be complete orthonormal bases in $\mathfrak{H}^{\mathrm{I}}$ and $\mathfrak{H}^{\mathrm{II}}$ respectively. Define

$$\Phi_{m,n} = \varphi^{\mathrm{I}}_m \otimes \varphi^{\mathrm{II}}_n.$$

Then $\{\Phi_{m,n}\}$ is a complete orthonormal basis in $\mathfrak{H} = \mathfrak{H}^{\mathrm{I}} \otimes \mathfrak{H}^{\mathrm{II}}$. The matrix representations of operators will be defined in the standard manner, $O_{mn} = (f_m, Of_n)$ for the operator O on the Hilbert space $\mathfrak{H}$. We shall denote the matrix elements of the density matrix ρ on $\mathfrak{H}$ by a pair of double indices:[24]

Definition 8.6

$$\rho_{mn,m'n'} = (\Phi_{m,n}, \rho\,\Phi_{m',n'}).$$

[22] The principles **I**, **II** are as follows: **I**: 'If the quantity $\mathfrak{R}$ has the operator R, then the quantity $f(\mathfrak{R})$ has the operator $f(R)$.' **II**: 'If the quantities $\mathfrak{R}, \mathfrak{S} \ldots$ have the operators $R, S \ldots$, then the quantities $\mathfrak{R} + \mathfrak{S} + \cdots$ have the operators $R + S + \cdots$. It is not assumed that $\mathfrak{R}, \mathfrak{S} \ldots$ are simultaneously measurable.'

[23] *Footnote* 210 *in the original*: 'For **I** this is clear, and for **II** also, so long as only polynomials are concerned. For general functions, it can be inferred from the fact that the correspondence of a resolution of the identity and a Hermitian operator is not disturbed in the transition $A_{\mathrm{I}} \to A$.'

[24] Density matrices were discussed in Subsection 7.1.2; see Definition 7.1.

The density matrices ρ^{I} on $\mathfrak{H}^{\mathrm{I}}$ and ρ^{II} on $\mathfrak{H}^{\mathrm{II}}$ are defined as partial traces:

$$\rho^{\mathrm{I}} = \mathrm{Tr}^{\mathrm{II}}\rho \text{ and } \rho^{\mathrm{II}} = \mathrm{Tr}^{\mathrm{I}}\rho. \tag{8.4}$$

It follows that

$$\begin{aligned} \rho^{\mathrm{I}}_{mm'} &= \sum_{n=1}^{\infty} \rho_{mn,m'n}, \\ \rho^{\mathrm{II}}_{nn'} &= \sum_{m=1}^{\infty} \rho_{mn,mn'}. \end{aligned} \tag{8.5}$$

Obviously, a given ρ determines ρ^{I} and ρ^{II} uniquely.

At this point von Neumann remarks (p. 425 of his book):

> We have thus established the rules of correspondence for the statistical operators of I, II, I+II, i.e., ρ^{I}, ρ^{II}, ρ. They proved to be essentially different from those [formulae (8.3)] which control the correspondence between the operators $A^{\mathrm{I}}, A^{\mathrm{II}}, A$ of physical quantities.

Note that, in general,

$$\rho \neq \rho^{\mathrm{I}} \otimes \rho^{\mathrm{II}}. \tag{8.6}$$

That being the case, the following is a natural question:

3. When do ρ^{I} and ρ^{II} determine ρ uniquely?

Equations (8.5) show that ρ determines ρ^{I} and ρ^{II} uniquely. Owing to (8.6), one would not expect the converse to be generally true. Von Neumann investigated the conditions under which the converse was true, and concluded that:

Theorem 8.7 *The density matrices ρ^{I} and ρ^{II} on $\mathfrak{H}^{\mathrm{I}}$ and $\mathfrak{H}^{\mathrm{II}}$ determine a unique density matrix ρ on $\mathfrak{H}^{\mathrm{I}} \otimes \mathfrak{H}^{\mathrm{II}}$ if and only if at least one of ρ^{I} or ρ^{II} represents a pure state.*

The theorem is stated on p. 426 of von Neumann's book. The proof (which is lengthy and will not be given here) consists of two parts. First, it is shown that if both ρ^{I} and ρ^{II} represent mixtures, then there are infinitely many ρ that satisfy (8.5), i.e., the condition of Theorem 8.7 is necessary. Then it is shown that the condition is sufficient, i.e., if one of ρ^{I} or ρ^{II} represents a pure state, then ρ is uniquely determined. In this case, von Neumann calls ρ^{I} and ρ^{II} *projections* of ρ; we shall make use of this terminology later.

Theorem 8.7 is used in only one place in von Neumann's measurement theory: proof that the initial state of the apparatus cannot be a mixture (page 155). The main burden falls upon the determination of ρ^{I} and ρ^{II} for a given pure state $\Phi \in \mathfrak{H}$, which is our next task.

4. The maps F and $F^\star$; determination of ρ^{I} and ρ^{II}

Now comes the critical step. Observe that if $\{\varphi_n^{\mathrm{II}}(r)\}$ is a complete orthonormal basis for $\mathfrak{H}^{\mathrm{II}}$, then so is $\{\bar{\varphi}_n^{\mathrm{II}}(r)\}$, the bar above denoting complex conjugation.[25] Let

$$\Phi(q,r) = \sum_{m,n=1}^{\infty} f_{mn}\varphi_m^{\mathrm{I}}(q)\bar{\varphi}_n^{\mathrm{II}}(r) \tag{8.7}$$

be a vector of unit norm in $\mathfrak{H}^{\mathrm{I}} \otimes \mathfrak{H}^{\mathrm{II}}$. The coefficients f_{mn} are restricted (only) by the condition

$$\sum_{m,n=1}^{\infty} |f_{mn}|^2 = ||\Phi||^2 = 1. \tag{8.8}$$

Remarks 8.8

(i) If f_{mn} does not factorize, i.e., $f_{mn} \neq a_m b_n$, then the systems I and II are entangled. In the mathematical analysis that follows, von Neumann aims at establishing a canonical form for the vector $\Phi(q,r)$ which describes an entangled state of the systems I and II.

(ii) According to London and Bauer, the state Φ of (8.7), which is clearly entangled, is the state of the object–apparatus system *after* the measurement interaction has taken place; the initial state, *before* the measurement, is separated (equations (1) and (2), and the text in between, on p. 246 of (Wheeler and Zurek, 1983)). This is physically reasonable, but the present author has not been able to find this assertion in von Neumann's book; indeed, the discussion on pp. 437–439 of his book, given in our Subsection 8.3.3.1, appears to contradict London and Bauer. It will become clear in the course of the discussion that von Neumann's mathematical analysis depends only on the form of $\Phi(q,r)$, and not upon its genesis.

Define now two linear transformations

$$\begin{aligned} F &: \mathfrak{H}^{\mathrm{I}} \to \mathfrak{H}^{\mathrm{II}}, \\ F^\star &: \mathfrak{H}^{\mathrm{II}} \to \mathfrak{H}^{\mathrm{I}} \end{aligned} \tag{8.9}$$

(determined by the state Φ of the composite system) by

$$\begin{aligned} F\varphi^{\mathrm{I}}(q) &= \int \overline{\Phi(q,r)}\varphi^{\mathrm{I}}(q)\mathrm{d}q, \\ F^\star\varphi^{\mathrm{II}}(r) &= \int \Phi(q,r)\varphi^{\mathrm{II}}(r)\mathrm{d}r. \end{aligned} \tag{8.10}$$

[25] This means that $\{\bar{\varphi}_n^{\mathrm{II}}\}$ is obtained from $\{\varphi_n^{\mathrm{II}}\}$ by an antiunitary transformation.

The transformations F and $F^\star$ are bounded, by definition. Substituting (8.7) into (8.10), we find that

$$\begin{aligned} F\varphi_m^{\mathrm{I}}(q) &= \sum_{j=1}^{\infty} \bar{f}_{mj}\varphi_j^{\mathrm{II}}(r), \\ F^\star\varphi_n^{\mathrm{II}}(r) &= \sum_{i=1}^{\infty} f_{in}\varphi_i^{\mathrm{I}}(q), \end{aligned} \tag{8.11}$$

so that

$$\begin{aligned} (\varphi_k^{\mathrm{II}}, F\varphi_m^{\mathrm{I}}) &= \bar{f}_{mk}, \\ (\varphi_l^{\mathrm{I}}, F^\star\varphi_n^{\mathrm{II}}) &= f_{ln}, \end{aligned}$$

i.e.,

$$\begin{aligned} F_{km} &= \bar{f}_{mk}, \\ (F^\star)_{ln} &= f_{ln}. \end{aligned} \tag{8.12}$$

Formulae (8.12) show that F and $F^\star$ are adjoints of each other, justifying the notation. The compositions

$$\begin{aligned} &\mathfrak{H}^{\mathrm{I}} \xrightarrow{F} \mathfrak{H}^{\mathrm{II}} \xrightarrow{F^\star} \mathfrak{H}^{\mathrm{I}}, \\ &\mathfrak{H}^{\mathrm{II}} \xrightarrow{F^\star} \mathfrak{H}^{\mathrm{I}} \xrightarrow{F} \mathfrak{H}^{\mathrm{II}} \end{aligned}$$

show that $F^\star F$ and $FF^\star$ are operators on $\mathfrak{H}^{\mathrm{I}}$ and $\mathfrak{H}^{\mathrm{II}}$ respectively.[26] They are clearly positive (Definition A6.14, page 327). Their matrix elements are, in the bases we are using,

$$\begin{aligned} (F^\star F)_{mn} &= \sum_{k=1}^{\infty} f_{mk}\bar{f}_{nk}, \\ (FF^\star)_{mn} &= \sum_{k=1}^{\infty} \bar{f}_{km} f_{kn}, \end{aligned} \tag{8.13}$$

so that

$$\operatorname{Tr}(FF^\star) = \operatorname{Tr}(F^\star F) = 1. \tag{8.14}$$

It follows that the operators $FF^\star$ and $F^\star F$ are compact (Theorem A6.18, page 327).

[26] We are using the fact that the composition of two linear transformations may be denoted multiplicatively.

On the other hand, the matrix elements of ρ for the pure state Φ defined by (8.7) are

$$\rho_{mn,m'n'} = \bar{f}_{mn} f_{m'n'}. \tag{8.15}$$

Combining (8.5), (8.15) and (8.13) we find that[27]

$$\begin{aligned} F^\star F &= \rho^{\rm I}, \\ FF^\star &= \rho^{\rm II}. \end{aligned} \tag{8.16}$$

The operators $\rho^{\rm I}$ and $\rho^{\rm II}$ so defined are positive and have unit trace. Von Neumann emphasizes that (8.16) are independent of the orthonormal bases that were used to derive them. He then draws attention to the fact that, as first noticed by Landau in 1927, the projections of the density matrix of a pure state in $\Phi \in \mathfrak{H}$ may be density matrices of *mixed states* in $\mathfrak{H}^{\rm I}$ and $\mathfrak{H}^{\rm II}$.

5. The structure of $\rho^{\rm I}$ and $\rho^{\rm II}$

Since the operators $\rho^{\rm I}$ and $\rho^{\rm II}$ are positive, self-adjoint and compact, their spectra are discrete and lie on $[0, \infty)$. Since the eigenvalue 0 cannot be ruled out for either of them, we begin by separating out their null spaces:

$$\begin{aligned} \mathfrak{H}^{\rm I} &= \mathfrak{H}_0^{\rm I} \oplus (\mathfrak{H}_0^{\rm I})^\perp, \\ \mathfrak{H}^{\rm II} &= \mathfrak{H}_0^{\rm II} \oplus (\mathfrak{H}_0^{\rm II})^\perp. \end{aligned} \tag{8.17}$$

In the above, $\mathfrak{H}_0^{\rm I,II}$ are the null spaces of $\rho^{\rm I,II}$ respectively. Their orthogonal complements $\mathfrak{H}_+^{\rm I,II} = (\mathfrak{H}_0^{\rm I,II})^\perp$ are spanned by eigenvectors of $\rho^{\rm I,II}$ with positive eigenvalues. Denoting the restrictions of $\rho^{\rm I,II}$ to $\mathfrak{H}_+^{\rm I,II}$ by $\rho_+^{\rm I,II}$ respectively, we may write the spectral resolution of $\rho_+^{\rm I}$ in the form (A6.12):

$$\rho_+^{\rm I} = \sum_{k=1}^{\infty} \lambda_k \, E_{\lambda_k}^{\rm I}, \tag{8.18}$$

where $\lambda_k > \lambda_{k'} > 0$ for $k < k'$ and $E_{\lambda_k}^{\rm I}$ is the projection operator on the subspace of eigenvectors with eigenvalue λ_k; recall that this subspace is finite-dimensional.[28] The same formula holds for the spectral decomposition of $\rho_+^{\rm II}$

[27] *Footnote 213 in the original*: 'The mathematical discussion is based on a paper by E. Schmidt, Math. Ann. **63**, 433-476 (1907).' [The volume number was printed incorrectly in the English translation. Schmidt's expansion formula for the asymmetric kernel of an integral equation is very similar to (8.30), and may be found in (Courant and Hilbert, 1953, p. 159).]

[28] The argument, based on the fact that the identity operator on an infinite-dimensional Hilbert space is not compact, is given on page 330.

on $\mathfrak{H}_+^{\mathrm{II}}$; one replaces I by II in (8.18), and λ_k by μ_k, where μ_k are the eigenvalues of ρ^{II}, $\mu_0 = 0$ and $\mu_k > \mu_{k'} > 0$ for $k < k'$.[29]

The operators ρ^{I} and ρ^{II} are closely related. Let ψ_j^{I} be a normalized eigenvector of ρ^{I} with eigenvalue λ_j. From $\rho^{\mathrm{I}} = F^\star F$ we have

$$F^\star F\psi_j^{\mathrm{I}} = \rho^{\mathrm{I}}\psi_j^{\mathrm{I}} = \lambda_j\psi_j^{\mathrm{I}}. \tag{8.19}$$

Applying F from the left and rearranging the left-hand side, we obtain

$$FF^\star(F\psi_j^{\mathrm{I}}) = \lambda_j(F\psi_j^{\mathrm{I}}), \tag{8.20}$$

which shows that $F\psi_j^{\mathrm{I}}$ is an eigenvector of $FF^\star = \rho^{\mathrm{II}}$ with eigenvalue λ_j. Since the argument is symmetric in the indices I and II, we conclude that: (i) ρ^{I} and ρ^{II} have the same eigenvalues, with the same multiplicity, and (ii) there is a one-to-one correspondence between the eigenvectors of ρ_+^{I} and ρ_+^{II}. We may thus write the spectral decomposition of ρ_+^{II} as

$$\rho_+^{\mathrm{II}} = \sum_{k=1}^{\infty} \lambda_k\, E_{\lambda_k}^{\mathrm{II}}, \tag{8.21}$$

where the λ_k are the same as in (8.18), and

$$\dim E_{\lambda_k}^{\mathrm{I}} = \dim E_{\lambda_k}^{\mathrm{II}}. \tag{8.22}$$

Equations (8.18), (8.21) and (8.22) show that the density matrices ρ^{I} and ρ^{II} differ only in the multiplicity of the eigenvalue zero.

6. Reduction of the matrix F

We continue to denote the eigenvalues of $\rho_+^{\mathrm{I,II}}$ by λ_k – but then, as they may be degenerate, we have to drop the convention $\lambda_k \neq \lambda_l$ for $k \neq l$. With this understanding, the set of eigenvectors $\{\psi_k^{\mathrm{I}} | k \in \mathbf{N}\}$ of ρ_+^{I} form an orthonormal basis for $\mathfrak{H}_+^{\mathrm{I}}$, with

$$\rho_+^{\mathrm{I}}\psi_k^{\mathrm{I}} = \lambda_k\psi_k^{\mathrm{I}}.$$

Then it is true that $\rho^{\mathrm{I}}\psi_k^{\mathrm{I}} = \lambda_k\psi_k^{\mathrm{I}}$, i.e., $F^\star F\psi_k^{\mathrm{I}} = \lambda_k\psi_k^{\mathrm{I}}$, so that

$$(\psi_k^{\mathrm{I}}, F^\star F\psi_k^{\mathrm{I}}) = (F\psi_k^{\mathrm{I}}, F\psi_k^{\mathrm{I}}) = \lambda_k. \tag{8.23}$$

We have seen in (8.20) that $F\psi_k^{\mathrm{I}}$ is an eigenvector of ρ^{II} with eigenvalue λ_k. Equation (8.23) now shows that

$$\psi_k^{\mathrm{II}} = \frac{1}{\sqrt{\lambda_k}} F\psi_k^{\mathrm{I}} \tag{8.24}$$

[29] For greater symmetry, we could have written λ_k as λ_k^{I} and μ_k as λ_k^{II}, but then the right-hand sides of (8.18) and (8.21) would have looked more cluttered.

is a normalized eigenvector of ρ^{II}, so that $\{\psi_k^{\mathrm{II}}\}$, $k \in \mathbf{N}$ is an orthonormal basis for $\mathfrak{H}_+^{\mathrm{II}}$. We also have the symmetric relation

$$\psi_k^{\mathrm{I}} = \frac{1}{\sqrt{\lambda_k}} F^\star \psi_k^{\mathrm{II}}. \tag{8.25}$$

Equations (8.24) and (8.25) show that the maps $F_+ : \mathfrak{H}_+^{\mathrm{I}} \to \mathfrak{H}_+^{\mathrm{II}}$ and $F_+^\star : \mathfrak{H}_+^{\mathrm{II}} \to \mathfrak{H}_+^{\mathrm{I}}$, which are restrictions of F to $\mathfrak{H}_+^{\mathrm{I}}$ and $F^\star$ to $\mathfrak{H}_+^{\mathrm{II}}$ respectively, are diagonal in the bases $\{\psi_k^{\mathrm{I}}\}$ and $\{\psi_k^{\mathrm{II}}\}$.

Now let $\{\xi_m^{\mathrm{I}}\}$, $m \in \mathbf{N}$, be an orthonormal basis for $\mathfrak{H}^{\mathrm{I}}$ such that[30]

$$\psi_k^{\mathrm{I}} = \xi_{\mu_k}^{\mathrm{I}}, \tag{8.26}$$

where the set $\{\mu_k\}$ is a countable subset of $\mathbf{N}$, and for $m \notin \{\mu_k\}$,

$$\xi_m^{\mathrm{I}} \in \mathfrak{H}_0^{\mathrm{I}}.$$

Similarly, let $\{\xi_n^{\mathrm{II}}\}$, $n \in \mathbf{N}$, be an orthonormal basis for $\mathfrak{H}^{\mathrm{II}}$ such that

$$\psi_k^{\mathrm{II}} = \xi_{\nu_k}^{\mathrm{II}}, \tag{8.27}$$

where $\{\nu_k\}$ is also a countable subset of $\mathbf{N}$, and for $n \notin \{\nu_k\}$,

$$\xi_n^{\mathrm{II}} \in \mathfrak{H}_0^{\mathrm{II}}.$$

We rewrite (8.24) as

$$\begin{aligned} F\,\xi_{\mu_k}^{\mathrm{I}} &= \sqrt{\lambda_k}\,\xi_{\nu_k}^{\mathrm{II}}, \\ F\,\xi_m^{\mathrm{I}} &= 0 \text{ for } m \notin \{\mu_k\}. \end{aligned} \tag{8.28}$$

That is, in these bases

$$f_{mn} = \begin{cases} \sqrt{\lambda_k}, & m = \mu_k, n = \nu_k, k \in \mathbf{N}, \\ 0, & \text{otherwise.} \end{cases} \tag{8.29}$$

Equivalently, the vector Φ can be written as

$$\Phi(q,r) = \sum_{k=1}^{\infty} \sqrt{\lambda_k}\,\xi_{\mu_k}^{\mathrm{I}}(q)\,\xi_{\nu_k}^{\mathrm{II}}(r). \tag{8.30}$$

At this point von Neumann writes (p. 434 of his book):

> By suitable choice of the complete orthonormal sets $\xi_m^{\mathrm{I}}(q)$ and $\xi_n^{\mathrm{II}}(r)$ we have established that each column of the matrix $[f_{mn}]$ contains at most one

[30] The orthonormal bases $\{\xi_m^{\mathrm{I}}\}$ and $\{\xi_n^{\mathrm{II}}\}$ are being introduced to cope with the possibility that $\dim \mathfrak{H}_0^{\mathrm{I}} \neq \dim \mathfrak{H}_0^{\mathrm{II}}$.

element $\neq 0$ (that this is real and > 0, namely $\sqrt{\lambda_k}$, is unimportant for what follows). What is the physical meaning of this mathematical statement?

Remark 8.9 It is clear from the derivation that the decomposition (8.30) of $\Phi(q,r)$ into a sum of mirrored pairs is unique. The coefficients $\sqrt{\lambda_k}$ are square roots of the common (nonzero) eigenvalues of ρ^{I}_{+} and ρ^{II}_{+}, and the vectors $\xi^{\mathrm{I}}_{\mu_k}(q)$, $\xi^{\mathrm{II}}_{\nu_k}(r)$ are eigenvectors of ρ^{I}_{+}, ρ^{II}_{+} respectively. We shall call (8.30) the *Schmidt–von Neumann canonical form* (or briefly, the canonical form) of the entangled state $\Phi(q,r)$.

7. Physical significance

We shall call I the *object*, II the *apparatus* and I + II the *combined system*. The discussion of Subsection 8.2.2 shows that an observable A^{I} may be assumed, without loss of generality, to be a compact self-adjoint operator. Von Neumann continues (p. 434 of his book):

> Let A^{I} be an operator with the eigenfunctions $\xi^{\mathrm{I}}_1, \xi^{\mathrm{I}}_2, \ldots$ and with only distinct eigenvalues, say $a_1, a_2, \ldots$; likewise A^{II} with $\xi^{\mathrm{I}}_1, \xi^{\mathrm{I}}_2, \ldots$ and $b_1, b_2, \ldots$. A^{I} corresponds to a physical quantity in I, A^{II} to one in II. They are therefore simultaneously measurable. It is easily seen that the statement 'A^{I} has the value a_m and A^{II} has the value b_n' determines the state $\Phi_{mn}(q,r) = \xi^{\mathrm{I}}_m(q)\,\xi^{\mathrm{II}}_n(r)$, and that this state has the probability
>
> $$(\Phi,\, P_{[\Phi_{mn}]}\Phi) = |(\Phi,\, \Phi_{mn}) = |f_{mn}|^2$$
>
> in the state $\Phi(q,r)$. Consequently, our statement means that $A^{\mathrm{I}}, A^{\mathrm{II}}$ are simultaneously measurable, and that if one of them was measured in Φ, then the value of the other is determined by it uniquely.

Then, after dealing with some minor technicalites (including redefining the $\sqrt{\lambda_k}$ of (8.30) as c_k), von Neumann arrives at the formulae

$$\rho^{\mathrm{I}}_{+} = \sum_{k=1}^{M} |c_k|^2 P_{[\xi^{\mathrm{I}}_{\mu_k}]} \tag{8.31}$$

and

$$\rho^{\mathrm{II}}_{+} = \sum_{k=1}^{M} |c_k|^2 P_{[\xi^{\mathrm{II}}_{\nu_k}]}. \tag{8.32}$$

He goes on to say (p. 436 of his book):

> Hence, when Φ is projected in I or II, it in general becomes a mixture, while it is a state[31] in I+II only. Indeed, it involves certain information regarding I+II *which cannot be made use of in* I *alone or in* II *alone, namely the one-to-one correspondence of the* A^{I} *and* A^{II} *values with each other* [emphasis added].

He concludes the section with the following:

> On the basis of the above results, we note: If I is in the state $\xi^{\mathrm{I}}(q)$ and II is in the state $\xi^{\mathrm{II}}(r)$, then I+II is in the state $\xi^{\mathrm{I}}(q)\,\xi^{\mathrm{II}}(r)$. If on the other hand I+II is in a state $\Phi(q,r)$ which is not a product $\xi^{\mathrm{I}}(q)\,\xi^{\mathrm{II}}(r)$, then I and II are mixtures and not states, but Φ establishes a one-to-one correspondence between possible values of certain quantities in I and in II.

8.3.3 Discussion of the measuring process

Von Neumann's Section VI.3 contains several pages devoted to proving that his measurement theory does not violate the principle of psycho-physical parallelism. As we shall not make use of the result, we have omitted the argument. For ease of reference, we have divided the remaining material, presented below, into two sections.

8.3.3.1 The initial state cannot be a mixture

Von Neumann argues that *the initial state of the apparatus cannot be a mixture* if a measurement is to be effected. The argument is as follows. We give as much of it as possible in his own words (pp. 437–438 of his book).

> Let I be the observed system, II the observer. If I is in the [pure] state $\rho^{\mathrm{I}} = P_{[\varphi^{\mathrm{I}}]}$ while II on the other hand is a mixture
>
> $$\rho^{\mathrm{II}} = \sum_{n=1}^{\infty} w_n P_{[\xi_n^{\mathrm{II}}]}, \tag{8.33}$$
>
> then I + II is a uniquely determined mixture
>
> $$\rho = \sum_{n=1}^{\infty} w_n P_{[\Phi_n]}, \quad \Phi_n(q,r) = \varphi^{\mathrm{I}}(q) \otimes \xi_n^{\mathrm{II}}(r). \tag{8.34}$$
>
> [The assertion follows from Theorem 8.7; ρ^{I} and ρ^{II} are clearly the projections of ρ defined by (8.34).] If now a measurement of a quantity A takes

31 Von Neumann uses the terms *state* and *mixture* to denote pure and mixed states respectively. We are using the terminology in current use, in which a state can be either pure or mixed.

place in I, then this [intervention] is to be regarded as an interaction of I and II. This is a process... [of unitary evolution under] an energy operator[32] H. If it has the time duration t, then we obtain

$$\rho(t) = \mathrm{e}^{-\mathrm{i}tH} \rho\, \mathrm{e}^{\mathrm{i}tH} \tag{8.35}$$

from ρ, and in fact,

$$\rho(t) = \sum_{n=1}^{\infty} w_n P_{[\mathrm{e}^{-\mathrm{i}tH} \Phi_n(q,r)]}. \tag{8.36}$$

At this stage von Neumann makes an assumption, which we have separated from the body of the text for emphasis:

Assumption 8.10 (von Neumann) For each n, the evolute

$$\mathrm{e}^{-\mathrm{i}tH} \Phi_n(q,r) \tag{8.37}$$

has the form $= \phi_n^{\mathrm{I}}(q)\, \eta_n^{\mathrm{II}}(r)$, where the ϕ_n^{I} are eigenfunctions of A and the η_n^{II} any fixed orthonormal set in $\mathfrak{H}^{\mathrm{II}}$.

He continues:

> [If Assumption 8.10 holds], then this intervention will have the character of a measurement. For it transforms each [pure] state φ of I into a mixture of the eigenfunctions ϕ_n^{I} of A. The statistical character therefore arises in this way: Before the measurement I was in a (unique) [pure] state, but II was a mixture – and the mixture character of II has, in the course of the interaction, associated itself with I + II, and, in particular, it has made a mixture of the projection in I...
>
> At this point, the attempted explanation breaks down. For quantum mechanics requires that $w_n = (\varphi^{\mathrm{I}}, P_{[\phi_n^{\mathrm{I}}]}\varphi^{\mathrm{I}}) = |(\varphi^{\mathrm{I}}, \phi_n^{\mathrm{I}})|^2$, i.e., w_n dependent on φ^{I}!...

Von Neumann concludes that (p. 439 of his book):

> Therefore the non-causal nature of the [collapse] process is not produced by any incomplete knowledge of the state of the observer...

Remark 8.11 The situation may be described, in words, as follows. Von Neumann is trying to demonstrate that, if the initial state of the apparatus is a mixture, then the principle that the initial states of object and apparatus

[32] It is this assertion of von Neumann that appears to justify the statement by London and Bauer, Remarks 8.8 (ii).

are independent of each other is violated if Assumption 8.10 holds. That is, if Assumption 8.10 is invalid, then the result too may cease to be valid.

Assumption 8.10 is a very specific assumption about the object–apparatus interaction; it asserts that eigenfunctions of A are mapped one-to-one onto vector states of the apparatus after the measurement interaction has run its course. Von Neumann seems to have accepted it as generic, which does not need justification. We shall find that the object–apparatus interaction that is at the heart of Sewell's theory (Chapter 10) clearly violates this assumption; nondegenerate eigenstates of A are mapped to *subspaces of* dim $\gg 1$ of the Hilbert space of the apparatus. From the point of view of physics, Assumption 8.10 may be generic for entanglement, but it surely is not for a measurement.

8.3.3.2 Apparatus with a classical display

For the last part, we again take up the story in von Neumann's own words (p. 439 of his book):

> Let us now apply ourselves again to the problem formulated at the end of [Section 8.3.1]. I, II and III shall have the meanings given there, and, for the quantum-mechanical investigation of I, II, we shall use the notation of [Section 8.3.2], while III remains outside of the calculation (cf. the discussion of this in [Section 8.3.1]). Let A be a quantity (in I) actually to be measured, $\varphi_1^{\mathrm{I}}, \varphi_2^{\mathrm{I}}, \ldots$ its eigenfunctions. Let I be in the state $\varphi^{\mathrm{I}}(q)$.
>
> If I is the observed system, II + III the observer, then we must apply the [collapse] process, and we find that the measurement transforms I from the state φ^{I} into one of the states φ_n^{I}, $(n = 1, 2, \ldots)$, the probabilities for which are respectively $|(\varphi^{\mathrm{I}}, \varphi_n^{\mathrm{I}})|^2$, $(n = 1, 2, \ldots)$.

The paragraph that follows is a direct continuation of the above. We have interrupted the continuation at this point as a way of emphasizing it.

> Now, what is the method of description if I + II is the observed system, and only III the observer? In this case we must say that II *is a measuring instrument which shows on a scale the value of* A (*in* I) : *the position of the pointer on this scale is a physical quantity* B (*in* II) *which is actually observed by* III... [emphasis added].

With this quotation, we end our summary of this section of von Neumann's book.

8.4 Wigner's reservations

Wigner's major reservations about von Neumann's measurement theory were condensed into one paragraph in his 1963 review, in the section entitled *Problems of the orthodox view* (Wigner, 1970, p. 167). Most of this paragraph is reproduced below. We have broken up the paragraph into a list of separate items. We shall consider these reservations in the sections and pages indicated at the end of each:

(i) The principal conceptual weakness of the orthodox view is, in my opinion, that it merely abstractly postulates interactions which have the effect of [reduction of the wave packet, Section 10.7, page 190].

(ii) For some observables, in fact for the majority of them (such as xyp_z), nobody seriously believes that a measuring apparatus exists [Section 9.3, pages 170–172].

(iii) It can even be shown that no observable which does not commute with the additive conserved quantities (such as linear or angular momentum or electric charge) can be measured precisely, and in order to increase the accuracy of the measurement, one has to use a very large measuring apparatus... [Section 11.8, pages 205–206].

(iv) On the other hand, most quantities which we believe to be able to measure, and surely all the very important quantities such as position, momentum, fail to commute with all the conserved quantities, so that their measurement cannot be possible with microscopic apparatus [Section 11.8, pages 205-206].

(v) This raises the suspicion that the macroscopic nature of the apparatus is necessary in principle... The joint state vector... resulting from a measurement with a very large apparatus, surely *cannot be distinguished as simply from a mixture* [emphasis in the original] as was the state vector obtained in the Stern-Gerlach experiment... [This experiment had been discussed by Wigner earlier in his paper. See Section 11.8, pages 205–206.]

In the third paragraph of the same section of the same article, Wigner wrote:

> The simplest... summary of the conclusions... is that... [the quantum] laws merely provide probability connections between the results of several consecutive observations on a system... However, there is a certain weakness in the word "consecutive", as this is not a relativistic concept. Most observations are not local and one will assume, similarly, that they have an irreducible extension in time, that is, duration...

This problem will be fully addresed in Section 9.3.

8.5 Reconsideration of von Neumann's theory

In the above, we have presented a fairly detailed account of the mathematical part of von Neumann's measurement theory, and a brief but adequate summary of Wigner's critique of it. To conclude this chapter, we shall rephrase some of von Neumann's his results in the language of entanglement, and shall draw attention to a statement of his which suggests that his 'physical' notion of the measuring apparatus may have differed essentially from the one he analysed mathematically.

8.5.1 Entanglement

The state Φ of the two-component system defined by (8.7) is clearly an entangled state. Von Neumann's analysis of the structure of composite systems is an analysis of the structure of entangled states of two-component systems, which may be extended to n-component systems. We shall therefore name the result summarized in paragraph 7 of Subsection 8.3.2 (pages 154–155) *von Neumann's entanglement theorem* and restate it as follows (recall the definition of projections of density matrices on page 148):

Theorem 8.12 (Von Neumann's entanglement theorem)

(1) *Projections of separated states of a two-component system are pure states of the component systems; projections of entangled states of a two-component system are mixed states of the component systems.*

(2) *In an entangled state, there is a* $(1,1)$ *correspondence between the values of an observable* A^{I} *with nondegenerate eigenvalues that commutes with* ρ^{I} *and the values of a corresponding observable* A^{II}*, so that the value of* A^{I} *may be inferred by observing the value of* A^{II}*.*

Remark 8.13 Von Neumann's remark on page 148 may now be restated as follows: The rules of correspondence for (density matrix) states reflect the phenomenon of entanglement, which does not apply to observables.

Let us return, briefly, to the analysis that transformed the state $\Phi(q,r)$ defined by (8.7) into the form (8.30). Suppose that the systems I and II were interacting with each other for $t < -|\tau|$, but are moving freely, each under its own Hamiltonian, for $t \geq 0$. Then the coefficients f_{mn} in (8.7) are functions of time. To avoid ambiguity we shall denote the time-dependent coefficients by $a_{mn}(t)$ and set $f_{mn} = a_{mn}(0)$. That is, the quantities λ_k in (8.30) are calculated with the coefficients $a_{mn}(0)$.

However, the λ_k are the common eigenvalues of the density matrices $\rho^{\mathrm{I}}(0)$ and $\rho^{\mathrm{II}}(0)$ at $t = 0$. If the systems I and II are not interacting with each other, then *these eigenvalues do not change with time.* That is, *the decomposition* (8.30) *holds for all* $t > 0$. We restate this as follows, for emphasis:

Conclusion 8.14 (Persistence of entanglement) Equation (8.30) establishes the phenomenon of *persistence of entanglement* in nonrelativistic physics.

8.5.2 Description of the measuring apparatus

In Section 8.3.3, von Neumann considers two possibilities: (i) I is the observed system, and II + III the observer, and (ii) I + II is the observed system, and only III the observer. He says (we reproduce this part of the quotation from page 157):

> Now, what is the method of description if I + II is the observed system, and only III the observer? In this case we must say that II *is a measuring instrument which shows on a scale the value of A (in* I)*: the position of the pointer on this scale is a physical quantity B (in* II) *which is actually observed by* III... [emphasis added].

To the present author, the above quotation suggests that von Neumann regarded the display of the apparatus as a classical display; it is difficult to see how the terms 'scale' and 'position of the pointer' can be interpreted otherwise. In this case, we may remark that III could well be a recording device, rather than a human observer. However, *von Neumann offers no explanation of how this physical quantity B (in* II) *– which must be a self-adjoint operator on* $\mathfrak{H}^{\rm II}$ *– could arise in the scheme of Subsection* 8.3.2; it may have been suggested (to him) by the analysis in Section V.4 of his book, entitled *The macroscopic measurement.*

Wigner thought differently (see the quotation on page 137):

> ...the state of the apparatus has no classical description...,

but, after reviewing the work of Yanase (1961),[33] he softened somewhat: 'This raises the suspicion that the macroscopic nature of the apparatus is necessary in principle...' (item (v) of the quotation on page 158). He did not offer any suggestion as to how a 'very large measuring apparatus' could differ qualitatively from a small one; he only noted that, in such a situation, it would not be easy to distinguish, experimentally, between pure and mixed states of the coupled system. Neither did he – or so it seems to the present author – take into account the developments that established that the term *classical observables* (as defined on page 141) was mathematically well defined and physically meaningful.

[33] Yanase's work dealt with increasing the accuracy of measurement of a class of operators which could only be measured approximately (Wigner, 1952; Araki and Yanase, 1960). The word *measurement* is used here in the sense that 'the state of the apparatus reflects the state of the object', nothing more.

9

Macroscopic observables in quantum physics

In a certain sense, object and apparatus were essentially on the same footing in von Neumann's measurement theory. They were systems with k and l degrees of freedom respectively (page 146). There were no constraints on the numbers k and l, which could, for example, be of the same order of magnitude. Quite possibly, it was this lack of differentiation between object and apparatus that led Wigner to assert that 'the state of the apparatus has no classical description' (page 137), a state of affairs that produced the infinite von Neumann chain which only ended in the observer's consciousness.[1]

In this and the following chapter we shall break with von Neumann and assume that k is *small*, i.e., of the order of unity, and that l is *large*, i.e., within a few orders of magnitude of Avogadro's number. The room for manoeuvre that this provides will allow us to break with Wigner and explore systems with states that *do* have classical descriptions. It will not surprise the informed reader that the room for manoeuvre created by our assumption will be filled, very substantially, by von Neumann's own work.

This chapter is divided into three sections. Section 9.1 is devoted to a theorem of von Neumann on observables that commute with each other. This prepares the way to our treatment of macroscopic observables, which is based on the commuting approximations to P and Q devised by von Neumann; these results, together with their antecedents, are presented in Section 9.2. In Section 9.3, an attempt is made to resolve some of Wigner's doubts.

We remind the reader that an adequate account of the theory of single operators is provided in Appendix A6.

9.1 Commuting self-adjoint operators

Let $\mathfrak{H}$ be a Hilbert space and T a bounded self-adjoint operator with the spectral decomposition (see page 336)

$$T = \int \lambda \, \mathrm{d}E_\lambda. \tag{9.1}$$

[1] See, however, Remark 8.3.

Define two bounded self-adjoint operators A and B by

$$A = \int f(\lambda)\,\mathrm{d}E_\lambda, \quad B = \int g(\lambda)\,\mathrm{d}E_\lambda, \tag{9.2}$$

where f and g are measurable functions. Clearly, the operators A and B commute with each other, and with T. This result has a converse, which we shall state in a more general form, i.e., for a finite or countable family of bounded or unbounded self-adjoint operators. However, we have first to clarify the notion of commutativity for operators that are not necessarily bounded.

An unbounded self-adjoint operator G on $\mathfrak{H}$ cannot be defined everywhere on $\mathfrak{H}$. Let $\xi \in \mathfrak{H}$ be a vector which is outside the domain of G. Then the equation $0{\cdot}G\xi = G{\cdot}0\xi$ has no meaning, because $G\xi$ on the left-hand side is undefined. We would like the relation $0{\cdot}G - G{\cdot}0 = 0$ to hold unconditionally. This can be achieved by redefining the notion of commutativity as follows.

Let R and S be two self-adjoint operators on $\mathfrak{H}$, bounded or unbounded, and let their spectral decompositions be

$$R = \int r(\lambda)\,\mathrm{d}R_\lambda, \quad S = \int s(\lambda)\,\mathrm{d}S_\lambda. \tag{9.3}$$

We define

$$[R, S] = 0 \text{ iff } [R_\lambda, S_\mu] = 0 \text{ for all } \lambda, \mu. \tag{9.4}$$

If R and S are both bounded, this definition agrees with the usual definition. If one of them is unbounded and the other, say S, is the constant operator $cI, c \in \mathbb{C}$, then $S_\mu = 0$ or I; the latter commute with every R_λ, so that the definition (9.4) achieves the desired result.[2]

We are now ready to state (somewhat loosely) the following result, which is a special case of a theorem of von Neumann (von Neumann, 1955, pp. 173–174).

Theorem 9.1 *Let $\{A_k | k \in \mathbf{N}\}$ be a family of commuting self-adjoint operators on $\mathfrak{H}$. Then there exists a self-adjoint operator T (possibly unbounded) on $\mathfrak{H}$ with a spectral resolution $T = \int \lambda\,\mathrm{d}E_\lambda$, and a family of real-valued measurable functions $\{f_k(\lambda)\}$, indexed by $k \in \mathbf{N}$, such that*

$$A_k = \int f_k(\lambda)\,\mathrm{d}E_\lambda.$$

This result shows that a 'complete set of commuting observables' can, in principle, be reduced to a single observable (and a family of real-valued measurable functions). This is convenient for measurement theory, but the success or failure of a practical calculation in quantum mechanics depends more often upon the choice of a suitable basis conforming to a set of commuting observables.

[2] Note that our general assumption that all observables be defined on a common dense domain does not address the problem that we have just settled.

9.1.1 Commuting observables with discrete spectra

For the special case of a finite set $\{A^{(k)}|k = 1, 2, \ldots, K\}$ of commuting observables with purely discrete spectra, Theorem 9.1 can easily be established. It is the result we need, and we shall give an informal sketch of the proof, because of the insight it provides.[3]

The observables $A^{(k)}$ have the spectral decompositions

$$A^{(k)} = \sum_j \lambda_j^{(k)} E_j^{(k)}. \tag{9.5}$$

For clarity, we write $A^{(1)} = \sum_p \lambda_p E_p$ and $A^{(2)} = \sum_q \mu_q F_q$. Let

$$G_{p,q} = E_p \cap F_q. \tag{9.6}$$

Then

$$E_p = \sum_q G_{p,q}, \quad F_q = \sum_p G_{p,q} \tag{9.7}$$

and

$$\sum_{p,q} G_{p,q} = I,$$

so that $A^{(1)}, A^{(2)}$ can be expressed in the form

$$A^{(1,2)} = \sum_{p,q} g_{p,q}^{(1,2)} G_{p,q}, \tag{9.8}$$

where $g_{p,q}^{(1,2)}$ are suitably chosen real numbers (many of which will be zero). However, *it will no longer be true that*

$$(p, q) \neq (r, s) \Rightarrow g_{p,q}^{(1,2)} \neq g_{r,s}^{(1,2)}.$$

We now relabel the set $\{G_{p,q}\}$ with a single index, say $a^{(1)}$ (note that $a^{(1)}$ is a running index; the superscript (1) denotes that this is the first step of a process), and rewrite (9.5) in terms of the $\{G_{a^{(1)}}\}$. Since the process (9.6) of subdividing a pair of projection operators into a set of finer projection operators does not depend on the eigenvalues, we may repeat the steps (9.7) and (9.8) with $\{G_{a^{(1)}}\}$ and $A^{(3)}$ and obtain a still finer set of projection operators $\{G_{a^{(2)}}\}$. After $k-1$ steps, when all the observables $A^{(k)}$ will have been so dealt with, we shall arrive at a set $\{G_{a^{(k-1)}}\} = \{\Pi_\alpha|\alpha = 1, 2, \ldots, \nu\}$ of pairwise-orthogonal projection operators which are fine enough to permit the reconstruction, by formulae like (9.8), of each $A^{(k)}$. This set $\{\Pi_\alpha\}$ will be the coarsest such set of projection operators, 'coarsest' here meaning the following. If $\{G'_z\} \neq \{\Pi_\alpha\}$ is another set

[3] A full proof, and some further discussion, is given in von Neumann's book, pp. 173–178.

of projection operators out of which one may reconstruct the set $\{A^{(k)}\}$, then one may reconstruct the $\{\Pi_\alpha\}$ out of the $\{G'_z\}$, but not the $\{G'_z\}$ out of the $\{\Pi_\alpha\}$.

We shall call the set of projection operators $\{\Pi_\alpha\}$ the *cellular decomposition*[4] of the set of operators $A^{(k)}$. Under appropriate physical conditions, the subspaces $\mathfrak{H}_\alpha = \Pi_\alpha \mathfrak{H}$ may be regarded as the quantum analogues of classical phase cells. Mathematically, the subspaces $\mathfrak{H}_\alpha$ may be defined by a single *cellular operator*

$$T = \sum_{\alpha=1}^{\nu} \lambda_\alpha \Pi_\alpha, \quad \lambda_\alpha \in \mathbb{R}, \; \lambda_\alpha \neq \lambda_\beta \text{ for } \alpha \neq \beta, \tag{9.9}$$

and the classical observables on $\mathfrak{H}$ will turn out to be the observables that commute with a suitable T.

Remark 9.2 The emphasis in the above sentence is on the word *suitable.* As we shall see in subsection 9.2.5, there are a vast number of observables that commute with each other, even on a single-particle Hilbert space.

9.2 The uncertainty principle and commuting observables

We begin with a summary, adapted from von Neumann's account (1955, pp. 230–237) of the mathematical theory of the uncertainty principle and of minimum uncertainty product states.[5] We remind the reader that we are using units in which $\hbar = 1$.

9.2.1 The uncertainty principle

The formalism of quantum mechanics is based on the canonical commutation relations

$$[P, Q] = -\mathrm{i} I_{\mathfrak{D}}. \tag{9.10}$$

Of the two operators P, Q, at least one must be unbounded (Theorem A6.8, page 324); $\mathfrak{D}$ is their common domain which is dense in $\mathfrak{H}$, and $I_{\mathfrak{D}}$ the identity operator[6] on $\mathfrak{D}$.

For any vector $\phi \in \mathfrak{D}$ with $||\phi|| = 1$, the identity

$$2\,\mathrm{Im}\,(P\phi, Q\phi) = -1 \tag{9.11}$$

[4] The terms *cellular decomposition* and *cellular operator* are ad hoc terms; the first has been borrowed from an entirely different branch of mathematics.

[5] References to the original sources are given in footnote 131 on p. 233 of von Neumann's book. The papers (Heisenberg, 1927) (in English translation) and (Robertson, 1929) may be found in (Wheeler and Zurek, 1983).

[6] In the following, we shall omit the subscript on $I_{\mathfrak{D}}$; this should not cause any confusion.

is established by straightforward computation; Im z denotes the imaginary part of z. The quantity Im $(P\phi, Q\phi)$ is antisymmetric in its arguments,

$$\text{Im}\,(P\phi, Q\phi) = -\text{Im}\,(Q\phi, P\phi),$$

and therefore we shall replace (9.11) by the equality of absolute values

$$2\,|\text{Im}\,(P\phi, Q\phi)| = 1, \tag{9.12}$$

which is symmetric in P and Q. We then have the inequalities

$$|\text{Im}\,(P\phi, Q\phi)| \leq |(P\phi, Q\phi)| \leq ||P\phi|| \cdot ||Q\phi||, \tag{9.13}$$

of which the second follows from the Schwarz inequality $|f, g)| \leq ||f|| \cdot ||g||$ (page 317). Combining (9.12) with (9.13), we obtain the uncertainty principle bound

$$||P\phi|| \cdot ||Q\phi|| \geq \tfrac{1}{2}. \tag{9.14}$$

We introduce the following notations for the means (or expectation values) and dispersions of P and Q in the (normalized) state ϕ:

$$\begin{aligned} \rho &= (\phi, P\phi), & \epsilon^2 &= ||(P - \rho I)\phi||^2, \\ \sigma &= (\phi, Q\phi), & \eta^2 &= ||(Q - \sigma I)\phi||^2. \end{aligned} \tag{9.15}$$

Define now the mean-shifted operators

$$\begin{aligned} P' &= P - \rho I, \\ Q' &= Q - \sigma I. \end{aligned} \tag{9.16}$$

Then, firstly,

$$(\phi, P'\phi) = (\phi, Q'\phi) = 0, \tag{9.17}$$

and secondly,

$$[P', Q'] = [P, Q] = -\mathrm{i}I. \tag{9.18}$$

The uncertainty principle inequality (9.14) was derived solely from the commutation relation (9.10). Therefore it continues to hold for P', Q', i.e.,

$$||P'\phi|| \cdot ||Q'\phi|| \geq \tfrac{1}{2}; \tag{9.19}$$

using the definitions (9.15) of ϵ and η, this inequality becomes

$$\epsilon \cdot \eta \geq \tfrac{1}{2}. \tag{9.20}$$

9.2.2 Minimum uncertainty product states

We now investigate whether there exist vectors $\psi \in \mathfrak{D}$ such that the equality holds in (9.20). Note first that the argument which led to the inequality (9.14) proceeded via the relations (9.12) and (9.13) – and the latter also remain true with P, Q replaced by P', Q' respectively. Therefore the equality will hold in (9.20) if and only if the following equalities hold:

$$\tfrac{1}{2} = |\mathrm{Im}\,(P'\psi, Q'\psi)| = |(P'\psi, Q'\psi)| = ||P'\psi||\cdot||Q'\psi||. \tag{9.21}$$

The last of these will be true if and only if $P'\psi$ is a constant multiple of $Q'\psi$, i.e., iff

$$P'\psi = (\beta + \mathrm{i}\gamma)Q'\psi, \quad \beta, \gamma \in \mathbb{R}. \tag{9.22}$$

If this condition is satisfied, then the second equality in (9.21) becomes

$$\mathrm{Im}\,[((\beta + \mathrm{i}\gamma)Q'\psi, Q'\psi,)] = |\beta + \mathrm{i}\gamma||(Q'\psi, Q'\psi)|,$$

which can be satisfied if and only if $\beta = 0$. In this case,

$$P'\psi = \mathrm{i}\gamma Q'\psi. \tag{9.23}$$

If (9.23) holds, then the first of (9.21) gives

$$1 = 2\gamma||Q'\psi||^2, \tag{9.24}$$

i.e., $\gamma > 0$. Then from (9.19), (9.20) and (9.24) we obtain

$$\epsilon = \sqrt{\frac{\gamma}{2}}, \quad \eta = \sqrt{\frac{1}{2\gamma}}. \tag{9.25}$$

9.2.3 Determination of ψ

The discussion of the last subsection does not prove the existence of ψ. This, however, can be done by direct calculation. Let $\mathfrak{H} = L^2(\mathbb{R}, \mathrm{d}q)$, and let Q be the operator of multiplication by q on it. Then

$$P = -\mathrm{i}\frac{\mathrm{d}}{\mathrm{d}q},$$

and (9.23) becomes

$$\left(-\mathrm{i}\frac{\mathrm{d}}{\mathrm{d}q} - \rho\right)\psi = \mathrm{i}\gamma(q - \sigma)\psi,$$

or

$$\frac{\mathrm{d}\psi}{\mathrm{d}q} = [-\gamma(q - \sigma) + \mathrm{i}\rho]\psi. \tag{9.26}$$

This is a first-order linear differential equation. The constant of integration in its solution is determined by the normalization condition $||\psi|| = 1$. The solution is

$$\psi_{\gamma;\rho,\sigma}(q) = \left(\frac{2\gamma}{\pi}\right)^{\frac{1}{4}} \exp\left[-\gamma(q-\sigma)^2/2\right] \cdot \exp\left(\mathrm{i}\rho q\right), \tag{9.27}$$

which is a damped oscillation centred around $q = \sigma$. This shows that minimum uncertainty product states exist, and form a three-parameter family. For the special case of equal uncertainties $\epsilon = \eta$, the parameter γ equals unity, and the solution (9.27) assumes the form

$$\psi_{1;\rho,\sigma}(q) = \left(\frac{2}{\pi}\right)^{\frac{1}{4}} \exp\left[-(q-\sigma)^2/2\right] \cdot \exp\left(\mathrm{i}\rho q\right). \tag{9.28}$$

9.2.4 Von Neumann's commuting operators $\hat{P}, \hat{Q}$

The self-adjoint operators P, Q considered above are unbounded and their spectra are strictly continuous. Therefore they have no eigenvectors, and can only be measured approximately (page 142). That being the case, von Neumann asked the following question: is it possible to find two *commuting* observables $\hat{P}, \hat{Q}$ which are reasonable approximations to P and Q respectively? If such operators exist, then one could claim to measure P and Q, albeit approximately, but *simultaneously.*[7] His answer was in the affirmative, and his construction is sketched below.

We see from the right-hand sides of (9.15) that the vector ψ of (9.27) satisfies the equations

$$\begin{aligned} ||(P-\rho I)\psi_{\gamma;\rho,\sigma}|| &= \epsilon, \\ ||(Q-\sigma I)\psi_{\gamma;\rho,\sigma}|| &= \eta. \end{aligned} \tag{9.29}$$

It may therefore be called a 'simultaneous approximate eigenvector' of P and Q corresponding to the spectral values ρ of P and σ of Q respectively. The goodness of the approximations is determined by ϵ and η respectively. The equality signs in (9.29) show that, given the uncertainty principle, they are the best possible.

Define now

$$\begin{aligned} \rho_\mu &= \sqrt{2\pi\gamma}\cdot\mu = \sqrt{4\pi}\cdot\epsilon\mu, \\ \sigma_\nu &= \sqrt{2\pi/\gamma}\cdot\nu = \sqrt{4\pi}\cdot\eta\nu, \end{aligned} \tag{9.30}$$

[7] Equation (9.20) merely shows the impossibility of a simultaneous measurement with $\epsilon \cdot \eta < 1/2$; it does not establish the *possibility* of a simultaneous measurement with $\epsilon{\cdot}\eta = K \geq 1/2$.

where $\mu, \nu \in \mathbb{Z}$. Then

$$\rho_\mu - \rho_{\mu-1} = \sqrt{4\pi}\cdot\epsilon,$$
$$\sigma_\nu - \sigma_{\nu-1} = \sqrt{4\pi}\cdot\eta.$$

We now relabel the states $\psi_{\gamma;\rho_\mu,\sigma_\nu}$ as $\psi_{\mu,\nu}$. Then (9.29) become

$$||(P - \rho_\mu I)\psi_{\mu,\nu}|| = \epsilon, \qquad ||(Q - \sigma_\nu I)\psi_{\mu,\nu}|| = \eta. \tag{9.31}$$

The vectors of $\{\psi_{\mu,\nu}|\mu,\nu \in \mathbb{Z}\}$ are normalized, but not orthogonal to each other. Von Neumann used the Gram–Schmidt process to obtain an orthonormalized set, and verified, 'without any particular difficulties', that it was complete. He denoted the set he obtained by $\{\psi'_{\mu,\nu}|\mu,\nu \in \mathbb{Z}\}$, and established the estimates

$$||(P - \rho_\mu I)\psi'_{\mu,\nu}|| \le C\epsilon, \qquad ||(Q - \sigma_\nu I)\psi'_{\mu,\nu}|| \le C\eta \tag{9.32}$$

for a certain fixed C. He added (von Neumann, 1955, p. 407) :

> A value $C \sim 60$ has been obtained in this way, and it could probably be reduced. The proof of this fact leads to rather tedious calculations, which require no new concepts, and we shall omit them. The factors $C \sim 60$ are not important, since $\epsilon\eta\ldots$ measured in macroscopic (CGS) units is exceedingly small ($c\cdot 10^{-28}$).

Denote now by $P_{[\psi'_{\mu,\nu}]}$ the projection operator onto $\psi'_{\mu,\nu}$, and define

$$\hat{P} = \sum_{\mu,\nu\in\mathbb{Z}} \rho_\mu \, P_{[\psi'_{\mu,\nu}]} \tag{9.33}$$

and

$$\hat{Q} = \sum_{\mu,\nu\in\mathbb{Z}} \sigma_\nu \, P_{[\psi'_{\mu,\nu}]}. \tag{9.34}$$

Then $\hat{P}$ and $\hat{Q}$ are von Neumann approximants to P and Q respectively (page 143), and

$$[\hat{P}, \hat{Q}] = 0. \tag{9.35}$$

The eigenvalues ρ_μ of $\hat{P}$ and σ_ν of $\hat{Q}$ are infinitely degenerate. Let

$$E_\mu = \sum_{\nu\in\mathbb{Z}} P_{[\psi'_{\mu,\nu}]}$$

and

$$F_\nu = \sum_{\mu \in \mathbb{Z}} P_{[\psi'_{\mu,\nu}]}.$$

Then we may write (9.33) and (9.34) as

$$\hat{P} = \sum_{\mu \in \mathbb{Z}} \rho_\mu E_\mu \tag{9.36}$$

and

$$\hat{Q} = \sum_{\nu \in \mathbb{Z}} \sigma_\nu F_\nu. \tag{9.37}$$

The E_μ and F_ν are projection operators onto infinite-dimensional subspaces of $\mathfrak{H}$. The numbers ρ_μ and σ_ν which characterize these subspaces are measurable with absolute precision in the von Neumann theory, and determine 'values' of P and Q up to errors ϵ and η respectively.

The above considerations may readily be extended to systems with N degrees of freedom, subject to the canonical commutation relations (7.46). Every well-defined operator function $F(P_i, Q_j)$ that has a continuous spectrum can be approximated by a function $F(\hat{P}_i, \hat{Q}_j)$ which commutes with every $\hat{P}_i, \hat{Q}_j$.

9.2.5 The von Neumann–Wigner characterization of macroscopic measurements

Section V.4 of von Neumann's book (pp. 398–416) is entitled *The macroscopic measurement.* Although the subject is discussed extensively, the term 'macroscopic measurement' itself does not appear to be defined explicitly. The following 'definition', so to speak, has been distilled by the present author from his discursive treatment:

Definition 9.3 (Macroscopic measurements)

(i) A macroscopic measurement is never infinitely precise; one has therefore to specify the accuracy of measurement for the quantity being measured.

(ii) A macroscopic measurement can be regarded as a yes–no experiment. Consequently, its possible results can be represented by a projection operator E on the Hilbert space $\mathfrak{H}$. The eigenvalues 1 and 0 of E correspond to the answers 'yes' and 'no' respectively. If the answer is yes, then the state of the system is to be found in the subspace $E\mathfrak{H}$; if the answer is no, it is to be found in the subspace $(I - E)\mathfrak{H}$.
(In footnote 203 on p. 402 of his book, von Neumann attributes this characterization of the macroscopic observer to Wigner.)

(iii) All macroscopic measurements that can be performed can be performed simultaneously; i.e., the corresponding projection operators $E_1, E_2, \ldots$ commute with each other.

(iv) The simultaneous eigenspaces $\mathfrak{H}_j, j \in \mathbf{N}$ of all the macroscopic projection operators $\{E_j\}$ correspond to classical phase cells.
(According to Boltzmann's formula ($S = k \log W$), the dimension of an eigenspace $\mathfrak{H}_k$ is proportional to the exponential of the entropy of the corresponding phase cell; it is ordinarily a large number.)

Further on, von Neumann remarks (footnote 204, pp. 410) that

> ... all macroscopically observable quantities are in no way all commutative with H [the Hamiltonian]. Indeed, many such quantities, for example the center of gravity of a gas in diffusion, change appreciably with t... Since all macroscopic quantities do commute, H is never a macroscopic quantity, i.e., the energy is not measured macroscopically with complete precision. This is plausible without additional comment.

Definition 9.3(i) suggests that von Neumann regarded the spectra of all macroscopic observables to be continuous. It followed that they could only be measured via approximants which had discrete spectra. If the macroscopic observables are assumed to be the quantum counterparts of functions of dynamical variables p_i, q_j of classical mechanics, then the substitutions

$$p_i \rightarrow \hat{P}_i, \quad q_j \rightarrow \hat{Q}_j \tag{9.38}$$

will produce, in one fell swoop, a family of operators which (a) have discrete spectra, (b) commute with each other, and (c) have a useful physical interpretation if the corresponding classical expression has one. This family will consist of n-particle operators for $n = 1, 2, \ldots, N$, that is, the projections in its cellular operator (9.9) will be one-dimensional (and the upper limit of the summation will be ∞).[8]

In current usage, a macroscopic operator is an n-particle operator with large n, meaning that n is within a few orders of magnitude of Avogadro's number. Definition 9.3 does not place any restriction on n. *Our subsequent use of the term* macroscopic *will conform to the current usage.*

9.3 Answers to Wigner, I

While the term measurement encompasses a wide variety of processes, measurement theory in quantum mechanics is concerned exclusively with those that

[8] The procedure described above is designed to be applicable simultaneously to all N. In practice, N is considered fixed and averaging (coarse-graining) precedes discretization.

lead to the reduction of a wave packet. The wave packet of a small quantum-mechanical system is reduced, as we shall see in the next chapter, when it interacts (in a manner which will be specified) with a macroscopic one *as a whole*. This has the effect of replacing the infinite von Neumann chain by a *bridge over the Heisenberg cut*, and renders invalid Wigner's remark that 'the state of the system has no classical description' (page 137) once the bridge is crossed.

But what exactly does a classical description achieve?

The final (metaphysical) stage of von Neumann's measurement theory was anthropocentric, and its anthropocentrism was accepted by Wigner for lack of a better alternative. This aspect was strongly controverted by Jauch, who began by pointing out a key difference between classical and quantum-mechanical measurements. In a classical measurement, the apparatus did not have a 'back effect' (Jauch's term) upon the object. He then went on to say (Jauch, 1968, p. 164):

> It is essential in the construction of an objective science that it be freed from anthropomorphic elements. This requirement of objectivity... can actually be satisfied [in microphysics] because... the last stages of observation, [namely] 'reading the scale' [are] on the classical level...
>
> [This] has the consequence... that the 'scale can be read' by a number of different observers who can communicate and establish that they read concurrent results... The individual observer, although necessary for completing an actual observation, can now fade into the background...

We may sum up the discussion as follows: if the state of the apparatus has a classical description, then the measurement may be regarded as completed when the object and the apparatus have ceased to interact.

Another of Wigner's objections was the following (page 158) :

> For some observables, in fact for the majority of them (such as xyp_z), nobody seriously believes that a measuring apparatus exists.

This objection too is anthropocentric.

There is considerable evidence – or so the present author contends – that microscopic quantum systems can interact with large quantum systems *as a whole*.[9] Examples include the Mössbauer effect and (in the nonrelativistic domain of finite-mass systems) the creation of elementary excitations by the inelastic

[9] We remind the reader once again that our subject-matter is nonrelativistic quantum mechanics, in which there is no upper bound to the signal velocity. Therefore this notion makes perfect sense. The fact that global effects can often be approximated very well by local calculations is a source of confusion regarding inferences that can be made from observations, which only strengthens the case for a clear, unambiguous statement of the theoretical framework.

scattering of neutrons by solids and by liquid helium. The 'measurement interaction' appears to be a subclass of such interactions. What is wholly lacking is *knowledge of the general principles – if indeed there be any – that govern such asymmetric interactions*, except for the realization that they may violate time-reversal invariance. These are major physical problems which, in the opinion of the present author, have been disregarded owing to the anthropocentric bias of quantum measurement theory. The problem, surely, is not the construction of a device for measuring xyp_z, but understanding how microscopic observables such as xyp_z can interact with macroscopic observables of a large, nonrelativistic quantum-mechanical system.

10

Sewell's theory of measurement

If the probability interpretation of quantum mechanics is to be maintained,[1] then an act of measurement – of a self-adjoint operator A with eigenvalues λ_i and eigenvectors ψ_i – should have the following effects:

(i) Interaction with the measuring apparatus should cause an initial pure state $\sum_i c_i \psi_i$ of the observed system to collapse to the mixed state $\sum_i |c_i|^2 P_{[\psi_i]}$.
(ii) The measuring apparatus should reflect an eigenvalue of A *upon a classical output device.*

The problem of quantum-mechanical measurement theory – as we shall understand it – is to establish that effects (i) and (ii) above can be caused by the interaction between the observed system and the measuring apparatus, where the interaction is described by standard quantum mechanics:[2] unitary evolution of the coupled object–apparatus system under a time-independent interaction Hamiltonian.[3] It should be emphasized that the object–apparatus system is assumed to be totally isolated from external influences.

As we saw in Chapter 8, von Neumann's measurement theory proper[4] did not meet these aims. Analysing the theory, Wigner concluded that '[its] principal conceptual weakness... is... that it merely abstractly postulates interactions which have the effect of [reduction of the wave packet]' (page 158). Von Neumann made up for this weakness by invoking the observer's conscious ego. In 1971–2 Klaus Hepp, using the newly developed algebraic quantum theory for systems with infinitely many degrees of freedom, showed that the extra generality afforded by this framework[5] made it possible to avoid the conscious ego hypothesis (Hepp, 1972).

[1] The standard form of the probability interpretation that was described on pages 117–119 tacitly assumed that measurements were instantaneous. In Sewell's theory, measurements are not instantaneous but require small but nonzero time τ – enough for a cat to be killed by a cyanide capsule – to be completed. As a result, assertion (i) on page 118 cannot be maintained in its naive form but must be modified. The modification required will emerge from the theory itself, and will be described in Subsection 10.5.2.

[2] The *quantum measurement problem* formulated on page 143 is item (i) of the above.

[3] Time-independent, i.e., apart from switching the interaction on and off.

[4] By *proper* we mean the theory excluding the conscious ego hypothesis. We shall use this term again, in Chapter 11.

[5] Namely, the existence of the *observables at infinity* of Lanford and Ruelle (1967); they will be briefly discussed in Section 12.5.

In Hepp's scheme, the measuring device was a system with infinitely many degrees of freedom, and different pointer positions corresponded to inequivalent representations of the canonical commutation and anticommutation relations. A unitary dynamical evolution could not cause the system to jump from one representation to an inequivalent one, and therefore the pointer position could change and the state vector collapse only in the limit $t \to \infty$. This result was controverted (among others) by Bell, who modified an example of Hepp by introducing observables which came into effect only at large values of t, and undid the temporal evolution of earlier times. This led him to 'insist, however, that $t = \infty$ never comes, so that the wave packet reduction never happens' (Bell, 1975). However, the conventional Schrödinger picture – in which all observables are time-independent – did not exist for Bell's model.[6] A little later, Whitten-Wolfe and Emch developed a model, again based on the $C^{\star}$-algebra framework, in which the state vector collapsed instantaneously, but in the infinite-volume limit of the object–apparatus interaction (Whitten-Wolfe and Emch, 1976). While the need to invoke the limit $t \to \infty$ or $V \to \infty$ could be considered a weakness of these schemes, the authors cited did succeed in establishing a fact of paramount importance: that von Neumann's metaphysical hypotheses of psychophysical parallelism and the abstract ego – staunchly defended by Wigner – could be replaced by physical hypotheses which admitted precise mathematical formulation.

Once the problem had been 'downgraded' from metaphysics to physics, it would have been natural for physicists to enquire whether the above results could be reproduced, in their essentials, in ordinary (linear, closed-system) quantum mechanics. But to the best of the present author's knowledge this problem was not addressed in a mathematically rigorous fashion for nearly 30 years until Sewell took it up in 2005.[7] Sewell's theory, in the words of its author, was 'designed to obtain conditions on the [object–apparatus] coupling that lead to [the effects (i) and (ii)]' in a finite time, and in a finite volume (Sewell, 2005). As in von Neumann's theory, the measuring apparatus is an N-particle quantum-mechanical system, N being of the order of magnitude of Avogadro's number. But, whereas von Neumann's theory results in the entanglement of mirrored pairs of vector states of object and apparatus (see (8.30) and the ensuing discussion), Sewell's theory seeks to pair an energy eigenstate of the object with a subspace of the Hilbert space of the apparatus characterized by values of certain macroscopic observables of the latter; the problem is how to bring this physical picture under effective mathematical control.

[6] Bell's critique will be discussed in greater detail in Chapter 11, pages 192–195.

[7] Plausibility arguments, within standard quantum mechanics, that the state vector could collapse without the intervention of the human observer had been advanced earlier in (Peres, 1980, 1986; van Kampen, 1988; Allahverdyan, Balian and Nieuwenhuizen, 2003). The author apologizes to those whose works he may have overlooked.

In this chapter, we shall provide a detailed exposition of Sewell's general theory (called by him the *generic model*), taking care to point out its differences with von Neumann's theory, and to make explicit the hypotheses invoked at each stage. We start with a discussion of the points (i) and (ii) mentioned at the beginning of this chapter.

10.1 Preliminary discussion

The problem of measurement *theory* is not just the determination of conditions on the object–apparatus interaction that lead to the effects (i) and (ii) described on page 173. The moment the conscious observer is removed from the equation, these effects have to be captured by mathematical formulae rather than verbal descriptions. The tool that is needed may be sensed by considering an imaginary variant of the classical Stern–Gerlach experiment.[8]

Figure 10.1 shows the scheme for a Stern–Gerlach experiment on atoms such as lithium or silver. The magnetic moments of these atoms are due entirely to the spin of the outermost electron, and can therefore assume only two values. We shall denote the two spin states by $1, 2$. The atoms in the beam β prepared by the source S are assumed to be in the state $c_1\psi_1 + c_2\psi_2$, with $|c_1|^2 + |c_2|^2 = 1$. The inhomogeneous magnetic field produced by the magnet MM is assumed to split the beam β into two beams β_1 and β_2, the atoms in β_1 being in the state ψ_1, and those in β_2 in the state ψ_2.

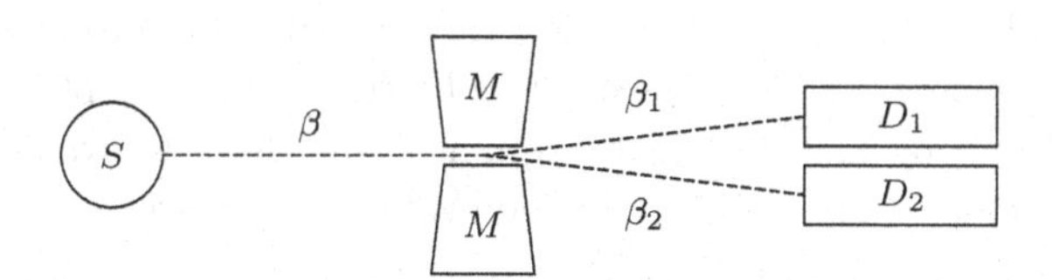

Fig. 10.1. Imaginary Stern–Gerlach experiment

A detector, D_1 or D_2, 'fires' when an atom traverses it. The experiment shows that there are only two possible outcomes: one, which we shall denote by Ω_1, is that D_1 has fired, but D_2 has not; the other, which we shall denote by Ω_2, is that D_2 has fired, but D_1 has not. The source S is assumed to emit atoms singly, and the interval between the emission of two successive atoms is assumed to be greater than the relaxation time of the detectors. Under these

[8] As Wigner said (Wigner, 1970, p. 159), in the Stern–Gerlach experiment 'the "apparatus" is [a] positional coordinate of the particle'; the object is part of the apparatus! Owing to this special circumstance – or so the present author contends – analysis of this experiment cannot yield much insight into the general problem of measurement. Our imaginary variant is an artifice, designed only to motivate a mathematical concept; it plays no role in the deductive development of the theory.

circumstances, the event that the detectors D_1 and D_2 fire simultaneously *never* occurs.

Denote by $p_{rs} = p(\psi_r|\Omega_s)$, $r, s = 1, 2$ the probability that if the outcome is Ω_s, the atom that traversed the apparatus is in the state ψ_r. In the experiment described above,

$$p(\psi_r|\Omega_s) = \delta_{rs}, \tag{10.1}$$

where δ_{rs} is the Kronecker delta. Equation (10.1) expresses the correlation – rather simple, in the present case – between the state of the apparatus and the state of the object. The quantities p_{rs} are *conditional probabilities*, but not in the sense of elementary probability theory, because one cannot assign physically meaningful a priori probabilities to the events Ω_1 and Ω_2. The mathematical object that we shall use to obtain the required information will be a certain conditional expectation. A brief but self-contained account of the topic, tailored to our needs, is given in Appendix A7. For the special case of measurement theory, the relevant results were established by Sewell by elementary means, and will be recounted in Section 10.4.

10.2 The object–apparatus interaction

We begin by defining our terms and setting up the basic notations. The system on which the measurement is carried out will be called the *object*. The measuring device will be called the *apparatus*. The object will be denoted by O and the apparatus by A. They will be considered to be two distinct quantum-mechanical systems, defined on the Hilbert spaces $\mathfrak{H}_\mathsf{o}$ and $\mathfrak{H}_\mathsf{a}$, with Hamiltonians H_o and H_a respectively. H_o and H_a will be assumed to be time-independent.

We shall denote by $\mathsf{S} = \mathsf{O} + \mathsf{A}$ the quantum-mechanical system defined on the Hilbert space $\mathfrak{H} = \mathfrak{H}_\mathsf{o} \otimes \mathfrak{H}_\mathsf{a}$. The time evolution of S will be assumed to be determined by a Hamiltonian

$$H = H_\mathsf{o} \otimes I_\mathsf{a} + I_\mathsf{o} \otimes H_\mathsf{a} + V, \tag{10.2}$$

where I_o and I_a are the identity operators on $\mathfrak{H}_\mathsf{o}$ and $\mathfrak{H}_\mathsf{a}$ respectively, and the operator V is the interaction between the object and the apparatus, which is switched on adiabatically at $t = 0$. The aim of Sewell's theory – which, at this point, begins to depart essentially from von Neumann's – is to determine conditions on V that lead to the effects (i) and (ii) described on page 173.

It follows from (10.2) that the dynamics of the composite system S under the Hamiltonian H will be given by the one-parameter group of unitary transformations $U(t)$ of $\mathfrak{H}_\mathsf{o} \otimes \mathfrak{H}_\mathsf{a}$ which is generated by H:

$$U(t) = \exp(\mathrm{i}Ht) \ \text{ for all } \ t \in \mathbb{R}. \tag{10.3}$$

We now make the following assumptions.

Assumption 10.1 (Initial states of object and apparatus) The object O and apparatus A are prepared independently of each other in the initial states represented by the density matrices $P_{[\psi]}$ and Ω on $\mathfrak{H}_o$ and $\mathfrak{H}_a$ respectively, where $P_{[\psi]}$ is the projection operator (7.18) onto a pure state ψ of O, and the interaction V is switched on adiabatically at $t = 0$.

Then the initial state (at $t = 0$) of S is

$$\Phi = \Phi(0) = P_{[\psi]} \otimes \Omega. \tag{10.4}$$

Assumption 10.2 (Dimensionalities of the Hilbert spaces) The Hilbert spaces $\mathfrak{H}_o$ and $\mathfrak{H}_a$ are finite-dimensional,[9] with

$$n = \dim \mathfrak{H}_o \ll \dim \mathfrak{H}_a.$$

The dimension of $\mathfrak{H}_a$ is not specified. The justification for the assumption $\dim \mathfrak{H}_a < \infty$ is provided by the physics of the situation. The microscopic states of the coupled object–apparatus system lie in an energy shell of finite thickness – before, during and after the measurement interaction.

The assumption $\dim \mathfrak{H}_o = n$ is far from trivial, and will be discussed in detail at later stages. Observe that Assumptions 10.1 and 10.2 are radical departures from von Neumann's theory.

Since $\mathfrak{H}_o$ is n-dimensional, it has an orthonormal basis consisting of eigenvectors of H_o; we denote this basis by $\{u_1, \ldots, u_n\}$, and the corresponding eigenvalues of H_o by ε_r:

$$H_o u_r = \varepsilon_r u_r. \tag{10.5}$$

The initial state ψ of O may be expressed as a linear combination of the vectors u_r,

$$\psi = \sum_{r=1}^{n} c_r u_r, \tag{10.6}$$

with

$$\sum_{r=1}^{n} |c_r|^2 = 1. \tag{10.7}$$

We now impose the following essential condition on the object–apparatus interaction V:

[9] The assumption $\mathfrak{H}_a < \infty$ is implicit in (Sewell, 2005). I would like to thank Professor G L Sewell for this clarification, and for the physical reasoning behind it.

Assumption 10.3 (Object–apparatus interaction) The object–apparatus interaction V is of the form

$$V = \sum_{r=1}^{n} P_{[u_r]} \otimes V_r, \tag{10.8}$$

where $P_{[u_r]}$ is the projection operator onto the state u_r and the V_r are observables of the apparatus, which are self-adjoint operators on $\mathfrak{H}_{\mathsf{a}}$.

This assumption (which has no analogue in von Neumann's theory) has the following consequence, which in fact is the reason for making it. Let $\xi, \zeta \in \mathfrak{H}_{\mathsf{a}}$. An easy computation shows that, with V given by (10.8),

$$(u_i \otimes \xi,\, V(u_j \otimes \zeta)) = \delta_{ij}(\xi,\, V_j\zeta) \quad \text{(no summation)},$$

i.e., V does not induce transitions between different eigenstates of H_{o}.

Since $H_{\mathsf{o}} = \sum_r \varepsilon_r P_{[u_r]}$ and $I_{\mathsf{o}} = \sum_r P_{[u_r]}$, (10.8) allows us to recast the Hamiltonian (10.2) of the composite system S as

$$H = \sum_{r=1}^{n} P_{[u_r]} \otimes K_r, \tag{10.9}$$

where K_r is defined by

$$K_r = H_{\mathsf{a}} + V_r + \varepsilon_r I_{\mathsf{a}}. \tag{10.10}$$

In order to proceed, we need a computational identity. Let $\mathfrak{H}_1$ and $\mathfrak{H}_2$ be any two Hilbert spaces, $E_1, \ldots, E_k$ a set of pairwise-orthogonal projection operators on $\mathfrak{H}_1$, and $B_1, \ldots, B_k$ a set of operators on $\mathfrak{H}_2$ that are defined on a common dense domain (they need not commute with each other). It is easily seen that

$$\left(\sum_{i=1}^{k} E_i \otimes B_i \right)^n = \sum_{i=1}^{k} E_i \otimes B_i^n.$$

We use this identity to exponentiate the right-hand side of (10.9), and obtain

$$U(t) = \exp(\mathrm{i}Ht) = \sum_{r=1}^{n} P_{[u_r]} \otimes U_r(t), \tag{10.11}$$

where

$$U_r(t) = \exp(\mathrm{i}K_r t). \tag{10.12}$$

Introducing (10.11) into the right-hand side of

$$\Phi(t) = U^\star(t)\Phi(0)U(t),$$

where $\Phi(0) = \Phi$ is the initial state given by (10.4), we find that

$$\Phi(t) = \sum_{r,s} \left[P_{[u_r]} P_{[\psi]} P_{[u_s]}\right] \otimes \left[U_r^\star(t) \Omega U_s(t)\right]. \tag{10.13}$$

Let $f \in \mathfrak{H}_\mathrm{o}$. Then, using (10.6) and (10.7), we find that

$$\begin{aligned} P_{[u_r]} P_{[\psi]} P_{[u_s]} f &= P_{[u_r]} P_{[\psi]} (u_s, f) u_s \\ &= (u_s, f)(\psi, u_s)(u_r, \psi) u_r \\ &= \bar{c}_s c_r (u_s, f) u_r. \end{aligned} \tag{10.14}$$

Define now an operator $R_{r,s}$ on $\mathfrak{H}_\mathrm{o}$ by

$$R_{r,s} f = (u_s, f) u_r \tag{10.15}$$

and an operator $\Omega_{r,s}(t)$ on $\mathfrak{H}_\mathrm{a}$ by

$$\Omega_{r,s}(t) = U_r^\star(t) \Omega U_s(t). \tag{10.16}$$

Since Ω has unit trace, so do the $\Omega_{r,r}(t)$:

$$\mathrm{Tr}\, \Omega_{r,r}(t) = \mathrm{Tr}\, \Omega = 1 \quad \text{for all} \quad r. \tag{10.17}$$

Furthermore,

$$(\Omega_{r,s}(t))^\star = \Omega_{s,r}(t), \tag{10.18}$$

as is evident from (10.16). These elementary facts will be called upon later. Next, using (10.15) and (10.16), we may write (10.13) as

$$\Phi(t) = \sum_{r,s=1}^{n} c_r \bar{c}_s R_{r,s} \otimes \Omega_{r,s}(t). \tag{10.19}$$

Note carefully that (10.19) describes the unitary temporal evolution of a coupled conservative system that is standard in quantum mechanics, subject to Assumption 10.2, namely that $\mathfrak{H}_\mathrm{o}$ and $\mathfrak{H}_\mathrm{a}$ are finite-dimensional. We remark in passing that this assumption is standard in the theoretical analysis of the Stern–Gerlach effect; see, e.g., (Wigner, 1983). The presence of the object–apparatus interaction V manifests itself in the subscripts r, s on $\Omega_{r,s}(t)$ in the right-hand side of (10.19). It is easy to verify that if $V = 0$, i.e., if object and apparatus do not interact, then (10.16) reduces to $\Omega_{r,s}(t) = U^\star(t) \Omega U(t)$, the right-hand side being independent of r, s. Then (10.19) becomes

$$\Phi(t) = \sum_{r,s=1}^{n} c_r \bar{c}_s R_{r,s} \otimes \Omega(t);$$

as expected, object and apparatus evolve independently of each other. However, we should also pay attention to the fact that the indices r, s that appear in (10.19) pertain *exclusively* to the states of the object; we shall therefore call these indices *object indices.*

We now make the assumption that the measurement is completed within a (short) time τ:

Assumption 10.4 (Duration of the measurement interaction) The measurement is completed within an interval τ, i.e., for $t > \tau$, object and apparatus no longer interact. The time τ is smaller, by orders of magnitude, than the Poincaré recurrence time for the object–apparatus system.

This assumption means that we may set $V = 0$ for $t > \tau$, so that, for $t > \tau$, the Hamiltonian (10.2) may be written as

$$H = H_{\mathsf{o}} \otimes I_{\mathsf{a}} + I_{\mathsf{o}} \otimes H_{\mathsf{a}}. \tag{10.20}$$

Then for $t > \tau$ the quantity K_r defined by (10.10) may be written as

$$K'_r = H_{\mathsf{a}} + \varepsilon_r I_{\mathsf{a}}. \tag{10.21}$$

Setting $t = t' + \tau$ for $t > \tau$, we may rewrite (10.16) as

$$\begin{aligned} \Omega_{r,s}(t' + \tau) &= U_r^\star(t')U_r^\star(\tau)\Omega U_s(\tau)U_s(t') \\ &= U_r^\star(t')\Omega_{r,s}(\tau)U_s(t'), \end{aligned} \tag{10.22}$$

where $U_r(t')$ and $U_s(t')$ are calculated with K'_r and K'_s, given by (10.21). Therefore, for $t = t' + \tau$, $t' > 0$, we have

$$\Omega_{r,s}(t' + \tau) = \exp(-\mathrm{i}H_{\mathsf{a}}t')\Omega_{r,s}(\tau)\exp(\mathrm{i}H_{\mathsf{a}}t'). \tag{10.23}$$

We set this equation aside for later use.

10.3 The macroscopic observables of A

We shall now make explicit use of an assumption that has only been implicit so far: that the apparatus A is a macroscopic N-particle system, where N is of the order of Avogadro's number or larger. That is, it is large enough to have macroscopic observables which commute with each other, as discussed in detail in Chapter 9. We now make the following assumption.

Assumption 10.5 (Characterization of the apparatus) The measuring apparatus is characterized by a *finite set* of macroscopic observables $\mathcal{M} = \{M_1, M_2, \ldots, M_K\}$, which is a subset of the set $\mathcal{B}_{\mathsf{a}}$ of all observables of A. Every observable in $\mathcal{M}$ commutes with H_{a}.

Since $\mathfrak{H}_{\mathsf{a}}$ is finite-dimensional, every observable $M_k \in \mathcal{M}$ has a purely discrete spectrum, consisting of a finite number of distinct eigenvalues. Note that Assumption 10.5 represents yet another radical departure from von Neumann's theory.

As discussed in Subsection 9.1.1, there exists a cellular operator T,

$$T = \sum_{\alpha=1}^{\nu} \lambda_\alpha \Pi_\alpha, \quad \lambda_\alpha \in \mathbb{R}, \ \lambda_\alpha \neq \lambda_\beta \text{ for } \alpha \neq \beta, \tag{9.9}$$

such that each operator $M_k \in \mathcal{M}$ can be expressed in terms of the spectral projections Π_α of T, and T is the coarsest operator with this property.[10] It follows easily from Assumption 10.5 that the Π_α, and T, commute with H_{a}. The *cellular decomposition* of $M_k \in \mathcal{M}$ is

$$M_k = \sum_{\alpha=0}^{\nu} m_{k;\alpha} \Pi_\alpha, \tag{10.24}$$

where the $m_{k;\alpha}$ are eigenvalues of M_k; the difference between the cellular decomposition of M_k and its spectral decomposition is that in the cellular decomposition two or more eigenvalues $m_{k;\alpha}$ may be equal, whereas in the spectral decomposition they are, by definition, distinct, which requires the projection operators to be modified accordingly. The projection operators Π_α satisfy

$$\Pi_\alpha \Pi_\beta = \Pi_\alpha \delta_{\alpha\beta} \tag{10.25}$$

and

$$\sum_{\alpha=0}^{\nu} \Pi_\alpha = I_{\mathsf{a}}. \tag{10.26}$$

Now define

$$\mathfrak{K}_\alpha = \Pi_\alpha \mathfrak{H}_{\mathsf{a}}. \tag{10.27}$$

The subspaces $\mathfrak{K}_\alpha$ are quantum analogues of classical phase cells, and represent a macroscopic state of the apparatus. The dimensions of these subspaces are very large; as remarked in Definition 9.3(iv), page 169, $\dim \mathfrak{K}_\alpha \sim \exp(cN)$, where $c \sim 1$ and $N \sim 10^{24}$. In the following, we shall refer to the subspace $\mathfrak{K}_\alpha$ of $\mathfrak{H}_{\mathsf{a}}$ as a *cell*, or *a state of the apparatus* A. The index α itself will be called the *apparatus index*.

10.4 Expectations and conditional expectations of observables

We shall denote the set of observables on $\mathfrak{H}_{\mathsf{o}}$ by $\mathcal{A}$. Note that $\mathcal{A}$ is not an algebra – the product of two self-adjoint operators is not self-adjoint unless the

[10] The term *coarsest* was defined, in the present context, on page 164.

two commute – but it is a linear space over $\mathbb{R}$. The observables of S of interest to us will be the tensor products $A \otimes M$, where $A \in \mathcal{A}$ and $M \in \mathcal{M}$. By contrast, the von Neumann theory appears to deal with all observables of the form $A \otimes B$, where A is as before but B is any bounded self-adjoint operator on $\mathfrak{H}_\mathsf{a}$.

The expectation value $E(A \otimes M)$ of the observable $A \otimes M \in \mathcal{A} \otimes \mathcal{M}$ in the time-dependent state $\Phi(t)$ is defined by the standard formula

$$E(A \otimes M) = \mathrm{Tr}\,(\Phi(t)[A \otimes M]) \text{ for all } A \in \mathcal{A},\ M \in \mathcal{M}. \tag{10.28}$$

In the following, we shall use the shorthand notations

$$E(A) = E(A \otimes I_\mathsf{a}) \tag{10.29}$$

and

$$E(M) = E(I_\mathsf{o} \otimes M). \tag{10.30}$$

$E(A)$ is the *unconditional* time-dependent expectation value of the observables A of the object. Since Π_α is the projection operator onto the cell $\mathfrak{K}_\alpha$, the time-dependent probability that the macroscopic state of the apparatus A is defined by the cell $\mathfrak{K}_\alpha$ is given by

$$w_\alpha(t) = E(I_\mathsf{o} \otimes \Pi_\alpha). \tag{10.31}$$

We single out the index 0 to denote the *rest-state*, or the state at $t = 0$, of the apparatus. Then

$$w_\alpha(0) = \delta_{\alpha,0}. \tag{10.32}$$

Since the apparatus was prepared in the initial state Ω, (10.31) and (10.32) mean that

$$\alpha \neq 0 \Longrightarrow \Pi_\alpha(\Omega\mathfrak{H}_\mathsf{a}) = 0, \tag{10.33}$$

which we shall state in words as follows: the mixed state Ω lies entirely in the cell $\mathfrak{K}_0 \subset \mathfrak{H}_\mathsf{a}$.

The time evolution of the state Ω can be pictured as follows. Initially, under the action of the Hamiltonian (10.9), the state Ω may 'spread' to cells other than $\mathfrak{K}_0$; however, the spreading process stops by $t = \tau$. Since the cellular projections Π_α commute with H_a, the motions induced by H_a after the interval τ remain strictly within the cells for all $\phi \in \mathfrak{H}_\mathsf{a}$. We express this conclusion in words as follows:

Conclusion 10.6 After a time τ, temporal evolution no longer transports any vector of $\mathfrak{H}_\mathsf{a}$ from one cell $\mathfrak{K}_\alpha$ to another; the cellular decomposition of $\mathfrak{H}_\mathsf{a}$ is stable under forward time translations for $t > \tau$.

To set up a measurement theory without von Neumann's conscious ego hypothesis, we have to be able to calculate the following quantities for times $t > \tau$, i.e., after the measurement interaction has run its course:

(i) The unconditional expectation value $E(A)$ of an observable A. This would give us the statistical distribution of the measured values of A for a large number of single measurements. We may also describe this quantity as the state of the object in the absence of a pointer reading.
(ii) The *conditional* expectation value $E(A|\mathfrak{K}_\alpha)$ of A, given that the macroscopic state of the apparatus is defined by the cell $\mathfrak{K}_\alpha$. This would give us the result of a single measurement.

As the preliminary discussion of Section 10.1 made clear, one cannot assign physically meaningful a priori probabilities on the set of cells $\{\mathfrak{K}_\alpha\}$. The existence of the conditional expectation value $E(A|\mathfrak{K}_\alpha)$ is therefore contingent upon the existence of a certain conditional expectation functional upon the set of observables $\mathcal{A}$. That this functional exists, and is unique, is the content of the following crucial lemma, due to Sewell:[11]

Lemma 10.7 (Sewell) *There exists a unique linear functional* $\boldsymbol{E}(\cdot|\mathcal{M}) : \mathcal{A} \to \mathcal{M}$ *that preserves positivity and normalization and satisfies the condition*

$$E(\boldsymbol{E}(A|\mathcal{M})M)) = E(A \otimes M) \text{ for all } A \in \mathcal{A},\ M \in \mathcal{M}, \tag{10.34}$$

where the left-hand side is understood in the sense of (10.30).[12]

Proof The proof consists of showing that (10.34), regarded as an equation for the unknown $\boldsymbol{E}(A|\mathcal{M})$, has a unique solution with the required properties.

Since $\boldsymbol{E}(A|\mathcal{M})$ is an element of $\mathcal{M}$, it can be written uniquely as

$$\boldsymbol{E}(A|\mathcal{M}) = \sum_{\alpha=0}^{\nu} f_\alpha(A)\Pi_\alpha.$$

Therefore

$$\begin{aligned} I_o \otimes (\boldsymbol{E}(A|\mathcal{M})M) &= I_o \otimes \left(\sum_{\alpha=0}^{\nu} f_\alpha \Pi_\alpha M\right) \\ &= \sum_{\alpha=0}^{\nu} m_\alpha f_\alpha(A)(I_o \otimes \Pi_\alpha), \end{aligned}$$

[11] The proof shows that the lemma is also valid if $\dim \mathfrak{H}_a = \infty$.

[12] This result is a noncommutative analogue of (A7.23), namely $E(\boldsymbol{E}(\boldsymbol{X}|\mathcal{G})\boldsymbol{Z}) = E(\boldsymbol{X}\boldsymbol{Z})$ of classical probability theory derived in Appendix A7. Let $\mathcal{F} = \mathcal{A} \otimes \mathcal{M}$ and $\mathcal{G} = I_o \otimes \mathcal{M}$. The quantity $\boldsymbol{E}(\cdot|\mathcal{M})$ is a functional $\mathcal{A} \otimes \mathcal{M} \to I_o \otimes \mathcal{M}$ restricted to $\mathcal{A} \otimes I_a$. Take $\boldsymbol{X} = A \otimes I_a$, $\boldsymbol{Z} = I_o \otimes M$ and $\boldsymbol{E}(\boldsymbol{X}|\mathcal{G}) = \boldsymbol{E}(A|\mathcal{M})$ in (A7.23). Equation (10.34) follows immediately.

so that, using (10.30) and (10.31), we obtain

$$E(\boldsymbol{E}(A|\mathcal{M})M) = \sum_{\alpha=0}^{\nu} m_\alpha f_\alpha(A) E(I_\circ \otimes \Pi_\alpha)$$
$$= \sum_{\alpha=0}^{\nu} f_\alpha(A) m_\alpha w_\alpha. \tag{10.35}$$

On the other hand,

$$E(A \otimes M) = \sum_{\alpha=0}^{\nu} m_\alpha E(A \otimes \Pi_\alpha), \tag{10.36}$$

and therefore (10.34) can hold if and only if

$$\sum_{\alpha=0}^{\nu} m_\alpha \left[f_\alpha(A) w_\alpha - E(A \otimes \Pi_\alpha)\right] = 0.$$

Since this has to hold for every $M \in \mathcal{M}$, it follows that

$$f_\alpha(A) w_\alpha = E(A \otimes \Pi_\alpha), \tag{10.37}$$

so that, for $w_\alpha \neq 0$,

$$f_\alpha(A) = E(A \otimes \Pi_\alpha)/w_\alpha \tag{10.38}$$

and

$$\boldsymbol{E}(A|\mathcal{M}) = \sum_{\alpha}{}' \frac{E(A \otimes \Pi_\alpha)}{w_\alpha} \Pi_\alpha, \tag{10.39}$$

where the prime on the summation sign indicates that the sum extends only over those values of α for which w_α is nonzero. This solution of (10.34) is unique; it clearly preserves positivity and normalization. □

Remark 10.8 In the following, we shall confine ourselves to the case in which the state of the object is described by a *single* macroscopic observable. It will then suffice to consider subalgebras $\mathcal{M}$ that consist of multiples of a single observable M. We shall then have $M = T$, and the cellular decomposition of M will be the same as its spectral decomposition.

Since the set of projectors $\{\Pi_\alpha\}$ define the cellular decomposition of $\mathcal{M}$, the conditional expectation values $E(A|\mathfrak{K}_\alpha)$ are determined by the operator $\boldsymbol{E}(A|\mathcal{M})$ as follows:

$$\boldsymbol{E}(A|\mathcal{M}) = \sum_{\alpha}{}' E(A|\mathfrak{K}_\alpha) \cdot \Pi_\alpha. \tag{10.40}$$

Comparing (10.40) with (10.39), we obtain

$$E(A|\mathfrak{K}_\alpha) = \frac{E(A \otimes \Pi_\alpha)}{w_\alpha} \tag{10.41}$$

for $w_\alpha \neq 0$.

10.4.1 Explicit expressions

Lemma 10.7 shows that all the information we require is contained in $E(A\otimes M)$. We therefore begin with an explicit calculation of this quantity.

By definition,

$$E(A \otimes M) = \mathrm{Tr}\,(\Phi(t)[A \otimes M]). \tag{10.42}$$

We introduce the expression (10.19) for $\Phi(t)$ and (10.24) for M into the above. Since the trace is a linear map and the sums are finite, we may take the constants and the summations outside the Tr sign. We then obtain

$$\begin{aligned} E(A \otimes M) &= \sum_{r,s=1}^{n} \sum_{\alpha=0}^{\nu} c_r \bar{c}_s m_\alpha \mathrm{Tr}\left([R_{r,s} \otimes \Omega_{r,s}(t)][A \otimes \Pi_\alpha)]\right) \\ &= \sum_{r,s=1}^{n} \sum_{\alpha=0}^{\nu} c_r \bar{c}_s m_\alpha \mathrm{Tr}\left([R_{r,s} A] \otimes [\Omega_{r,s}(t)\Pi_\alpha]\right). \end{aligned} \tag{10.43}$$

Let now $\{\xi_k\}$ be any orthonormal basis for $\mathfrak{H}_\mathsf{a}$. Using the orthonormal basis $\{u_i \otimes \xi_k\}$ for $\mathfrak{H}$, we may write the trace in (10.43) explicitly as

$$\begin{aligned} &\mathrm{Tr}\left([R_{r,s} \otimes \Omega_{r,s}(t)][A \otimes \Pi_\alpha)]\right) \\ &\qquad = \sum_{i=1}^{n} \sum_{k=1}^{\infty} \left(u_i \otimes \xi_k,\, [R_{r,s} A \otimes \Omega_{r,s}(t)\Pi_\alpha] \cdot [u_i \otimes \xi_k]\right) \\ &\qquad = \sum_{i=1}^{n} \left(u_i,\, R_{r,s} A\, u_i\right) \sum_{k=1}^{\infty} \left(\xi_k,\, \Omega_{r,s}(t)\Pi_\alpha\, \xi_k\right), \end{aligned} \tag{10.44}$$

where we have used formula (A6.5) for the inner product in tensor product spaces to reduce the first equation to the second. Using the definition (10.15) of $R_{r,s}$, we easily find that the first sum on the last line is

$$\sum_{i=1}^{n} \left(u_i,\, R_{r,s} A\, u_i\right) = (u_s,\, A u_r), \tag{10.45}$$

while the second sum is clearly just the trace of $\Omega_{r,s}\Pi_\alpha$. Lastly, we define

$$F_{r,s;\alpha}(t) = \mathrm{Tr}\,(\Omega_{r,s}\Pi_\alpha). \tag{10.46}$$

Introducing (10.45) and (10.46) into (10.44) and then the resulting expression into (10.43), we arrive at the formula

$$E(A \otimes M) = \sum_{r,s=1}^{n} \sum_{\alpha=0}^{\nu} c_r \bar{c}_s (u_s, Au_r) m_\alpha F_{r,s;\alpha}. \tag{10.47}$$

Next, comparing (10.47) with (10.36), we find that

$$E(A \otimes \Pi_\alpha) = \sum_{r,s=1}^{n} c_r \bar{c}_s (u_s, Au_r) F_{r,s;\alpha}. \tag{10.48}$$

Lastly, setting $A = I_o$ in (10.48) and using the definition (10.31) of w_α, we obtain

$$w_\alpha = \sum_{r=1}^{n} |c_r|^2 F_{r,r;\alpha}. \tag{10.49}$$

The state of the coupled system for $t > 0$ is described *completely* by the $F_{r,s;\alpha}$.

10.4.2 Properties of $F_{r,s;\alpha}$

The following properties of $F_{r,s;\alpha}(t)$ follow almost immediately from the definition (10.46) and the properties (10.17) and (10.18) of $\Omega_{r,s}(t)$:

$$0 \leq F_{r,r;\alpha} \leq 1, \tag{10.50}$$

$$\sum_{\alpha=0}^{\nu} F_{r,r;\alpha} = 1 \tag{10.51}$$

and

$$F_{r,s;\alpha} = \bar{F}_{s,r;\alpha}. \tag{10.52}$$

It also follows from (10.50) and (10.52) that, for $z_1, \ldots, z_n \in \mathbb{C}$, the sesquilinear form

$$\sum_{r,s=1}^{n} z_r \bar{z}_s F_{r,s;\alpha}$$

is positive for each α, from which follows the inequality

$$F_{r,r;\alpha} F_{s,s;\alpha} \geq |F_{r,s;\alpha}|^2, \tag{10.53}$$

which holds for each α.

10.5 Ideal measurements

In an ideal measurement, the final state of the apparatus should be readable unambiguously, and should point to a unique state of the object. These aims will be achieved if, for $t > \tau$, the object–apparatus dynamics meets the following conditions:

(i) The density matrix $\Omega_{r,r}(t)$ lies entirely in a single cell $\mathfrak{K}_\alpha$; i.e., vectors in $\mathfrak{H}_\mathrm{a}$ that lie outside $\mathfrak{K}_\alpha$ do not contribute to $\Omega_{r,r}(t)$.
(ii) The observed state of the apparatus (for $\alpha \neq 0$) indicates a unique state of the object.

It is not necessary for every nonnull state of the apparatus to correspond to a state of the object, but those that do not will never result from an ideal measurement. Therefore (in view of Remark 10.8) there is no loss of generality in assuming that every nonnull value of the apparatus index corresponds to a unique value of the object index, and vice versa. That is, there exists an invertible map γ

$$r = \gamma(\alpha) \quad \text{for} \quad \alpha \neq 0 \tag{10.54}$$

from the set of nonnull apparatus indices to the set of object indices. Then, if condition (i) is satisfied, we shall have

$$\mathrm{Tr}\,(\Omega_{r,r}\Pi_\alpha) = \delta_{\gamma^{-1}(r),\alpha}, \tag{10.55}$$

i.e.,

$$F_{r,r;\alpha} = \delta_{\gamma^{-1}(r),\alpha}. \tag{10.56}$$

That means that for $r \neq s$ and fixed α, at least one of $F_{r,r;\alpha}$ and $F_{s,s;\alpha}$ must vanish. Then, from (10.53), it follows that

$$F_{r,s;\alpha} = 0 \quad \text{for} \quad r \neq s, \quad \alpha \neq 0, \tag{10.57}$$

which shows that, for times $t > \tau$, the density matrix for the state of the apparatus *lies in a single cell.* This evolution of the object–apparatus system has taken place *under the ordinary Schrödinger–von Neumann equations.* What is observed in a single experiment is a macroscopic 'pointer position' that describes the cell. What quantum mechanics cannot tell us is in *which* cell will the final state of the apparatus be in a particular experiment.

10.5.1 Interpretation

Setting $M = I_\mathrm{a}$ and using the shorthand notation (10.29), we may rewrite (10.47) as

$$E(A) = \sum_{r=1}^{n} |c_r|^2 (u_r, Au_r) + \sum_{\substack{r,s=1 \\ r\neq s}}^{n} \sum_{\alpha=0}^{\nu} F_{r,s;\alpha} c_r \bar{c}_s (u_r, Au_s). \tag{10.58}$$

Then, for $\tau > 0$ and $\alpha \neq 0$ we have, using (10.57),

$$E(A) = \sum_{r=1}^{n} |c_r|^2 (u_r, Au_r). \tag{10.59}$$

This is the classical expectation value of a random variable which assumes the values (u_r, Au_r) with probabilities $p_r = |c_r|^2$. *The quantity being measured will therefore be the diagonal matrix element of A in the basis $\{u_r\}$.*

Let us now inject the information that the pointer has been read at some time $t > \tau$ and the apparatus found in the macroscopic state $\alpha \neq 0$. Then the relevant expectation value of the observable A is the conditional expectation value $E(A|\mathfrak{K}_\alpha)$. Using (10.48) and (10.49), the expression (10.41) for $E(A|\mathfrak{K}_\alpha)$ becomes

$$E(A|\mathfrak{K}_\alpha) = (u_{\gamma(\alpha)}, Au_{\gamma(\alpha)}). \tag{10.60}$$

This shows that the object–apparatus interaction has sent the initial state ψ of the object to the eigenstate $u_{\gamma(\alpha)}$ of H_o; the wave packet has collapsed to one of the eigenstates of H_o. Combining (10.59) and (10.60), we conclude that the final state of the object is

$$\rho_o(t' + \tau) = \sum_{r=1}^{n} |c_r|^2 P_{[u_r]}. \tag{10.61}$$

10.5.2 The probability interpretation in Sewell's theory

Equations (10.59) and (10.60) show that the expectation and conditional expectation values are defined in terms of the diagonal matrix elements of the observable A in a basis consisting of eigenstates of H_o. These diagonal matrix elements reduce to eigenvalues of A if and only if A commutes with H_o. Sewell's theory is therefore consistent with an extension of the naive probability interpretation in which observables that do not commute with the Hamiltonian may also be measured. In this case the measured value will not be an eigenvalue, but rather *a diagonal matrix element of the measured observable.*

10.6 Consistency and robustness of Sewell's generic model

The reader who is familiar with Wigner's most serious reservation on von Neumann's measurement theory may be puzzled by the assertion that the observable A is not required to commute with H_o. We shall set aside this question for the next chapter, and begin by addressing two questions that, in our opinion, have priority: the questions of consistency and of robustness, or stability, of Sewell's generic model.

10.6.1 The question of consistency

The theory developed so far was based on a number of assumptions: Assumptions 10.1–10.5. No matter how reasonable physically or mathematically, they remain just that: assumptions, and it is pertinent to ask whether they are *mathematically consistent with each other.*

Consistency problems in mathematics are at best tricky, at worst undecidable. Since this situation is not really tolerable in day-to-day life, logicians have developed a strategy to deal with it. This strategy consists of eschewing maximum generality and accepting a specific theory as consistent if there is a model which satisfies the axioms of the theory in a nontrivial manner. We shall not define the mathematical meaning of the term *model,* but shall accept the thinking behind the strategy. That is, we shall accept the assumptions of Sewell's theory as being consistent with each other if there is an exactly soluble model – as physicists understand this term – in which the assumptions of the theory are fulfilled nontrivially.

Such a model has been developed by Sewell, who named it the *finite Coleman–Hepp model.* The existence of this model provides an affirmative answer to the nontriviality problem. We shall omit details of this model, referring the reader to the original article (Sewell, 2005). Sewell employs '...the phase cell representation of van Kampen and Emch for the description of macroscopic observables...',[13] which leads to the same final results as our treatment of macroscopic observables of large quantum systems in Chapter 9, based on the work of von Neumann: the full algebra of observables $\mathcal{B}_{\mathsf{a}}$ of the apparatus A has a subalgebra $\mathcal{M}$ of macroscopic observables that commute with each other, and this is the formal structure that is employed by Sewell.[14]

10.6.2 The question of robustness

As we have seen, (10.56) implies that the measurement is perfect; there is no possibility for the pointer reading to give a result other than the correct state of the object. In a finite system – no matter how large N is – thermodynamic fluctuations are not smoothed out completely, and a reasonable theory of measurement ought to be robust enough to cope with the resulting errors of measurement. This robustness is generic to Sewell's scheme, and has been sketched in (Sewell, 2007).

According to its definition (10.46), the quantity $F_{r,r;\alpha}$ is the probability that,[15] for $t > \tau$, the object is in the state r when the apparatus is in the state α. The

[13] The references cited by Sewell are (van Kampen, 1954; Emch, 1964).

[14] It is not necessary to be conversant with the mathematical definition of algebras of observables in order to appreciate the physical content of the above statement. See footnote 8 on page 117.

[15] The complete statement is: the object is in the state u_r when the apparatus is in the macroscopic state determined by the cell $\mathfrak{K}_\alpha$.

requirement (10.56) of a bijective correspondence between object and apparatus states forces these probabilities to be either 1 or 0. There exists, therefore, a natural way to weaken this requirement, and that is by replacing it by

$$F_{r,r;\alpha} = p_r(\alpha; N), \tag{10.62}$$

where p_r is a probability distribution on r for given α, depending on the parameter N, which has a high peak at $r = \gamma(\alpha)$ but also a small but nonzero variance. Then a correspondence (r', α), where $r' \neq r(\alpha)$, will be a rare event which can be analysed by *the theory of large deviations.*[16] Such an analysis has been carried out by Sewell,[17] who obtained the following estimates:

$$|1 - F_{r,r;\alpha}| \sim \exp(-cN/n), \tag{10.63}$$

and, for $r \neq s$,

$$|F_{r,s;\alpha}| \sim \exp(-cN/2n), \tag{10.64}$$

where c is a positive constant of order unity. The factor n in the denominators of the exponents is dim $\mathfrak{H}_o$, and equals 2 for the finite Coleman–Hepp model solved by Sewell (Sewell, 2005). If, as we have assumed, N is of the order of Avogadro's number and n of order unity, then the right-hand sides of (10.63) and (10.64) are indeed small; one may consider the measuring apparatus A to be a reliable instrument. If, on the other hand, n is within a few orders of magnitude of N, the measuring apparatus will no longer be a reliable instrument.

It may, however, be argued that the experimentalist has some control over n; otherwise even the assumption that n is finite becomes untenable. The problem of the size of n belongs properly to the domain of design of the experiment; as we have seen in Section 6.4, one may expect the experimentalist to make n as small as possible, to increase the accuracy of measurement. Unexpectedly, in quantum-mechanical measurement theory more accurate instruments also appear to be the more reliable ones.

10.7 Answers to Wigner, II

The first of Wigner's objections to von Neumann's measurement theory, listed on page 158, was:

> The principal conceptual weakness of the orthodox view is, in my opinion, that it merely abstractly postulates interactions which have the effect of [reduction of the wave packet].

[16] The theory of large deviations, as it is understood today, was initiated by Varadhan (Varadhan, 1966). A standard reference for physicists is (Ellis, 1985).

[17] Private communication.

In the opinion of the present author, this criticism goes to the heart of the matter. The root of the problem is Wigner's own insistence that the state of the apparatus have no classical description. It should be recalled that Hepp, Whitten-Wolfe and Emch developed explicitly soluble models in which the object–apparatus interaction was specified precisely.[18] Sewell, in his generic model, attempted – successfully – to determine the key qualitative features of object–apparatus interactions that lead to the effects (i) and (ii) described on page 173 (Sewell, 2005). It may thus be claimed that Sewell's theory is not subject to Wigner's principal objection. As we shall see in Section 11.8, Wigner's remaining objections (page 158) do not apply to Sewell's theory, precisely because it takes cognizance of his principal objection.

[18] The references are (Hepp, 1972; Whitten-Wolfe and Emch, 1976).

11
Summing-up

In this chapter we shall discuss a variety of topics that are not covered by Sewell's scheme. Our main conclusion will be that Sewell's scheme can be extended to cover these topics, and the extension provides adequate answers to the problems of measurement theory in quantum mechanics. We shall then attempt to meet the last three of Wigner's objections listed on page 158, and the problem arising from the failure of localizability in relativistic physics that was stressed by him. The material will be arranged as follows.

In Section 11.1 we shall deal with an example of Bell that challenged the notion of quantum mechanics as it has been used so far in this book. In Section 11.2 we shall discuss a few extensions of Sewell's scheme, leaving aside the crucial extension to continuous spectra. That discussion, provided in Section 11.6, will be preceded by short accounts of the results of Araki and Yanase in Section 11.3, the impossibility theorems of Shimony and Busch in Section 11.4, and the Heisenberg cut in Section 11.5. The results of Araki and Yanase were obtained within von Neumann's measurement theory proper,[1] whereas Shimony and others based their attempts on a class of modifications of it. Section 11.7 will be devoted to establishing the adequacy of Sewell's scheme, and Section 11.8 to meeting the objections of Wigner and providing our answer to the question that has induced the writing of Part II of this book.

11.1 Bell's example

We have already referred to Hepp's 1972 paper, which (to the best of the present author's knowledge) was the first to demonstrate that the metaphysical assumptions of von Neumann could be replaced by mathematical ones to achieve reduction of the wave packet. This paper was strongly criticized by Bell. The following quotations are from his critique (Bell, 1975; reprinted in Bell, 2004) of Hepp's paper, which was mentioned in passing in the previous chapter:

> K. Hepp has discussed quantum measurement theory [using]...the $C^\star$ algebra description of infinite quantum systems...Many people not familiar

[1] We remind the reader that the adjective *proper* means that the intervention of the observer's conscious ego has been excluded from the theory. See footnote 4 on page 173.

with the $C^\star$ algebra approach... have been intrigued by the following statement in Hepp's abstract:

> In several explicitly soluble models, the measurement leads to macroscopically different "pointer positions" and to a rigorous "reduction of the wave packet" with respect to *all* [emphasis added] local observables.

This looks like a clean solution at last to the infamous measurement problem.[2] But it is not so... Here we will take one[3] of his models and analyse it in elementary text-book terms...

Bell then does precisely that. After having done so, he continues:

> The result [quoted above]... shows that any *fixed* observable Q will eventually give a very poor (zero, in this case) measure of persisting coherence. But *nothing forbids the use of different observables as time goes on* [emphasis added]...

Then Bell constructs the following set of *explicitly time-dependent* self-adjoint operators:

$$z(t) = \sigma_0^1 \prod_{n=1}^{[t-r-w]} \sigma_n^2,$$

where $\sigma^1, \sigma^2, \sigma^3$ are the Pauli spin matrices and the subscript n denotes the particle number. The quantity $[x]$ in the upper limit denotes the integral part of $x, x > 0$ (Bell denotes it by $N(x)$). For our present purposes, we need not be concerned with r and w, except to note that they are positive numbers. By definition, $z(t) = 0$ for $[t - r - w] < 1$. In Bell's words, 'the increasing string of factors here serves to unflip the flipped spins [flipped by the object–apparatus interaction]'. From this he concludes that

> So long as nothing, in principle, forbids consideration of such arbitrarily complicated observables, it is not permitted to speak of wave packet reduction.

Bell's fallacy (or so the present author contends) was in not recognizing that he was interpreting the word *all* used by Hepp as a layman, whereas Hepp was using it as a mathematician. There *was* a principle that forbade consideration of such observables: von Neumann's axiomatization of quantum mechanics, in which observables were, by definition, time-independent; the unconditional existence

[2] *Footnote in the original*: For a general survey, see, for example, d'Espagnat (1971).

[3] *Footnote in the original*: Note that Hepp considers several other models, making points not presented here, in particular concerning the possibility of 'catastrophic' time evolutions.

of the Schrödinger picture (as defined by us in Subsection 7.1.5, page 116) was a *sine qua non.*

The last paragraph of Bell's paper reads as follows:

> The continuing dispute about quantum measurement theory is not between people who disagree on the results of simple mathematical manipulations. Nor is it between people with different ideas about the actual practicality of measuring arbitrarily complicated observables. It is between people who view with different degrees of concern or complacency the following fact: so long as the wave packet reduction is an essential component, and so long as we do not know exactly when and how it takes over from the Schrödinger equation, we do not have an exact and unambiguous formulation of our most fundamental physical theory.

The present author sees things differently:

(i) Wave packet reduction was indeed an essential component of von Neumann's theory.

(ii) When Bell was writing (1975) it could indeed be argued that 'we do not know exactly when and how [wave packet reduction] takes over from the Schrödinger equation' (since in Hepp's work it happened only as $t \to \infty$), and therefore 'we do not have an exact and unambiguous formulation of our most fundamental physical theory' – *provided* that 'this most fundamental physical theory' is understood as von Neumann's axiomatization of quantum mechanics.

(iii) However, what Bell did, in effect, was to depart radically from von Neumann's axiomatization. Therefore, while his counterexample did contradict the collapse postulate, one can only draw the following conclusions from this fact (or so the present author believes):

  (a) There are two theories under consideration; N (von Neumann's) and B (Bell's), and $\mathsf{N} \neq \mathsf{B}$. In N all observables are time-independent; in B this is not true, by definition.

  (b) Theory N has a lacuna; to fill it, it is necessary that a result, which we shall call theorem C (collapse of the wave packet), be proven within it. This had not yet been done in 1975, without invoking the limit $t \to \infty$.

  (c) Bell's counterexample proves that theorem C is false in theory B. It makes no statement about theory N.

After this hiatus, let us summarize what we have learnt from Sewell's theory so that we may continue.

(i) In Sewell's theory, the algebras of observables of both object and apparatus are fixed from the outset, and consist of time-independent operators.

(ii) Wave packet reduction 'takes over' from the Schrödinger equation *when* the observed microscopic system interacts with the macroscopic apparatus. The transition, which is probably not instantaneous, nevertheless takes place within a very short time (in the laboratory frame).
(iii) Except in special cases, the quantity measured is a diagonal matrix element and not an eigenvalue of the observable.

11.2 Extension of Sewell's scheme

We shall now consider a particular case and an extension of Sewell's generic model. The really important extension, to observables with continuous spectra, will be postponed till after a review of the results of Wigner, Araki and Yanase in Section 11.3, and of the results of Shimony and Busch in Section 11.4.

11.2.1 Degenerate eigenvalues

Let us now assume that the observable A commutes with the Hamiltonian but has degenerate eigenvalues. Then the u_r are eigenvectors of A as well:

$$Au_r = a_r u_r. \tag{11.1}$$

To simplify the discussion we assume that only one of the eigenvalues is degenerate:

$$\begin{aligned} a_i \neq a_j \quad &\text{for} \quad i, j = 1, \ldots, m+1, \\ a_i = a_j = b \quad &\text{for} \quad i, j > m. \end{aligned} \tag{11.2}$$

In this case we may expect the wave packet to collapse, but, as we shall see below, the apparatus will no longer be able to distinguish between different eigenstates u_j of the object for $j > m$.

From (11.1) it follows that

$$(u_s, Au_r) = a_r \delta_{rs} \quad \text{(no summation)},$$

so that (10.47) becomes, with $M = I_{\mathrm{a}}$,

$$E(A \otimes I_{\mathrm{a}}) = \sum_{r=1}^{n} a_r |c_r|^2 \sum_{\alpha=0}^{\nu} F_{r,r;\alpha} \tag{11.3}$$

and (10.48) becomes

$$E(A \otimes \Pi_\alpha) = \sum_{r=1}^{n} a_r |c_r|^2 F_{r,r;\alpha}. \tag{11.4}$$

After the time τ, the apparatus is in one of the states $\alpha = 1, \ldots, m, m+1$. We assume that if $1 \leq \alpha \leq m$, then the state of the apparatus corresponds to one of the nondegenerate eigenstates of the object. Under these conditions (10.51), which can be written as

$$\sum_{\alpha=1}^{m} F_{r,r;\alpha} + F_{r,r;m+1} = 1,$$

yields the information that, for $\alpha = r < m+1$,

$$F_{r,r;r} = 1,$$

whereas, for $\alpha = m+1$,

$$F_{r,r;m+1} = 1 \text{ for } r = m+1, \ldots, n.$$

Then (11.3) and (11.4) become, respectively,

$$E(A \otimes I_{\mathsf{a}}) = \sum_{r=1}^{m} a_r |c_r|^2 + (n-m)b \sum_{r=m+1}^{n} |c_r|^2 \tag{11.5}$$

and

$$E(A \otimes \Pi_\alpha) = \begin{cases} a_r |c_r|^2 & \text{for} \quad 1 \leq \alpha \leq m \\ b\left(\sum\limits_{m+1}^{n} |c_r|^2\right) & \text{for} \quad \alpha = m+1. \end{cases} \tag{11.6}$$

Equation (11.5) shows that the wave packet has collapsed, but (11.6) shows that the pointer reading $\alpha = m+1$ no longer corresponds to a unique vector state of the object, as von Neumann had asserted (von Neumann, 1955, p. 218).

11.2.2 The case of $\dim \mathfrak{H}_{\mathsf{o}} = \infty$

Sewell's generic model is based on the assumption that $\dim \mathfrak{H}_{\mathsf{o}} = n$, where n is finite. Von Neumann's theory assumes that $\mathfrak{H}_{\mathsf{o}}$ is an infinite-dimensional separable Hilbert space, and the Araki–Yanase theorem, which will be proven in the following section, is established in the framework of von Neumann's theory. It will therefore be useful, for purposes of comparison, to see what additional assumptions are required to extend Sewell's generic model to the case $\dim \mathfrak{H}_{\mathsf{o}} = \infty$. The first of these is $\dim \mathfrak{H}_{\mathsf{a}} = \infty$, which requires no further comment.

If $\mathfrak{H}_{\mathsf{o}}$ is infinite-dimensional, then one will have to assume that H_{o} has a purely discrete spectrum. Correspondingly, one will have to assume that the macroscopic operator M has a countable infinity of distinct eigenvalues. It is easy to see that the contents of Sections 10.2–10.4 remain valid in this case as well, *except* possibly for the very last inequality (10.53), which is $F_{r,r;\alpha} F_{s,s;\alpha} \geq |F_{r,s;\alpha}|^2$. This inequality cannot be asserted without further analysis. However, we have seen in Subsection 11.2.1 that the terms with $F_{r,s;\alpha}, r \neq s$ drop out of the expectation values if the operator A being measured commutes with the Hamiltonian. It follows that:

Conclusion 11.1 The conclusions of Sewell's generic model remain valid also for $\dim \mathfrak{H}_{\mathsf{o}} = \dim \mathfrak{H}_{\mathsf{a}} = \infty$ if the Hamiltonian H_{o} has a purely discrete spectrum, the cells $\mathfrak{K}_\alpha$ are finite-dimensional, and one of the following conditions holds:

(i) The operator A being measured commutes with H_{o}.
(ii) The quantities $F_{r,s;\alpha} = \mathrm{Tr}\,(\Omega_{r,s}\Pi_\alpha)$ vanish for $r \neq s$.

In these cases one has to replace the upper limits of summation n and ν in (11.5) and (11.6) by ∞.

11.3 The Araki–Yanase theorem

In 1952, Wigner considered the problem of measuring the spin components of a particle, and found that, if the z-component was conserved, then the x- and y-components could not be measured precisely (Wigner, 1952). The term measurement was understood by him in the sense of von Neumann,[4] and encapsulated as 'the state of the apparatus reflects the state of the object'. Wigner's result was generalized by Araki and Yanase, who proved that an observable A with a purely discrete spectrum cannot be measured exactly (in the von Neumann sense) unless it commutes with all conserved quantities (Araki and Yanase, 1960). They also proved that, under certain conditions on the conserved quantities, A could always be measured approximately, provided that the notion of approximate measurement was understood in the rather special sense of Definition 11.3; the larger the measuring apparatus, the better the approximation. All of these results were proven within a slightly sharpened form of von Neumann's measurement theory proper; whereas the actual interaction between object and apparatus was passed over in silence (except to establish the result reported in Section 8.3.3.1) by von Neumann, Araki and Yanase assumed that it could be represented as a unitary evolution of the combined state of the object and apparatus. We begin by setting up the notations.

As before, we denote by $\mathfrak{H}_{\mathsf{o}}$ and $\mathfrak{H}_{\mathsf{a}}$ the Hilbert spaces of object and apparatus respectively. The Hilbert space of the combined system will be $\mathfrak{H} = \mathfrak{H}_{\mathsf{o}} \otimes \mathfrak{H}_{\mathsf{a}}$. Vectors in $\mathfrak{H}_{\mathsf{o}}$ will be denoted by φ, with subscripts when necessary, and vectors in $\mathfrak{H}_{\mathsf{a}}$ by the letters ξ, ϑ and ζ, again with subscripts as necessary.

Araki and Yanase considered the measurement of an observable A. The initial state of the object–apparatus system was $\varphi_\mu \otimes \xi$, where φ_μ is an eigenvector of

[4] We have called the result *von Neumann's entanglement theorem*; see page 159.

A with eigenvalue μ: $A\varphi_\mu = \mu\varphi_\mu$. The measurement process is expressed by the equation

$$U(t)[\varphi_\mu \otimes \xi] = \varphi_\mu \otimes \zeta_\mu, \tag{11.7}$$

where $t > \tau$, τ being the time it takes for the apparatus to arrive at its final state ζ_μ. The vectors $\varphi_\mu \in \mathfrak{H}_{\mathsf{o}}$ and $\zeta_\mu \in \mathfrak{H}_{\mathsf{a}}$ satisfy the orthogonality conditions

$$\begin{aligned} (\varphi_\mu, \varphi_\nu) &= \delta_{\mu\nu}, \\ (\zeta_\mu, \zeta_\nu) &= 0 \quad \text{if} \quad \mu \neq \nu, \end{aligned} \tag{11.8}$$

the second orthogonality requirement in (11.8) being necessary for distinguishing between different states of the object. At the end of the process, 'the state of the apparatus reflects the state of the object', nothing more. Araki and Yanase allowed for the eigenvalues μ of A to be degenerate, but we shall assume that they are nondegenerate. This merely simplifies the appearance of the formulae; the argument remains the same.

Suppose now that there is a self-adjoint operator L on $\mathfrak{H}$ which is additive in the sense that

$$L = L_{\mathsf{o}} \otimes I_{\mathsf{a}} + I_{\mathsf{o}} \otimes L_{\mathsf{a}}, \tag{11.9}$$

where L_{o} and L_{a} are self-adjoint operators on $\mathfrak{H}_{\mathsf{o}}$ and $\mathfrak{H}_{\mathsf{a}}$ respectively, and that L obeys the conservation law

$$[U(t), L] = 0, \tag{11.10}$$

which is 'universal' in the sense that it holds for every possible choice of the apparatus. Then:

Theorem 11.2 (Araki–Yanase) *With A, L, L_{o} and L_{a} as above, the evolution equation* (11.7) *cannot hold unless*

$$[L_{\mathsf{o}}, A] = 0. \tag{11.11}$$

In words, unless A commutes with L_{o}, *the final state of the apparatus will not reflect the state of the object.*[5]

[5] The usual definition of commuting operators has to be modified to cover unbounded operators. This modification was given in (9.4), page 162.

Proof Let $\mu \neq \nu$. Using unitarity, the conservation law (11.10) and the measurement condition (11.7), we find that

$$\begin{aligned}(\varphi_\nu \otimes \xi,\, L[\varphi_\mu \otimes \xi]) &= (U(t)[\varphi_\nu \otimes \xi],\, U(t)L[\varphi_\mu \otimes \xi]) \\ &= (U(t)[\varphi_\nu \otimes \xi,]\, LU(t)[\varphi_\mu \otimes \xi]) \\ &= (\varphi_\nu \otimes \zeta_\nu,\, L[\varphi_\mu \otimes \zeta_\mu]).\end{aligned}$$

The last line may be simplified by using the additivity condition (11.9):

$$\begin{aligned}(\varphi_\nu \otimes \zeta_\nu,\, L[\varphi_\mu \otimes \zeta_\nu]) &= (\varphi_\nu \otimes \zeta_\nu,\, (L_o \otimes I_a + I_o \otimes L_a)[\varphi_\mu \otimes \zeta_\nu]) \\ &= (\varphi_\nu \otimes \zeta_\nu,\, \varphi_\mu \otimes L_a\zeta_\nu) + (\varphi_\nu \otimes \zeta_\nu,\, L_o\varphi_\mu \otimes \zeta_\mu) \\ &= (\varphi_\nu,\, \varphi_\mu)(\zeta_\nu,\, L_a\zeta_\mu) + (\varphi_\nu,\, L_o\varphi_\mu)(\zeta_\nu,\, \zeta_\mu) \\ &= 0, \text{ for } \mu \neq \nu,\end{aligned}$$

where we used the orthogonality relations (11.8). We therefore conclude that

$$(\varphi_\nu \otimes \xi,\, L[\varphi_\mu \otimes \xi]) = 0 \text{ for } \mu \neq \nu. \tag{11.12}$$

Next, using again the additivity condition (11.9), we may recast the left-hand side of (11.12) as

$$\begin{aligned}(\varphi_\nu \otimes \xi,\, L[\varphi_\mu \otimes \xi]) &= (\varphi_\nu \otimes \xi,\, (L_o \otimes I_a + I_o \otimes L_a)[\varphi_\mu \otimes \xi]) \\ &= (\varphi_\nu \otimes \xi,\, L_o\varphi_\mu \otimes \xi) + (\varphi_\nu \otimes \xi,\, \varphi_\mu \otimes L_a\xi) \\ &= (\varphi_\nu,\, L_o\varphi_\mu)(\xi,\, \xi) + (\varphi_\nu,\, \varphi_\mu)(\xi,\, L_a\xi) \\ &= (\varphi_\nu,\, L_o\varphi_\mu)(\xi,\, \xi) \text{ for } \mu \neq \nu.\end{aligned} \tag{11.13}$$

Combining (11.12) and (11.13), we obtain

$$(\varphi_\nu,\, L_o\varphi_\mu) = 0 \text{ for } \mu \neq \nu. \tag{11.14}$$

This result states that the off-diagonal matrix elements of L_o vanish in a basis consisting of the eigenvectors of A. To complete the proof of $[A, L_o] = 0$, we use the spectral representation of A (recall that we have assumed the eigenvalues of A to be nondegenerate):

$$A = \sum_\lambda \lambda P_{[\varphi_\lambda]}. \tag{11.15}$$

It suffices to prove that $[P_{[\varphi_\lambda]}, L_o] = 0$ for all λ. Now

$$(\varphi_\nu,\, P_{[\varphi_\lambda]}L_o\varphi_\mu) - (\varphi_\nu,\, L_oP_{[\varphi_\lambda]}\varphi_\mu) = \delta_{\lambda\nu}(\varphi_\nu, L_o\varphi_\mu) - \delta_{\lambda\mu}(\varphi_\nu, L_o\varphi_\mu).$$

Owing to (11.14), the right-hand side vanishes unless $\lambda = \mu = \nu$. But then if $\lambda = \mu = \nu$, the equation becomes the trivial identity $0 = 0$. This establishes that $U(t)[\varphi_\mu \otimes \xi] = \varphi_\mu \otimes \zeta_\mu$ implies that $[L_o, A] = 0$. The assertion of the theorem is the contrapositive of this result, assuming that the conservation law $[U(t), L] = 0$ continues to hold. $\square$

We have ignored domain questions for the operators L_o and L_a in the above proof. Araki and Yanase have shown that the result continues to hold even if L_a is unbounded, as long as L_o is bounded. Details may be found in footnote 4 of (Araki and Yanase, 1960).[6]

For purposes of comparison, we note that in Sewell's theory, (10.59) and (10.60) show that, if A does not commute with H_o, then the quantity measured is no longer an eigenvalue of A, but an approximation to one. The goodness of the approximation depends, among other factors, on the choice of $\mathfrak{H}_o$ and H_o. Indeed, Sewell's assumption $n < \infty$ may be interpreted as exhibiting the centrality of the notion of approximate measurements in his theory.

Since the position operator, a key observable, will seldom commute with the Hamiltonian (which is an absolutely conserved quantity), it is natural to ask if such a quantity can be measured approximately in von Neumann's theory proper. But the first problem is to define the notion of an approximate measurement in von Neumann's theory proper, in which states of the measuring apparatus do not have classical descriptions. Therefore the result of a measurement cannot be assumed to be a real number, and the statistical notions of mean and deviation are no longer available.

Araki and Yanase defined a notion of *malfunctioning of the apparatus* which could be quantified. It was based on the observation that, in von Neumann's theory, one could envisage a situation in which the state of the apparatus *did not* reflect the state of the object. Using this definition, they were able to prove that an approximate measurement is possible, provided that L has a discrete spectrum *and* L_o *has a finite number of eigenvalues.* As we shall not be using their result, we shall omit the proof, and content ourselves with stating their definition.

Definition 11.3 (Approximate measurement (Araki–Yanase)) Suppose that the observable A with the spectral representation (11.15) does not commute with the operator L_o of (11.9). Then A will be said to be *approximately measurable* if there exist states $\varphi_\mu, \psi \in \mathfrak{H}_o$ and $\zeta_\mu, \vartheta_\mu \in \mathfrak{H}_a$ with the following properties:

$$
\begin{aligned}
(\zeta_\mu, \zeta_\nu) &= 0 \quad \text{if} \qquad \mu \neq \nu, \\
(\zeta_\mu, \vartheta_\nu) &= 0 \quad \text{for all} \quad \mu, \nu,
\end{aligned}
$$

[6] The Araki–Yanase result has been examined in detail, and some of its limitations exposed, by Stein and Shimony in Appendix A of their paper (Stein and Shimony, 1971). As will be seen in Section 11.5, neither the results of Araki and Yanase nor those of Stein and Shimony will be sufficient for our purposes, and therefore we shall not go into the technicalities.

$$(\vartheta_\mu, \vartheta_\nu) = 0 \quad \text{for} \quad \mu \neq \nu,$$
$$||\psi \otimes \zeta_\mu||^2 < \varepsilon,$$

and temporal evolution of the combined system has the form, for $t > \tau$,

$$U(t)[\varphi_\mu \otimes \xi] = \varphi_\mu \otimes \zeta_\mu + \psi \otimes \vartheta_\mu. \tag{11.16}$$

Araki and Yanase proved the existence of such states for any $\varepsilon > 0$. However – and this is the important point – the states of the apparatus A were superpositions of eigenstates of L_a; to reduce ε, one had to go on increasing the number of eigenstates, in effect converting A into a large device, for which a pure state could hardly be distinguished from a mixed one.

This concludes our survey of the results of Araki and Yanase. We shall return to their notion of approximate measurements in Section 11.5.

11.4 Impossibility theorems of Shimony and Busch

As we have seen, von Neumann's measurement theory proper did not lead to the reduction of the wave packet. Several authors, starting with Fine, have investigated whether or not it was possible to arrive at a different result in a modified version of von Neumann's theory, with a somewhat different notion of measurement (Fine, 1969). The answer was uniformly negative, and led to a number of theorems on the 'insolubility of the quantum measurement problem'. Recall that von Neumann had argued that the initial state of the apparatus cannot be a mixture (pages 155–156). Shimony attempted to revive the possibility of the initial state of the apparatus being a mixture by relaxing the condition that the measurement be exact (Shimony, 1974; see also Fehrs and Shimony, 1974). His assumptions were as follows (the notation is obvious):

(a) $\{\mathfrak{K}_\mathsf{o}^m\}$ is a finite or countable family, indexed by m, of pairwise orthogonal subspaces of $\mathfrak{H}_\mathsf{o}$ that span $\mathfrak{H}_\mathsf{o}$.
(b) $\{\mathfrak{K}_\mathsf{a}^m\}$ is a family of pairwise orthogonal subspaces of $\mathfrak{H}_\mathsf{a}$.
(c) U is a unitary operator on $\mathfrak{H}_\mathsf{o} \otimes \mathfrak{H}_\mathsf{a}$.
(d) T is a statistical operator (density matrix) on $\mathfrak{H}_\mathsf{a}$ such that for every m and every $\varphi \in \mathfrak{K}_\mathsf{o}^m$,

$$U(P_{[\varphi]} \otimes T)U^{-1} = \sum_{r,n} a_{r,n} P_{[\zeta_{r,n}]}, \tag{11.17}$$

where $\zeta_{r,n} \in \mathfrak{H}_\mathsf{o} \otimes \mathfrak{K}^n$, and the $a_{r,n}$ are nonnegative real numbers such that

$$\sum_{r,n} a_{r,n} = 1,$$
$$\sum_r a_{r,n} \ll 1 \text{ for } n \neq m.$$

As (11.17) shows, the modification attempted by Shimony may be described as follows. The initial state of the object is a pure state, but that of the apparatus is a mixture. The final state of the object–apparatus system is a density matrix which has a sharp peak at $n = m$ (but it still does not have a 'classical description'). This is the sense in which the measurement is approximate. Shimony proved that under these conditions, if the number of subspaces $\mathfrak{K}_o^m$ is greater than 1, then 'there exist initial states of the object for which the final statistical state of the object plus apparatus is not expressible as a mixture of eigenstates of the apparatus observable'.

The proof of this result is by contradiction. Shimony assumed the contrary, and derived, by a detailed computational argument, a contradiction with the fact that the eigenvalues of a density matrix are invariant under unitary transformations. Some years later, Brown gave a much simpler proof of a similar result under slightly stronger conditions (Brown, 1986), but it turned out that this weakened the final result considerably. A critique of Brown's result may be found in (Shimony, 1993, pp. 46–47).

Subsequently Busch and Shimony, using the notion of unsharp observables defined by positive operator valued measures,[7] proved that Shimony's impossibility theorem held even for unsharp observables (Busch and Shimony, 1996). We should like to draw the reader's attention to the last sentence of their abstract, which is as follows: 'Both theorems show that the measurement problem is not the consequence of neglecting the ever-present imperfections of actual measurements.' A 'maximal extension' of Shimony's theorem was obtained by Stein in 1997, using arguments that were more conceptual than computational (Stein, 1997).

11.5 The Heisenberg cut

As already discussed in the Introduction to Part II, the 1930s saw two conceptions of measurement attempting to 'complete' Born's probability interpretation of quantum mechanics. One was that of Heisenberg and Bohr, in which the result of a measurement was supposed to be represented on a classical output device. As noone had any idea how an apparatus made up of quantum-mechanical subsystems could do so, Heisenberg invented the '*Heisenberg cut*', which would be better described as the Heisenberg chasm. Its merit, from the point of view that we have espoused, was that the notions of measurement error and approximate measurements were the same as they were in classical physics, and there was no difficulty in accepting that almost every measurement of a continuous spectrum had to be approximate. To recapitulate: in the notion of measurement advanced by von Neumann and Wigner, the state of the apparatus had no classical description.

[7] These measures, which we shall not define (they are defined in the article by Busch and Shimony), are used extensively in the quantum theory of open systems. See, for example, (Davies, 1976).

It followed immediately that there was no canonically defined notion of approximate measurements. Araki and Yanase therefore proposed to augment the notion of exact measurement by a quantifiable notion of malfunctioning of the apparatus, and defined an approximate measurement as one with a small but nonzero probability of apparatus malfunction. With this notion they concluded, within von Neumann's theory proper, that an approximate measurement was always possible. Recall that the wave packet did not collapse in von Neumann's theory proper.

Shimony, and then Busch and Shimony, modified von Neumann's theory proper to admit mixed initial states of the apparatus, but then, in view of von Neumann's result that the initial state of the apparatus could not be a mixture (Subsection 8.3.3.1), they had to abandon the notion of precise measurements altogether. They devised a notion of approximate measurements that was closer to intuition than the 'apparatus malfunction' of Araki and Yanase, but their conclusion was that, with these premises, even approximate measurements were impossible.

Sewell's theory modifies von Neumann's premises to permit the building of a bridge over the Heisenberg cut; collapse of the wave packet results, so to speak, from crossing this bridge.

11.6 Measurement of continuous spectra

An operator R with a strictly continuous spectrum does not have eigenvectors, and therefore cannot be measured in accordance with von Neumann's theory. In Subsection 8.2.2 we described how von Neumann formulated the problem of approximate measurement of such an operator, and reduced it to the problem of exact measurement of an operator $G(R) = G$ with a purely discrete spectrum which was an approximation to R. This operator G, which we called a von Neumann approximant to R, was not canonically defined; it had, presumably, to be chosen by the experimentalist, who then had to design an apparatus to measure it.

Choose $\varepsilon > 0$, and let $\{\varphi_n | n \in \mathbf{N}\}$ be a set of approximate eigenvectors of R,

$$||(R - \lambda_k I)\varphi_k|| < \varepsilon_k, \;\; k \in \mathbf{N},$$

such that every spectral value λ of R lies in one (or more) of the intervals $(\lambda_k - \varepsilon, \lambda_k - \varepsilon)$. We shall assume that the λ_k are equally spaced, i.e., the quantity

$$\Delta = \lambda_k - \lambda_{k-1}$$

is independent of k. Let G be the operator that has a purely discrete spectrum, with the eigenvalues and eigenvectors

$$G\varphi_k = \lambda_k \varphi_k.$$

Then G is a von Neumann approximant of R, and an exact measurement of G is equivalent, according to von Neumann, to an approximate measurement of R to an accuracy of $\pm\varepsilon$. The theory places no limitations on ε, except that it be greater than zero.

The problem of exact measurement of G has already been dealt with. If G does not commute with H_o, then a measurement of G will yield not one of its eigenvalues but one of its diagonal matrix elements in a basis consisting of eigenvectors of H_o. In principle, if G and H_o are known, then the matrix elements

$$g_{kk} = (u_k, Gu_k)$$

can be calculated. Set

$$\delta_k = \min_j |g_{kk} - \lambda_j|.$$

Then, if

$$\delta_k \ll \Delta, \tag{11.18}$$

a reading of g_{kk} will reflect, with near-certainty, one and only one eigenvalue of G, namely the one nearest to g_{kk}. In Sewell's theory, an approximate measurement of an operator with a continuous spectrum is possible (via a von Neumann approximant) *even if the operator in question does not commute with any quantity that obeys a universal additive conservation law.*

11.7 Adequacy of Sewell's scheme

Sewell's theory has not resulted in the complete elimination of anthropocentrism from the theory of measurement in quantum mechanics. But it has succeeded, or so we shall argue, in reducing the role of the experimental quantum physicist to roughly the same as that of the experimental classical physicist, which was discussed in Chapter 6. There the role of the experimentalist was explained in economic terms: the efficient allocation of scarce resources. The role of the physicist designing and performing quantum-mechanical measurements is exactly the same.

An observable A can be measured exactly only if it has a purely discrete spectrum, and the measurement can be recorded exactly only if the eigenvalue being measured is rational. An observable with a continuous spectrum can never be measured exactly, but one can devise a von Neumann approximant that can be measured exactly. These statements may have to be modified slightly if the observable being measured does not commute with the Hamiltonian, but the same would be true of classical measurements. In the final analysis – or so the present author claims – it is the limitation of physical resources that limits the accuracy of *ideal measurements*, both in classical and in quantum mechanics.[8]

Recall the equation

$$m = nr. \tag{6.3}$$

that we wrote down in Section 6.4. There m was the total memory available, n the number of observation points, and r the memory required to record each observation, which (possibly by a change of scale) is the same thing as the number of significant digits in the record of the observation. The only change required here is to interpret $n = \dim \mathfrak{H}_0$. That means that the experimentalist who wishes to obtain the most accurate results will confine himself or herself to as small an n as is feasible, i.e., to as small a range as possible of the variable concerned.

11.8 Answers to Wigner, III

We shall conclude this part of our enterprise by answering the remaining questions raised by Wigner. These fall into two parts: items (iii)–(v) that were quoted on page 158, and the 'tension with the special theory of relativity', to use a beautiful phrase coined by Shimony (Shimony, 2002).

The first set of questions share two common themes: (a) the impossibility of precise measurements of most observables, and (b) the need for using a large apparatus to make a reasonably accurate measurement.

Regarding the impossibility of precise measurements, it is a 'flaw', as first noted by Born, that quantum mechanics shares with classical mechanics. Even if one is willing to disregard all instrumental and human errors (and to work at absolute zero), there is little that one can do about the structure of the real number system. Quantum mechanics affords enough freedom to make quantum-mechanical errors smaller than any preassigned quantity in any single measurement, provided that measurements are understood in Bohr's sense as formulated mathematically by Sewell.

Regarding the need for using a large apparatus, it appears to the present author that Wigner could be described as the quintessential reductionist, which is why (i) he rejected Bohr's idea, and (ii) this possibility made him uneasy. The

[8] Real measurements are subject to random errors. Sewell has attempted, with some success (Sewell, 2007; private communication), to trace these random errors to thermodynamic fluctuations, and to estimate the probability of large fluctuations using the theory of large deviations. The same argument should hold true for classical measurements, but the present author is not aware of any study.

first paragraph of an essay he wrote honouring Einstein reads as follows (Wigner, 1970, p. 3):

> The world is very complicated and it is clearly impossible for the human mind to understand it completely. Man has therefore devised an artifice which permits the complicated nature of the world to be blamed on something which is called accidental and thus permits him to abstract a domain in which simple laws can be found. The complications are called initial conditions; the domain of regularities, laws of nature.[9] Unnatural as such a division of the world's structure may appear from a very detached point of view, and probable though it is that the possibility of such a division has its own limits, the underlying abstraction is probably one of the most fruitful ones the human mind has ever made. It has made the natural sciences possible.

At the Istanbul Summer School in 1962, Wigner told a group of students that this division into initial conditions and laws of nature 'was the essence of reductionism' (or words to that effect). The present author was one of the group.

It could be that Wigner did not believe in the seemingly antireductionist (and possibly antirelativistic) proposition that a microscopic system could interact with a macroscopic system *as a whole*; the appeal to the observer's conscious ego, about which he said that it was '... shrouded in mystery and no explanation has been given so far in terms of quantum mechanics, or in terms of any other theory' (page 137), could therefore be construed as the counsel of despair.

11.8.1 The tension with the theory of relativity

Two of the most important observables in nonrelativistic quantum mechanics are the position and momentum operators of a single particle. The momentum operators will be constants of motion in many situations, but exactly the opposite will be true for the position operators. The position operator in nonrelativistic quantum mechanics – if it exists at all – will, firstly, have a continuous spectrum and secondly, will seldom commute with the Hamiltonian. Observables with continuous spectra cannot be measured exactly in von Neumann's formulation of quantum mechanics. If one does not admit measuring instruments with states that have classical descriptions, they may not even be approximately measurable.

However, if one does admit measuring instruments with states that *do* have classical descriptions, then Sewell's theory and its extension provides an adequate solution within von Neumann's formulation of quantum mechanics. In the

[9] We shall refer to this as *Wigner's paradigm*, the word paradigm being used in the sense of Kuhn (Kuhn, 1957).

opinion of the present author, the situation is no worse in quantum mechanics than it is in classical mechanics.

In relativistic quantum theory, the impossibility of precise measurement of the position of a particle may be a blessing rather than a curse. In 1974, in a paper that is remarkably short and easy to read, Hegerfeldt proved that the notion of particle localizability is in conflict with the notion of causality (Hegerfeldt, 1974). We quote the abstract of his paper in full:

> We show that under quite general assumptions localization of particles in a finite space region at a given time is inconsistent with causality. The same holds for localization in a finite region of space-time. The derivation is short and very simple.

This result made Wigner very unhappy. We quote from his lecture notes (Wigner, 1983, p. 312):

> No matter how one defines the position, one has to conclude that the velocity, defined as the ratio of two subsequent position measurements divided by the time interval between them, has a finite probability of assuming an arbitrarily large value, exceeding c. One either has to accept this, or deny the possibility of measuring the position precisely or even giving significance to this concept: a very difficult choice!

If the impossibility of precise measurement of the position of a point-particle is reason enough to question the usefulness of the notion of a geometrical point in physics, then it surely is an argument that extends to Newtonian physics as well. Wigner is on record as having remarked that 'no one asks whether the outcome of the measurement is a rational or an irrational number' (Wigner, 1983, p. 274) when discussing the measurement of an observable with a continuous spectrum, but the present author does not know whether, unlike Max Born, he ever considered the analogous problem in classical physics.

The present author disagrees with Wigner's assessment quoted above, and believes that denying the possibility of precise measurements is the *only* rational choice.[10] Unfortunately, the rationality of this choice may have been masked by the fact that it is superficially similar to the one advocated by the school of thought that maintains that what is good enough 'for all practical purposes' does not require further analysis – what Bell described as the FAPP school of thought.

[10] As discussed earlier, this means that the notion of precise measurements has to be replaced by the notion of *arbitrarily precise* measurements, i.e., measurements with error $\varepsilon > 0$, where ε can be arbitrarily small.

12

Large quantum systems

The aim of this final chapter of Part II is rather different from that of the preceding chapters. It is to provide a glimpse into the generalizations of the formalism of quantum mechanics on Hilbert space that are required to describe the symmetries and dynamics of systems with infinitely many degrees of freedom, owing to the qualitative differences that arise when the number of degrees of freedom tends to infinity.

The burden of the preceding chapters was that the notion of a geometrical point is as meaningful in quantum physics as it is in classical physics. The argument involved a lengthy excursion into quantum-mechanical measurement theory. During this excursion, we found that the notion of *design of the experiment* played an essential role.

An experiment, no matter how ingenious, does not create laws of nature; it substantiates, or refutes, an assumed law. According to the dichotomy between laws of nature and initial conditions that was posited by Wigner (page 206), the role of human intervention in an experiment is to realize, under controlled and repeatable conditions, a conjunction of initial conditions that may be improbable in nature. Seen in this light:

(i) The failure of von Neumann's theory proper to explain wave function collapse is either a failure of the theory to establish control over complicated initial conditions or an indicator of as-yet-undiscovered laws of nature; the success of Sewell's theory suggests that it is the former, and not the latter.
(ii) The centrality of 'design of the experiment' in Sewell's theory falls short of anthropocentrism. A quantum-mechanical measurement *reveals consequences of laws of nature* that would seldom be manifested in nature owing to the very low probability of the required conjunction of initial conditions; human intervention merely increases this probability.

Let us try to interpret Sewell's theory in terms of Wigner's paradigm. In this theory the apparatus is an N-particle system subject to ordinary quantum mechanics, where N is large. The magnitude of N permits a *reduced description* of the apparatus by a set of *commuting macroscopic observables*, and it makes sense to talk about the macroscopic evolution of the apparatus. Wigner's paradigm should be applicable to this situation as well.[1] However, a macroscopic

[1] It also shows that Wigner's paradigm is not equivalent to reductionism; it does not always honour Francis Bacon's motto '*dissecare naturam*'. Precisely for that reason, it may be more significant scientifically.

measuring apparatus can, by definition, be *driven* by a microscopic system. It is therefore pertinent to ask:

Question 12.1 What are the principles that govern the interactions that permit a microscopic quantum-mechanical system to drive the macroscopic observables of a large system?

The above formulation is intended to suggest a paradigm shift: the problems of quantum-mechanical measurement theory lie, not in the intervention of the observer, but in the physics of the interaction between two qualitatively different entities, one a microscopic system and the other a macroscopic nonrelativistic quantum-mechanical system considered as a whole.[2] In the following, we shall try to prepare the ground for this study.

12.1 Elementary excitations in superfluid helium

The theory of symmetry in quantum mechanics developed by Wigner does not suffice for infinite systems which cannot be described adequately on a single Hilbert space. A striking example is provided by the elementary excitations in superfluid helium, which were postulated by Landau and detected later by inelastic neutron scattering experiments. The reason is that these excitations have *zero nonrelativistic mass*, and objects with zero nonrelativistic mass cannot be described on a Hilbert space.

We have seen, in Subsection 7.4.1, that particles of nonzero mass are described by projective representations of the Galilei group G. The true representations of G were analyzed by Inönü and Wigner (Inönü and Wigner, 1952).[3] They concluded that irreducible unitary representations of G did not contain states that were either localizable in space or in velocity space. The reason, briefly, is as follows. One of the invariants of the Lie algebra of G is the quantity $\boldsymbol{P}^2$ (page 123); in an irreducible representation, it will be a constant. In such a representation there are simply not enough $\boldsymbol{p}$-values available to form a δ-function in $\boldsymbol{x}$-space.[4] This naive argument was replaced by mathematically correct ones by Inönü and Wigner. The irreducible representations are defined on the spaces $L^2(S_2(r), \mathrm{d}S)$, where $S_2(r)$ is the 2-sphere $\boldsymbol{p}^2 = r > 0$ and $\mathrm{d}S$ is the rotation-invariant measure

[2] The Reeh–Schlieder theorem of relativistic quantum theory suggests the possibility that the adjective *nonrelativistic* used above may not be essential. We shall not stop to discuss this fascinating question. The interested reader is referred to (Reeh and Schlieder, 1961; Streater and Wightman, 1964; Haag, 1993), for a start.

[3] Although this paper was published two years before Bargmann's 1954 paper in which the latter showed that a nonrelativistic particle of mass $m \neq 0$ corresponded to a projective representation of G (discussed in Subsection 7.4.1), Bargmann's results were known to Inönü and Wigner. See p. 706 of their paper, particularly footnote 2.

[4] The reader who is not familiar with the true representations of the Galilei group is referred to the review article (Lévy-Leblond, 1972), or the article (Sen, 1972). These articles use the Dirac formalism. The quantities E and $\boldsymbol{p}$ are not eigenvalues but spectral values, but it is convenient to talk about the energy and momentum of states, and lack of mathematical rigour does not lead to serious errors in this instance.

on it.[5] One may, instead, try to work with a *reducible* representation, containing all $\boldsymbol{p}$-values, defined on the space $L^2(\mathbb{R}^3, \mathrm{d}\boldsymbol{p})$. In this case $\mathbb{R}^3$ may be considered as momentum space, with $\mathrm{d}\boldsymbol{p}$ the Lebesgue measure on it. If E and $\boldsymbol{p}$ are the spectral values of H and $\boldsymbol{P}$ (normally interpreted as energy and momenta), then their transformation properties under the boost $(0, 0, \boldsymbol{v}, 1)$ turn out to be

$$\begin{aligned} E &\to E + \boldsymbol{p}\cdot\boldsymbol{v}, \\ \boldsymbol{p} &\to \boldsymbol{p}. \end{aligned} \tag{12.1}$$

That is, the momentum is invariant under the boosts, but the energy is not. We have the following situation:

(i) According to (12.1), it is not possible to assign a definite E to a given $\boldsymbol{p}^2$; in the language of many-body problems, the existence of a *dispersion law* is precluded in a unitary representation.
(ii) According to Landau, it is precisely elementary excitations with the transformation properties (12.1) that are responsible for the phenomenon of superfluidity below a certain critical velocity.[6] These excitations have a dispersion law $E = E(\boldsymbol{p}^2)$.

From (12.1) we see that a value of $\boldsymbol{p}^2$ will never define a unique value of E as long as the boosts are implemented on the Hilbert space. To create room to define a dispersion law, we have somehow to ensure that the boosts are *not* implemented on the Hilbert space. One possibility is to use a Hilbert *bundle* as state space, in which the boosts are implemented not on the fibres, but only on the base space.[7] This will be done in the following. We begin with a definition of Hilbert bundles which will suffice for our present purposes.

12.2 Hilbert bundles

A **Hilbert bundle** $\mathfrak{B}$ is the topological product $\mathfrak{B} = X \times \mathfrak{H}$, where X is a second-countable, Hausdorff topological space and $\mathfrak{H}$ a Hilbert space over $\mathbb{C}$ with its metric topology. Thus a point $\mathfrak{b} \in \mathfrak{B}$ may be written uniquely as $\mathfrak{b} = (x, \phi)$, where $x \in X$ and $\phi \in \mathfrak{H}$. The map $\pi : \mathfrak{B} \to X$ defined by $\pi(x, \phi) = x$ is called the **projection** of the bundle. A map $\sigma : X \to \mathfrak{B}$ that satisfies $\pi(\sigma(x)) = x$ is called a **section**, or **cross-section** of X in $\mathfrak{B}$.

A Hilbert bundle may be pictured on a plane, with the X-axis representing the base space and the Y-axis the fibre, as in Fig. 12.1. The lines parallel to the Y-axis are some of the fibres, and s is a *continuous* cross-section.

[5] The case $r = 0$ is exceptional, and need not be considered in the present context.

[6] An account of Landau's theory is given in the book by Khalatnikov (Khalatnikov, 1965). This book also contains translations of two of Landau's original papers.

[7] A brief account of the topological spaces called *fibre bundles* is given in Section A8.1.

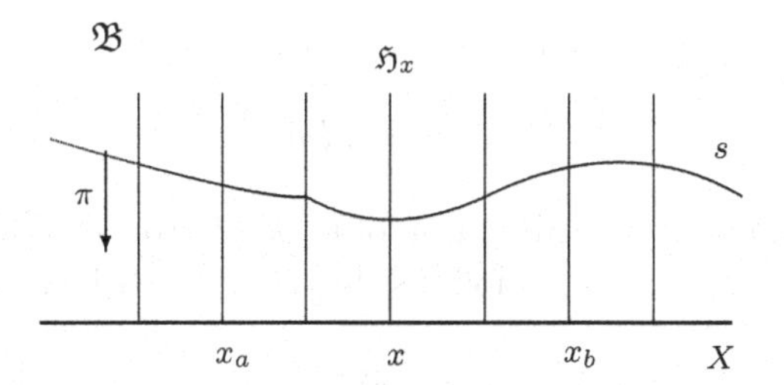

Fig. 12.1. A Hilbert bundle

Sections of a Hilbert bundle can be added, and multiplied by complex numbers; they form a linear space over $\mathbb{C}$. If the base space has a measure defined on it, then one can use this measure, together with the metric on the fibre, to define a metric on the space of sections. This construction defines **direct integrals** of Hilbert spaces, which one encounters in the theory of infinite-dimensional group representations.

A **bundle map** $h : \mathfrak{B} \to \mathfrak{B}$ is a homeomorphism that *preserves fibres*, i.e.,

$$\pi(\mathfrak{b}) = \pi(\mathfrak{b}') \Longrightarrow \pi(h(\mathfrak{b})) = \pi(h(\mathfrak{b}')).$$

This means that θ induces a **base map** $\bar{\theta} : X \to X$ according to the formula

$$\pi \circ h = \bar{h} \circ \pi$$

(see Fig. A8.2, page 352); the induced base map is a homeomorphism.

12.3 Bundle representations

In the following we shall assume that G is a topological group, H a closed subgroup of it, $X = G/F$ the space of left-cosets of F furnished with the quotient topology, and $p : G \to G/F$ the projection. The action of G upon itself by left-translations, i.e., the map $G \times G \to G$ defined by $(g_1, g_2) = g_1 g_2$, defines an action of G on X by left-translations which we shall denote by gx, where $x \in X$.

We shall represent G on the Hilbert bundle $\mathfrak{B} = X \times \mathfrak{H}$, where $\mathfrak{H}$ is a separable Hilbert space over $\mathbb{C}$. Let $x \in X$, $\phi \in \mathfrak{H}$. Then $\mathfrak{b} = (x, \phi) \in \mathfrak{B}$, and we may write

$$g\mathfrak{b} = g(x, \phi) = (gx, u(g, x)\phi), \tag{12.2}$$

where $u(g, x)$ is a unitary operator on $\mathfrak{H}$ which satisfies the condition

$$u(g'g, x) = u(g', gx)u(g, x) \tag{12.3}$$

that arises from the associativity of multiplication in G,

$$(g'g)\mathfrak{b} = g'(g\mathfrak{b}). \tag{12.4}$$

The problem of determining the action of G on $\mathfrak{B}$ reduces to solving (12.3).

These equations are solved as follows. Let $\eta : X \to G$ be a section of X in G, and set

$$k(g,x) = \eta(gx)^{-1} \cdot g \cdot \eta(x). \tag{12.5}$$

It is easily seen that the quantities $k(g,x)$ so defined satisfy (12.3). Although each factor on the right-hand side of (12.5) is an element of G, one may readily verify that $k(g,x)$ is in fact an element of H. It is called a (G,X,H)-**cocycle**. Let now D be a continuous unitary representation of H upon $\mathfrak{H}$, and set

$$u(g,x) = D(k(g,x)). \tag{12.6}$$

This $u(g,x)$ fulfils the cocycle condition (12.3), and solves the algebraic part of our problem.[8]

However, as G is a topological group, one would expect a group action to satisfy some continuity requirements before it can properly be called a representation. The continuity requirement that has proved itself is *strong continuity*, which we have already encountered in the statement of Stone's theorem (Subsection 7.5.3). The action of G on X is continuous, but the continuity of the action of $u(g,x)$ on $\mathfrak{H}$ remains to be addressed. According to a theorem of von Neumann, a unitary group representation is strongly continuous if it is weakly measurable (page 306). For the latter, it is sufficient that η be measurable, and such choices are always possible. We refer the reader to Mackey's Chicago lectures (Mackey, 1976) for full statements and proofs. Measurability of η implies the measurability of $k(g,x)$, which in turn implies the continuity of $u(g,x) = D(k(g,x))$. Finally, we observe that the representation itself is independent of the section η; a change of sections is equivalent to a unitary transformation on the Hilbert space $\mathfrak{H}$, a result which is not difficult to establish.

Note that we have not assumed that G is a Lie group. If it is, there may be further results of physical interest. We refer the interested reader to (Sen, 1978, 1986; Sen and Sewell, 2002) for details and for alternative constructions.

12.3.1 Landau excitations

We shall use the term *Landau excitations* for the excitations which were called phonons and rotons by Landau; the two are now believed to be the same, and

[8] This method of solving (12.3) was devised by Wigner (Wigner, 1939), who traced its antecedants to Frobenius and Schur. See footnote 15 on page 167 of his article (page 47 in (Dyson, 1966)). It seems to have been discovered independently by Whitney in the course of his pioneering investigations on fibre bundles. The references may be found in (Steenrod, 1972).

are called phonons. Our justification for using a different term is that phonons in superfluid helium are very different from phonons in crystal lattices.[9]

We shall apply the construction of the preceding section to the Galilei group. That is, we shall take (in the notations of Section 7.4):

(i) G to be the Galilei group, $g = (b, \boldsymbol{a}, \boldsymbol{v}, R) \in G$.
(ii) $H = T_0 \times E_3 \subset G$. Here T_0 and E_3 are respectively the subgroups of time translations and Euclidean transformations, and $H = \{(b, \boldsymbol{a}, 0, R)\}$.

Then:

(i) H is a normal subgroup of G.
(ii) The space X of left-cosets of H in G is a group: $X \sim \{(0, 0, \boldsymbol{w}, 1)\}$. We shall call X the *space of boosts* and denote an element of it by $\boldsymbol{w}$.
(iii) G has a natural action on X given by $g\boldsymbol{w} = (b, \boldsymbol{a}, \boldsymbol{v}, R)\boldsymbol{w} = \boldsymbol{v} + R\boldsymbol{w}$.
(iv) There exists a natural *continuous* cross-section of X in G given by $\eta(\boldsymbol{w}) = (0, 0, \boldsymbol{w}, 1)$.
(v) With $g = (b, \boldsymbol{a}, \boldsymbol{v}, R)$ and the above cross-section, the cocycle $k(g, \boldsymbol{w})$ is given by

$$k(g, \boldsymbol{w}) = (b, \boldsymbol{a} - (\boldsymbol{v} + R\boldsymbol{w})b, 0, R). \tag{12.7}$$

The Hilbert bundle on which we shall represent G will be $\mathfrak{B} = X \times L^2(\mathbb{R}^3, \mathrm{d}\boldsymbol{p})$, where $\mathbb{R}^3$ is the three-dimensional momentum space, $\mathrm{d}\boldsymbol{p}$ the Lebesgue measure on it and X as defined above. The only thing that remains to be done is to choose the representation D of H.

12.3.1.1 Unitary representations of E_3 and H

The group E_3 itself is a semidirect product[10] of space translations T_3 and rotations: $E_3 = T_3 \wedge O^+(3)$. The representation theory of such groups is based on two pillars:

(i) That of one-parameter groups, which are commutative; therefore their irreducible unitary representations are one-dimensional. They are complex numbers of modulus unity and are called *characters*. Their theory was developed by Pontrjagin (Pontrjagin, 1939). If the group is compact, it is

[9] However, the latter can also be described in terms of bundle representations, but with a space group replacing the Euclidean group (Borchers and Sen, 1975).

[10] We have denoted the semidirect product, for which there is no standard notation, by the symbol $\wedge$. It should be noted that, unlike direct products, semidirect products of two groups cannot always be defined; one of the groups has to be a group of automorphisms of the other, and the semidirect product structure is defined in terms of this automorphism (which need not be unique). In physics, the rotation group is the group of automorphisms of T_3 which leaves the Euclidean metric invariant, and the Lorentz and space-time translation groups are similarly related.

The groups $O(3)$ and $O(2)$ are the orthogonal groups in three and two dimensions respectively, which include reflections. The rotation groups are their connected components, and are denoted by $O^+(3)$ and $O^+(2)$ respectively.

isomorphic with the two-dimensional rotation group $O(2)$, and its characters are labelled by an integer; if it is not compact, it is isomorphic with $\mathbb{R}$, and its characters are labelled by a real number.

(ii) That of compact groups, of which $O(3)$ is one. The representation theory of compact groups has an extensive literature, of which we mention only the book by Weyl, the main architect of the subject (Weyl, 1946).

It should be added that the representation theory of $O(3)$ was developed in much greater detail, for application to atomic spectroscopy, by Wigner (Wigner, 1931; English translation, 1959). Wigner's main results will be known to every student of physics.

The representations of E_3 may be constructed by the method of induced representations that was developed by Wigner to deal with the Poincaré group (Wigner, 1939). A classification of the irreducible unitary representations of E_3 may be found in (Lévy-Leblond, 1972) or (Sen, 1972).

We shall denote an element of E_3 by $(\boldsymbol{a}, R)$. The group multiplication law in E_3 is $(\boldsymbol{a}', R')(\boldsymbol{a}, R) = (\boldsymbol{a}' + R'\boldsymbol{a}, R'R)$. Let $\phi \in L^2(\mathbb{R}^3, \mathrm{d}\boldsymbol{p})$. We choose the following representation of E_3 on $L^2(\mathbb{R}^3, \mathrm{d}\boldsymbol{p})$:

$$(U(\boldsymbol{a}, R)\phi)(\boldsymbol{p}) = \mathrm{e}^{-\mathrm{i}\boldsymbol{a}\cdot\boldsymbol{p}}(\phi \circ R^{-1})(\boldsymbol{p}), \tag{12.8}$$

where $\phi \in L^2(\mathbb{R}^3, \mathrm{d}\boldsymbol{p})$ (the sign of the exponent is immaterial; it has been chosen to achieve conformity with Landau's equations (12.1)). Note that on the right-hand side of (12.8) the character of T_3 is computed *before* the rotation is applied.[11] This representation is not irreducible, but is the direct integral of all irreducibles with helicity zero with the Lebesgue measure. Finally, we choose, arbitrarily, a dispersion law (function)

$$E = E(\boldsymbol{p}^2) \tag{12.9}$$

and construct with it the following representation of H:

$$(D(b, \boldsymbol{a}, R)\phi)(\boldsymbol{p}) = \mathrm{e}^{\mathrm{i}(bE - \boldsymbol{a}\cdot\boldsymbol{p})}(\phi \circ R^{-1})(\boldsymbol{p}). \tag{12.10}$$

This is the representation that we shall use.

12.3.1.2 Landau excitations as bundle representations

With $k(g, \boldsymbol{w})$ given by (12.7) and D by (12.10), the action of G on the bundle $\mathfrak{B}$ is given explicitly by the formula

$$(b, \boldsymbol{a}, \boldsymbol{v}, R)(\boldsymbol{w}, \phi) = (\boldsymbol{v} + R\boldsymbol{w}, \mathrm{e}^{\mathrm{i}b[E + (\boldsymbol{v} + R\boldsymbol{w})\cdot\boldsymbol{p}]}\, \mathrm{e}^{-\mathrm{i}\boldsymbol{a}\cdot\boldsymbol{p}}\; \phi(R^{-1}\boldsymbol{p})). \tag{12.11}$$

[11] The same rule has to be followed for computing $(U(\boldsymbol{a}', R')[U(\boldsymbol{a}, R)\phi])\boldsymbol{p}$ to verify the group multiplication law.

To make the result more transparent, let us look at the rest-fibre $\boldsymbol{w} = 0$ and the action of the group elements $(b, \boldsymbol{a}, 0, 1)$ and $(b, \boldsymbol{a}, \boldsymbol{v}, 1)$ on it. We find that

$$(b, \boldsymbol{a}, 0, 1)(0, \phi) = (0, \mathrm{e}^{\mathrm{i}bE}\, \mathrm{e}^{-\mathrm{i}\boldsymbol{a}\cdot\boldsymbol{p}}\, \phi(\boldsymbol{p})), \tag{12.12}$$

whereas

$$(b, \boldsymbol{a}, \boldsymbol{v}, 1)(0, \phi) = (\boldsymbol{v}, \mathrm{e}^{\mathrm{i}b(E+\boldsymbol{v}\cdot\boldsymbol{p})}\, \mathrm{e}^{-\mathrm{i}\boldsymbol{a}\cdot\boldsymbol{p}}\, \phi(\boldsymbol{p})). \tag{12.13}$$

This shows that E and $\boldsymbol{p}$ indeed transform under the boosts as required by Landau's formula (12.1). However, as a boost maps a fibre of $\mathfrak{B}$ to a wholly different fibre, there can be no superposition of states with same $\boldsymbol{p}^2$ but different E; in other words, the existence of a dispersion law is no longer contradicted by the requirement of Galilei invariance.

If one takes the transformation properties of the energy and momentum of a particle of mass m under a Galilei boost $\boldsymbol{v}$ and sets $m = 0$, one recovers Landau's relations (12.1). This justifies the statement made earlier that elementary excitations described by the bundle representation (12.11) are *nonrelativistic zero-mass systems.*

12.3.1.3 Unstable excitations

Clearly, there are enough momenta available in the representation (12.8) to form a δ-function in $\boldsymbol{x}$-space. This, however, is physically insignificant, as elementary excitations generally occupy fairly narrow momentum bands. They also tend to have finite lifetimes. This suggests that they be viewed as representations of the Galilei *semigroup* $G^{\uparrow}$, which is the same as the Galilei group, except that time translations are not invertible; $b \geq 0$ in $(b, \boldsymbol{a}, \boldsymbol{v}, R)$. However, $G^{\uparrow}$ has an identity which is the same as the identity of G, and therefore, although the generic element $(b, \boldsymbol{a}, \boldsymbol{v}, R)$ of $G^{\uparrow}$ does not have an inverse, the elements $(0, \boldsymbol{a}, \boldsymbol{v}, R)$ do:

$$(0, \boldsymbol{a}, \boldsymbol{v}, R)^{-1} = (0, -R^{-1}\boldsymbol{a},\ R^{-1}\boldsymbol{v},\ R^{-1}). \tag{12.14}$$

The existence of the identity element implies[12] that $G^{\uparrow}$ has infinitesimal generators (which are the same as those of G). It also means that, with suitable restrictions,[13] one can develop the theory of both Hilbert space and bundle representations for such semigroups along the same lines as that for groups. The

[12] We shall not go into details. The reader is referred to (Sen and Sewell, 2002) and the references cited there.

[13] The restrictions are that the semigroup have a coset decomposition with respect to a maximal subsemigroup, i.e., the base space be definable, and that the subsemigroup be represented on the fibres. These restrictions are sufficient, but their necessity has not been established. There appears to be no general theory of either linear or bundle representations of semigroups. Since decaying particles and excitations are part of nature, this may be considered as a gap in the literature.

analogue of the bundle representation (12.13) becomes

$$(b, \boldsymbol{a}, \boldsymbol{v}, 1)(0, \phi) = (\boldsymbol{v}, \mathrm{e}^{-b\Gamma}\, \mathrm{e}^{\mathrm{i}b(E+\boldsymbol{v}\cdot\boldsymbol{p})}\, \mathrm{e}^{-\mathrm{i}\boldsymbol{a}\cdot\boldsymbol{p}}\, \phi(\boldsymbol{p})) \ \text{ for } \ b \geq 0, \tag{12.15}$$

where $\Gamma > 0$ is the half-life of the excitation. We refer the reader to (Sen and Sewell, 2002) for further details.

12.4 Dynamics on Banach bundles

Banach spaces are norm-complete linear spaces in which the norm does not necessarily satisfy the polarization identity (page 294).[14] A brief account of normed spaces is given in Section A5.1. We shall define a **Banach bundle** to be the topological product $\mathfrak{C} = X \times \mathfrak{W}$, where the fibre $\mathfrak{W}$ is a Banach space. The following paragraph is taken from (Sen and Sewell, 2002):[15]

> Symmetries are usually studied via group rather than Lie algebra actions, because they avoid the problems associated with unbounded operators. By contrast, dynamics is usually studied via the infinitesimal generator of the one-parameter group or semigroup of time translations – the Hamiltonian – because the Hamiltonian represents the total energy of the system in question.[16] In quantum mechanics, the Hamiltonian is an operator on a Hilbert space. In classical mechanics, it is a complete vector field on a manifold of states. From physical considerations, one would expect a one-parameter group or semigroup of time translations of a Banach bundle to possess an infinitesimal generator which would act as a linear transformation on the fibres and a vector field on the base space. To the best of the authors' knowledge, such mathematical objects have not yet been studied...

The last sentence is still valid. Therefore we have to write separate equations of motion on the base space and on the fibres. The action of the group itself provides *solutions* of these equations of motion.

Let $\mathfrak{C}$ be a Banach bundle and T a one-parameter group or semigroup of fibre-preserving maps $T_t : \mathfrak{C} \to \mathfrak{C}$, where the parameter is denoted by the subscript t, which stands for time.[17] If T is a group, then $t \in \mathbb{R}$; if it is a semigroup, then $t \in \mathbb{R}^+ = [0, \infty)$. The pair $(\mathfrak{C}, T)$ will be called a *dynamical system.*

[14] The polarization identity expresses the inner product as a function of the norm.

[15] The material of this section, including its title, is taken from the same reference.

[16] Here and in the following the term semigroup will *always* denote a *semigroup with identity,* because only these semigroups have infinitesimal generators. The interested reader is referred to (Hille, 1965).

[17] By convention, a one-parameter group or semigroup is assumed to be at least once differentiable with respect to that parameter.

We shall denote a point in $\mathfrak{C}$ by $\mathfrak{c} = (x, \vartheta)$, where $x \in X$ and $\vartheta \in \mathfrak{W}$. The action of T on $\mathfrak{C}$ will be written explicitly as

$$T_t(x, \vartheta) = (S_t x, W(t, x)\vartheta), \tag{12.16}$$

where $S = \{S_t | t \in \mathbb{R} \text{ or } t \in \mathbb{R}^+\}$ is a one-parameter group or semigroup of base maps and $W(x, t)$ is a linear transformation on $\mathfrak{W}$. They have to satisfy the requirements of associativity of multiplication in T, namely

$$S_{t'}(S_t x) = S_{t'+t}\, x, \tag{12.17}$$

and

$$W(t' + t, x) = W(t', S_t x)W(t, x). \tag{12.18}$$

The entities x and ϑ may be regarded as functions of time, and written accordingly as $x(t)$ and $\vartheta(t)$. Then the x and ϑ that appear in (12.16) should be regarded as their initial values $x(0)$ and $\vartheta(0)$, and the expressions $S_t x$ and $W(t, x)\vartheta$ as solutions of the initial-value problems for the equations of motion for $x(t)$ and $\vartheta(t)$ respectively. Note that $S_0 = \mathrm{id}$ and $W(0, x) = I$, the identities on X and $\mathfrak{W}$ respectively. The equations of motion themselves may be written as

$$\frac{\mathrm{d}x}{\mathrm{d}t} = F(x) \tag{12.19}$$

and

$$\frac{\mathrm{d}\vartheta}{\mathrm{d}t} = L(x)\vartheta, \tag{12.20}$$

where

$$F(x) = \lim_{h \to 0} \frac{(S_h - \mathrm{id})x}{h} \quad \text{for all } x \in X \tag{12.21}$$

and

$$L(x)\vartheta = \lim_{h \to 0} \frac{(W(x, h) - I)\vartheta}{h} \quad \text{for all } x \in X,\ \vartheta \in \mathfrak{W}. \tag{12.22}$$

Equation (12.19) shows that time translations engender a flow on the base space which is autonomous; equation (12.20) then shows that this flow *drives* a flow on the fibres. If $\mathfrak{W}$ is a Hilbert space and X a manifold, then one may describe the situation as an autonomous classical system driving a quantum system. On the other hand, physics would suggest that it is the *motion of the composite system*, represented on the bundle, that defines the motion on the base space. Passing from the global to the local seems to interchange potter and pot![18]

The reader is referred to (Sen and Sewell, 2002) for further development of these ideas.

[18] The allusion is to the line from *The Rubayyat of Omar Khayyam* that was quoted on page 4.

12.5 Operator algebras and states

It is clear that, if the multiplication of operators is defined in the standard way, the bounded operators on a complex Hilbert space $\mathfrak{H}$ form an algebra over $\mathbb{C}$. We shall denote this set by $\mathcal{B}$. Recall that there is a norm defined on $\mathcal{B}$ (page 323):

$$||A|| = \sup_{||x||=1} ||Ax||,$$

where the supremum is taken over all vectors of unit length in $\mathfrak{H}$. The norm satisfies the conditions

$$\begin{aligned} ||A|| \geq 0, \ \ ||A|| = 0 \ \text{ iff } \ A = 0, \\ ||cA|| = |c| \cdot ||A|| \ \text{ for all } \ c \in \mathbb{C}, \\ ||A+B|| \leq ||A|| + ||B||, \\ ||AB|| \leq ||A|| \cdot ||B||. \end{aligned} \tag{12.23}$$

The third inequality is the triangle inequality, which shows that the norm is, *inter alia*, a metric on the algebra. Therefore the notions of Cauchy sequences and completeness make sense. However, the norm will *never* satisfy the polarization identity in any nontrivial case, i.e., there is no question of an inner product. The algebra is a Banach space which is almost never a Hilbert space.

The metric topology induced by the norm on $\mathcal{B}$ in the standard way is called the **uniform** or **norm** topology.The algebra is equipped with the operation of taking the adjoint – called **involution** in the mathematical literature – which satisfies

$$\begin{aligned} (cA)^\star = \bar{c}A^\star, \\ (AB)^\star = B^\star A^\star. \end{aligned} \tag{12.24}$$

An algebra with an involution is called a ***-algebra** (read: star-algebra). An algebra over $\mathbb{C}$ that satisfies conditions (12.23) and (12.24) is called a **normed *-algebra**. Equations (12.23) show that the algebraic operations are continuous in the norm topology.

12.5.1 $C^\star$-algebras

We now remove the underlying Hilbert space, and define an abstract algebra as follows:

Definition 12.2 ($C^\star$-algebras) A normed *-algebra $\mathcal{A}$ over $\mathbb{C}$ is called a *$C^\star$-algebra* if it satisfies the following conditions:

(a) It is complete in its norm.
(b) $||A^\star A|| = ||A||^2$ for all $A \in \mathcal{A}$.

It follows from the inequality $||AB|| \leq ||A|| \cdot ||B||$ and the second of these conditions that $||A^\star|| = ||A||$; the easy verification is left to the reader. (An algebra that is complete in the norm but does not necessarily satisfy condition (b) is called a **Banach algebra**.)

Examples 12.3 (Examples of $C^\star$-algebras)

(i) Let X be a compact Hausdorff space, and $\mathcal{C}(X)$ the space of complex-valued continuous functions on it. This space is complete in the sup norm (denoted by $||\cdot||_\infty$; see Subsection A5.2.1). Since the complex conjugate of a continuous function is a continuous function, $\mathcal{C}(X)$ is a commutative $C^\star$-algebra.

It is a remarkable theorem of Gelfand that every commutative $C^\star$-algebra is the algebra of continuous complex-valued functions on some compact Hausdorff space. The reader will find an excellent easy-to-read account in the text by Simmons (Simmons, 1963).

(ii) The set $\mathcal{B}$ of all bounded operators on a Hilbert space is a $C^\star$-algebra with the usual norm. In this case neither property of Definition 12.2 is obvious. Proofs may be found in (Bratteli and Robinson, 1979).

Note that an abstract $C^\star$-algebra is not required to have an identity. The examples given above do have identities, and we shall not have occasion to consider operator algebras that do not have identities.

12.5.2 States on $C^\star$-algebras

A **state** on a $C^\star$-algebra is defined to be a positive, normalized linear functional on the algebra.[19] The formal definition is as follows:

Definition 12.4 (States on $C^\star$-algebras) Let $\mathcal{A}$ be a $C^\star$-algebra with identity. A functional $\omega : \mathcal{A} \to \mathbb{C}$ is called a *state* on $\mathcal{A}$ if it fulfils the following conditions for all $c_1, c_2 \in \mathbb{C}$ and all $A_1, A_2 \in \mathcal{A}$:

$$\begin{aligned} \omega(c_1 A_1 + c_2 A_2) &= c_1\omega(A_1) + c_2\omega(A_2), \\ \overline{\omega(A)} &= \omega(A^\star), \\ \omega(A^\star A) &\geq 0, \\ \omega(I) &= 1. \end{aligned} \tag{12.25}$$

Crucially, states on $\mathcal{A}$ also satisfy the **Schwarz inequality**:

$$|\omega(AB)|^2 \leq \omega(A^\star A)\omega(B^\star B) \;\text{ for all }\; A, B \in \mathcal{A}. \tag{12.26}$$

We shall denote the set of states on $\mathcal{A}$ by $\mathfrak{S}(\mathcal{A})$, or briefly by $\mathfrak{S}$. It is an important fact that $\mathfrak{S}$ is a convex set; i.e., if $\omega_1, \omega_2 \in \mathfrak{S}$ and $0 \leq \lambda \leq 1$, then

[19] We remind the reader that a functional is a map from a linear space into the real or complex numbers. See the footnote on page 292. The term *state*, in use in the mathematical literature, has been taken from physics.

their convex combination

$$\lambda\omega_1 + (1-\lambda)\omega_2$$

also belongs to $\mathfrak{S}$. A state that cannot be expressed as a convex combination of two different states is called an **extremal state**. In the language of physics, extremal states correspond to pure states.

12.6 Representations of $C^\star$-algebras

Let $\mathcal{A}$ be a $C^\star$-algebra, and $\mathcal{B}$ an algebra of bounded operators, furnished with the norm topology, on a separable Hilbert space $\mathfrak{H}$. With these topologies,[20] a continuous map

$$\pi : \mathcal{A} \to \mathcal{B} \tag{12.27}$$

is called a **representation** of $\mathcal{A}$ on $\mathfrak{H}$ if the following conditions are satisfied:

$$\begin{aligned} \pi(aA + bB) &= a\pi(A) + b\pi(B), \\ \pi(AB) &= \pi(A)\pi(B), \\ \pi(A^\star) &= \pi(A)^\star, \\ \pi(I_{\mathcal{A}}) &= I_{\mathfrak{H}}. \end{aligned} \tag{12.28}$$

We are assuming that $\mathcal{A}$ has an identity; if it does not, the last condition will be inapplicable. A representation is called **faithful** if $\pi(A) = 0 \Leftrightarrow A = 0$.

A representation π of $\mathcal{A}$ as defined above has the following basic properties (we omit the proofs):

(i) π preserves positivity, i.e., $A \geq 0 \Rightarrow \pi(A) \geq 0$ for all $A \in \mathcal{A}$.
(ii) π preserves continuity of the algebraic operations.
(iii) $||\pi(A)|| \leq ||A||$ for all $A \in \mathcal{A}$, equality holding only if π is a faithful representation.

12.6.1 The Gelfand–Naimark–Segal theorem

The usefulness of $C^\star$-algebras in physics stems from the following theorem:

Theorem 12.5 (Gelfand–Naimark–Segal) *Let $\mathcal{A}$ be a $C^\star$-algebra and ω a state on it. Then there exists a Hilbert space $\mathfrak{H}_\omega$, a vector $\Omega_\omega \in \mathfrak{H}_\omega$ and a representation π_ω of $\mathcal{A}$ such that*

$$\omega(A) = (\Omega_\omega, \pi_\omega(A)\Omega_\omega).$$

[20] As there are several different and useful topologies on the algebra of bounded operators on a Hilbert space, it is essential to specify which topology is being used to define a representation.

Proof The full proof requires more machinery than we have developed. We shall therefore outline only the main steps, which will be enough to bring out the feature that is our chief concern.

(1) Recall that the algebra $\mathcal{A}$ is, *inter alia*, a linear space over $\mathbb{C}$. The state ω defines a Hermitian form on this space through the formula

$$(A|B) = \omega(A^\star B). \tag{12.29}$$

This form, linear in B and antilinear in A, is *semidefinite*; the equality in $\omega(A^\star A) \geq 0$ cannot be excluded.

(2) There is a standard algebraic procedure to deal with this deficiency. Let $\mathcal{I}_\omega$ be the subset of $\mathcal{A}$ defined by

$$\mathcal{I}_\omega = \{X|X \in \mathcal{A},\ (X|X) = 0\}. \tag{12.30}$$

Then $\mathcal{I}_\omega$ is a subspace of $\mathcal{A}$, i.e., an additive group. It can be shown that $X \in \mathcal{I}_\omega \Rightarrow AX \in \mathcal{I}_\omega$, i.e., $\mathcal{I}_\omega$ is, in algebraic terms, a **left-ideal** in $\mathcal{A}$. It is sometimes called the *Gelfand ideal of the state* ω.

(3) The left-ideal $\mathcal{I}_\omega$ partitions $\mathcal{A}$ into equivalence classes as follows: A and B belong to the same class iff $A - B \in \mathcal{I}_\omega$. The class containing X is the subset $\{X + J|J \in \mathcal{I}_\omega\}$. The set of equivalence classes is a linear space over $\mathbb{C}$. The zero vector in this space is the class $\mathcal{I}_\omega$ itself. We denote the equivalence class of A by $[A]$.

(4) Let $[X]$ and $[Y]$ be two distinct equivalence classes, neither of them being $\mathcal{I}_\omega$ itself. Let $X' \in [X]$ and $Y' \in [Y]$. One verifies that $(X|Y) = (X'|Y')$, and that $(X|X) > 0$.

(5) One therefore defines an inner product $([X], [Y])$ on the space of equivalence classes by the formula $([X], [Y]) = (X|Y)$; the particular choices of X and Y are immaterial. One verifies that $([X], [Y])$ have all the properties required of an inner product.

(6) The space of equivalence classes, furnished with this inner product, is a **pre-Hilbert space**, which is the name given to a linear space that has all the properties of Hilbert space *except* completeness. Its metric completion in the metric defined by the above inner product (Appendix A4) is the required Hilbert space $\mathfrak{H}_\omega$.

(7) The product in $\mathcal{A}$ defines an action of $\mathcal{A}$ on the vectors of the pre-Hilbert space. Let $\varphi = [B]$ and $\psi = [AB]$. Then this action is defined by the formula

$$\pi_\omega(A)\varphi = \psi. \tag{12.31}$$

The action of $\mathcal{A}$ on the pre-Hilbert space can be extended to the whole of $\mathfrak{H}_\omega$ by continuity (by a general procedure in *linear analysis*).

(8) Finally, one defines $\Omega_\omega = [I]$, the equivalence class containing the identity. Then one shows that

$$\omega(A) = (\Omega_\omega, \pi_\omega(A)\Omega_\omega).$$

□

It is clear that the left-ideal $\mathcal{I}_\omega$ depends strongly on the state ω, which need not be a pure state. The state Ω_ω, however, is a vector state in the Hilbert space $\mathfrak{H}_\omega$.

Since the proof of the Gelfand–Naimark–Segal theorem is strictly constructive, it is also known as the GNS-**construction**.

12.6.2 The weak operator topology on a representation

An abstract $C^\star$-algebra has a natural topology defined on it, namely the norm topology. It has been proved that the norm which makes a normed algebra into a $C^\star$-algebra is unique. The situation is very different with algebras of operators on a Hilbert space. As we have just seen, the GNS-construction produces, from any state on an abstract $C^\star$-algebra, an algebra of operators on a Hilbert space.[21] We have already encountered a topology other than the norm topology on an algebra of operators on Hilbert space in connection with Stone's theorem: the strong topology (Subsection 7.5.3 and Appendix A6). We shall now define another topology, called the *weak operator topology.*

Definition 12.6 (Weak operator topology) Let $\mathcal{B}$ be the set of bounded operators on the Hilbert space $\mathfrak{H}$, $\{A^{(k)}|k \in \mathbf{N}\}$ a sequence in $\mathcal{B}$ and $\{\psi_n|n \in \mathbf{N}\}$ an orthonormal basis in $\mathfrak{H}$. Denote the matrix elements of $A^{(k)}$ in this basis by $a_{mn}^{(k)} = (\psi_m, A^{(k)}\psi_n)$. If the sequences

$$\{a_{mn}^{(k)}|m, n \text{ fixed}, \ k \in \mathbf{N}\}$$

converge for all m, n, and

$$\lim_{k\to\infty} a_{mn}^{(k)} = a_{mn},$$

then they define an operator A with matrix elements a_{mn} in $\mathcal{B}$. This notion of convergence defines a topology on $\mathcal{B}$ which is called the weak operator topology.[22]

[21] An algebra of operators on a Hilbert space which is also a $C^\star$-algebra is called a **concrete $C^\star$-algebra**.

[22] For a proper definition, sequential convergence has to be replaced by filter convergence (defined in Subsection A3.8.1), but we shall gloss over this point.

Weak operator convergence is often written as follows:

$$A^{(k)} \xrightarrow{w} A.$$

Recall that we have already defined the notion of *strong operator convergence* and used it to formulate Stone's theorem (page 133). If we denote norm convergence by $A^{(k)} \to A$ and strong operator convergence by $A^{(k)} \xrightarrow{s} A$, then we have the implications

$$A^{(k)} \longrightarrow A \implies A^{(k)} \xrightarrow{s} A \implies A^{(k)} \xrightarrow{w} A,$$

but the implications cannot be reversed. Weakening (= coarsening) the topology shrinks the family of open sets and therefore enlarges the family of limit points; and that is why a sequence that converges in the weak operator topology may fail to converge in the norm topology. More generally, as there are several distinct (and useful) notions of convergence on a representation but only one on the algebra itself, a sequence that converges in a representation in one of these senses may fail to converge in the algebra.

12.7 Algebraic description of infinite systems

Our discussion of $C^\star$-algebras, states and representations involved neither space nor time. One needs both to describe a physical system. Typically, we would like to describe the temporal evolution of a physical system which contains a finite number of particles in a finite nonzero volume. We therefore associate with each bounded open region of space $O \subset \mathbb{R}^3$ a $C^\star$-algebra $\mathcal{A}_O$ of observables which satisfies the following conditions:[23]

(i) If O, O' are bounded open regions in $\mathbb{R}^3$, then

$$O \subset O' \Longrightarrow \mathcal{A}_O \subset \mathcal{A}_{O'}.$$

This property is called *isotony.*

(ii) If $O \cap O' = \emptyset$, then every $A \in \mathcal{A}_O$ commutes with every $A' \in \mathcal{A}_{O'}$.

We now define

$$\mathcal{A} = \overline{\bigcup_{O \subset \mathbb{R}^3} \mathcal{A}_O}, \tag{12.32}$$

where the overbar indicates completion in the norm. That is, we first form the algebra which is the union of the $\mathcal{A}_O$ over all bounded open regions O in space, and then complete this algebra in the norm. It has, of course, to be

[23] Recall that the term *algebra of observables* means the algebra *generated* by a given set of observables. The product of two self-adjoint operators is not self-adjoint unless the two commute, and therefore the algebra of observables contains elements that are not self-adjoint, and therefore cannot be called observables.

established – just as in the completion of $\mathbb{Q}$ to $\mathbb{R}$ – that new elements so introduced can be welded seamlessly into the algebra. The problem here is simpler than in the completion of $\mathbb{Q}$ to $\mathbb{R}$, because there is no operation of division that has to be dealt with.

The result is the algebra $\mathcal{A} = \mathcal{A}(\mathbb{R}^3)$, which is a $C^\star$-algebra. It is called an *algebra of quasi-local observables.*

12.7.1 $C^\star$-dynamical systems

Before defining the dynamics, let us state some important ways in which the algebraic approach differs from the Hilbert space approach.

(i) Observables as self-adjoint operators are associated with representations and not with the abstract algebra. If two representations are unitarily inequivalent, they will not have the same set of observables in the ordinary quantum-mechanical sense. It is perhaps unfortunate that the term *observable* is used in two senses when so much of the physics lies precisely in the difference between the two.

(ii) A Hilbert space description may be recovered by the GNS-construction from an arbitrarily chosen state ω. The result depends strongly upon ω, and this is the extra generality that one achieves in the algebraic description as compared with von Neumann's formalism. This extra generality is, in some sense, equivalent to the generality afforded by inequivalent representations of the canonical commutation and anticommutation relations for infinitely many degrees for freedom (CCR and CAR). It is effectively exploitable in the algebraic formalism,[24] e.g., to describe different thermodynamic phases and equilibrium states at different temperatures; states at different temperatures correspond to inequivalent representations.

(iii) The formalism of ordinary quantum mechanics devolves around the transformation theory of Dirac, i.e., the equivalence between the Schrödinger and the Heisenberg pictures. A similar situation does *not* obtain in the $C^\star$-algebra formulation. The action of a group of automorphisms on the algebra will define, uniquely, a group of automorphisms on the states, but *the converse may not be true.* For this reason, a $C^\star$-dynamical system is generally defined as a one-parameter group of automorphisms of the *algebra*; see, e.g., (Bratteli and Robinson, 1981) or (Sewell, 2002).

Definition 12.7 ($C^\star$-dynamical systems) A $C^\star$-dynamical system is defined to be a triple $(\mathcal{A}, \mathfrak{S}(\mathcal{A}), \alpha_t)$, where $\mathcal{A}$ is a $C^\star$-algebra, $\mathfrak{S}(\mathcal{A})$ the set of states on it, and $\{\alpha_t | t \in \mathbb{R}\}$ a one-parameter group of automorphisms of $\mathcal{A}$. This group induces a one-parameter group of automorphisms $\{\alpha_t^\star | t \in \mathbb{R}\}$ of $\mathfrak{S}(\mathcal{A})$ by the

[24] It is possible to associate $C^\star$-algebras with the CCR and CAR. See, for example, (Emch, 1972) for a general discussion, or (Sewell, 2002) for simple examples.

formula

$$(\alpha_t^\star \omega)(A) = \omega(\alpha_t A).$$

12.7.2 Observables at infinity

Let Λ be a fixed bounded open set in $\mathbb{R}^3$, and let O be any bounded open set that does not intersect Λ: $O \cap \Lambda = \emptyset$. Define

$$\tilde{\mathcal{A}}_\Lambda = \overline{\bigcup_{O \subset \mathbb{R}^3 \setminus \Lambda} \mathcal{A}_O}. \tag{12.33}$$

Then, if $\mathcal{A}_\Lambda$ is the algebra of observables *inside* Λ, $\tilde{\mathcal{A}}_\Lambda$ can clearly be interpreted as the algebra of observables *outside* Λ. Furthermore, $\tilde{\mathcal{A}}_\Lambda \subset \mathcal{A}$.

Choose now a state $\omega \in \mathfrak{S}(\mathcal{A})$ and let $\pi_\omega(\mathcal{A})$ be the representation of $\mathcal{A}$ on the Hilbert space $\mathfrak{H}_\omega$. Then $\pi_\omega(\tilde{\mathcal{A}}_\Lambda)$ is well defined, and so is the intersection of the $\pi_\omega(\tilde{\mathcal{A}}_\Lambda)$ for all Λ; call it Δ_ω:

$$\Delta_\omega = \bigcap_\Lambda \pi_\omega(\tilde{\mathcal{A}}_\Lambda).$$

This set is nonempty, because it contains at least the multiples of the identity. Now comes the essential part. Since Δ_ω is a set of operators on the Hilbert space $\mathfrak{H}_\omega$, *we can form its closure in the weak operator topology*:

$$\Delta_\omega^{\text{w-cl}} = \left[\bigcap_\Lambda \pi_\omega(\tilde{\mathcal{A}}_\Lambda)\right]^{\text{w-cl}}. \tag{12.34}$$

The weak closure operation may introduce limit points which *belong to the representation π_ω, but not to the algebra $\mathcal{A}$*. It is this feature which is not available in N-particle quantum mechanics on a Hilbert space. Of course, if the state ω is identifiable as an N-particle state with $N < \infty$, there will be no such limit point.

The set $\Delta_\omega^{\text{w-cl}}$ was introduced by Lanford and Ruelle (1969), who called it the set of *observables at infinity* of the representation π_ω. The phrase 'observables at infinity' means that these observables make their presence felt *outside* any bounded region of $\mathbb{R}^3$; they may be compared with intensive thermodynamic variables of infinitely extended systems.

The quantum theory of measurement devised by Hepp (Hepp, 1972) to account for wave function collapse without invoking the observer's conscious ego was based on these observables at infinity.

12.8 Temporal evolution of reduced descriptions

In Sewell's measurement theory, the Hilbert space of the apparatus was assumed to be finite-dimensional. If we wish to lift this restriction, we have to be able to accommodate the possibility that an eigenspace of a macroscopic observable is infinite-dimensional. This may be difficult to achieve with a density matrix; since a density matrix has unit trace, its nonzero eigenvalues cannot be infinitely degenerate. An alternative that suggests itself is a Hilbert bundle based upon the space of eigenvalues of the macroscopic operator, the fibres being infinite-dimensional Hilbert spaces.

Let us now leave the measurement problem aside, and consider a macroscopic observable O with a continuous spectrum. Let $\mathfrak{B} = \Lambda \times \mathfrak{H}$ be a Hilbert bundle based on the spectrum Λ of O; we shall assume the latter to be a connected subset of $\mathbb{R}$. The Hilbert bundle description of the system is useful only if the temporal evolution of the system is given by a one-parameter family $\{\Theta(t)\}$ of bundle maps $\mathfrak{B} \to \mathfrak{B}$ which induces a family $\{\bar{\Theta}(t)\}$ of base maps $\Lambda \to \Lambda$. The latter are the maps that will be of physical interest. Let us consider two cases:

(i) O is a constant of motion, i.e., it commutes with the Hamiltonian H. In this case all motion takes place on the fibres, and $\{\bar{\Theta}(t)\}$ is the identity map.
(ii) O does not commute with the Hamiltonian. In this case the map $\{\bar{\Theta}(t)\}$ is nontrivial. This is exactly the case considered earlier, in Section 12.4 (with a Banach space as fibre, but that makes no difference). There we noticed that, while the base map is induced by the bundle map, the (differential) equations of motion describe a flow on the base space that appears to be autonomous, while the flow on the fibres is driven by it. Looking at (12.19) and (12.20), it would not immediately be apparent that they derive from the bundle map (12.16).

The temporal evolution of Λ, a macroscopic quantity, is described by (12.19). This equation has some similarities to Pauli's master equation (Pauli, 1928), but in a different mathematical setting. We refer the reader to (van Kampen, 1954, 1962) for a more detailed analysis of macroscopic observables, and to (Emch, 1964) for generalized master equations.

The master equation is similar to the Boltzmann equation in that it describes the temporal evolution of probabilities. However, the Boltzmann equation may have solutions that are far from equilibrium, and cannot be captured by the dynamical variables of equilibrium hydrodynamics. In 1977, Wightman wrote that: 'In fact, the existence and uniqueness of solutions [of the Boltzmann equation] for all t is not known to this day except in special cases' (Wightman, 1977, p. 151).[25]

The problem was attacked by Hilbert (Hilbert, 1912).[26] Wightman continues:

> However, the problem that Hilbert chose to attack was a different one and, in some sense, more subtle: the problem of normal solutions. The

[25] Reference in the original: (Grad, 1969).

[26] A simplified version of some of Hilbert's work and that of Chapman and Enskog may be found in the textbook by Huang (Huang, 1963).

Boltzmann equation ought to have a distinguished class of solutions in which the probability distribution at each time can be characterized completely by a few macroscopic parameters such as the density $n(\mathbf{x}, t)$, velocity $\mathbf{v}(\mathbf{x}, t)$ and thermal energy density $\varepsilon(\mathbf{x}, t)$ mentioned above. The evolution of these parameters is governed by equations of a kind of macroscopic continuum mechanics. The important problem of kinetic theory is to show that general solutions of the Boltzmann equation are asymptotic for large t to normal solutions; that is the problem of *approach to equilibrium.*

In the classical case, existence of the density and velocity fields $n(\mathbf{x}, t)$ and $\mathbf{v}(\mathbf{x}, t)$ may be justifiable on kinetic theory grounds alone, but that of the thermal energy density field $\varepsilon(\mathbf{x}, t)$ requires an additional concept, that of *local thermodynamic equilibrium*, which was introduced by K. Schwarzschild (Schwarzschild, 1906). It has turned out to be rather difficult to give a rigorous definition of this concept, both classically and quantum-mechanically, and we refer the reader to the book (Sewell, 2002) and the articles (Roos, 1995; Buchholz, Ojima and Roos, 2002; Bahr, 2006), where further references will be found. Except in special situations, the stability of macroscopic observables would generally be contingent on some form of local thermodynamic equilibrium. The subject needs to be pursued.

Epilogue

This work has explored the implications of Einstein–Weyl causality for the mathematical structure of space-time using the concept of geometrical points, and has tried to justify the use of the latter in physics. In part, it may also be viewed as a mathematical analysis of aspects of (i) the gap between theoretical and experimental physics, (ii) Wigner's 'unreasonable effectiveness of mathematics in the natural sciences' and (iii) consequences of a partial abandonment of the philosophy of reductionism. From these perspectives, a variety of other problems suggest themselves – or so it seems to the present author. This Epilogue is devoted to brief discussions, partly speculative, of some of these problems.[1]

Causal automorphisms of an ordered space

In 1964, Zeeman published a paper with the title 'Causality implies the Lorentz group', in which he showed that the group of causal automorphisms of Minkowski space was the same as the inhomogeneous Lorentz group plus the dilatations (Zeeman, 1964). Zeeman included space reflections (but not time inversion) in the inhomogeneous Lorentz group. A generalization of Zeeman's theorem was given a few years later by Borchers and Hegerfeldt (Borchers and Hegerfeldt, 1972a,b).

Consider an ordered space M with cardinality $\aleph_0$ such that M itself is a D-set. Denote by G the group of causal automorphisms of M which do not reverse the direction of time, and by $\check{M}$ the order-completion of M, and assume that $\check{M}$ is locally compact. It should be possible to extend the action of G on M to an action on $\check{M}$ by a procedure analogous to completion. Having done that, it should be possible to make G into a topological group using the compact-open topology induced by the action[2] of G on $\check{M}$. The topological space G will be completely regular,[3] i.e., it will be uniformizable, and G should therefore admit a uniform completion, which we shall denote by $\widetilde{G}$. It should, in the first instance, be possible to extend the group structure of G to $\widetilde{G}$, which should then

[1] The speculative parts aim to raise scientific questions, without attempting philosophical analyses which are beyond the present author's competence. The dangers inherent in exceeding this competence have been spelled out brilliantly in (Stebbing, 1944).

[2] See (Steenrod, 1972, pp. 19–20) for a definition of the compact-open topology. Some mild restrictions on the group G may be necessary.

[3] This is a famous theorem of A. Weil. See (Hewitt and Ross, 1963, p. 70).

make $\widetilde{G}$ into a group of causal automorphisms of $\check{M}$. If $\check{M}$ is not plagued by the cushion problem, then Zeeman's result suggests that, under certain conditions, the group $\widetilde{G}$ should be the inhomogeneous Lorentz group. The problem is to determine these conditions. This problem may be addressed with existing tools, and its solution will provide one further clue to understanding the 'unreasonable effectiveness of mathematics in the natural sciences'.

Narrowing the von Neumann–Heisenberg gap in measurement theory

Part II of this book is devoted to measurement theory, but our treatment is incomplete 'by definition', as it were, owing to the limited nature of our objectives. However, it may be completed to an account of what may be described as 'conservative measurement theory', as opposed to the 'orthodox measurement theory' of von Neumann and Wigner.[4] To do so, one has – in the opinion of the present author – to begin with a thorough discussion of the impossibility theorems of Shimony *et al.* Then one should examine how these impossibility theorems hold up if the Hilbert spaces of object and apparatus are assumed to be finite-dimensional, as Sewell does, with dim $\mathfrak{H}_{\mathsf{o}}$ being many orders of magnitude smaller than dim $\mathfrak{H}_{\mathsf{a}}$. Such a study may narrow, significantly, the gap that separates the Bohr–Heisenberg and the von Neumann–Wigner conceptions of measurement. This hypothesis – or hope – is based on the following observations. (i) The random errors of physical measurements may be attributed to thermodynamic fluctuations in finite systems. Except perhaps near $T = 0$ K, these effects will be very similar for classical and quantum systems. (ii) As a result, in most quantum-mechanical measurements made away from $T = 0$ K, errors of measurement will completely dominate the uncertainty principle bounds, so that the latter may be ignored.[5] (iii) Consequently, the use of finite-dimensional Hilbert spaces becomes permissible. It is worth repeating that most of the observed consequences of the superposition principle in the nonrelativistic region do not depend on the Hilbert space being finite-dimensional. Von Neumann was only concerned with problems that arose in infinite-dimensional Hilbert spaces.

Sewell has initiated a programme of analysing errors of measurement using the large deviation principle (Sewell, 2007, personal communication).

Sets, real numbers and physics

As we saw earlier, Wigner's article on 'the unreasonable effectiveness of mathematics in the natural sciences' was based on Cantor's view that mathematics is a free creation of the human mind, and Cantor's view was based on his discovery of the power-set construction (page 245). What concerns us here is

[4] The term *orthodox measurement theory* was used by Wigner (Wigner, 1963).

[5] This is a point that van Kampen has made repeatedly (van Kampen, 1954, 1962, 1988).

that $|\mathcal{P}(S)| > |S|$ unless $S = \emptyset$. Cantor's set theory also contains an undecidable proposition, the continuum hypothesis, and both the continuum hypothesis (page 246) and its negation are consistent with the ZFC axioms (page 249).

So much for the mathematics. In physics, if one keeps experimental limitations in mind, the notion of a geometrical point (as emphasized by Borchers) cannot be divorced from its neighbourhoods. The theorist talks to the experimentalist not in terms of the points that constitute a set S, but only in terms of *covers* of S by sets that are open, connected and small in some sense. The power-set construction destroys the possibility of replacing points by neighbourhoods to make contact with experimental physics, and therefore – or so the present author believes – is not relevant to physics. Physics (for the moment, at least) seems to need only two infinities, $\aleph_0$ and $\aleph$, and loses nothing by adopting the continuum hypothesis. In the opinion of the present author, these assumptions are entirely admissible at the time of writing (2009), and seriously weaken Wigner's 'unreasonable effectiveness' case.

BenDaniel has made a suggestion which is iconoclastic, both mathematically and physically. He proposes, effectively, to replace the real number system by a nonstandard enlargement of the rationals, which should suffice to describe experimental data and contain the infinitesimals,[6] which would accord with intuition. Derivatives can be defined, but not infinite sums. This is accomplished by a modification of the ZFC axioms for set theory, resulting in a theory which BenDaniel calls 'theory T'. Cantor's power-set is no longer relevant to theory T, and the problem of convergence or divergence of the renormalized perturbation series becomes an undecidable proposition. The present author is not persuaded by the physical arguments that underlie BenDaniel's suggestions, but the mathematics implied by his theory T appears to be worth pursuing independently of these arguments. For further details, the reader is referred to (BenDaniel, 1998a,b, 2006); of these, the article (BenDaniel, 1998a) is the most easily readable.

Interaction of quantum and classical systems

The results obtained in experiments on the Mössbauer effect or the inelastic scattering of neutrons by superfluid helium can be interpreted in two different ways: (i) The incident particle interacts with a small group of particles in the target, but the result can be approximated *as if* the incident particle was interacting with the target as a whole. (ii) The incident particle indeed interacts with the target as a whole. However, the experiments on the holographic manipulation of cold atomic beams that were carried out by Morinaga and collaborators[7] may be interpreted more easily as single microscopic objects interacting with a

[6] The term is used here in the technical sense of Abraham Robinson's nonstandard analysis. See (Robinson, 1966).

[7] See (Morinaga *et al.*, 1996). A review of neutron interferometry may be found in (Rauch and Werner, 2000).

macroscopic system *as a whole*.[8] This raises the following question: *What are the principles that govern the interaction of a microscopic system with a macroscopic quantum system as a single entity*? One obvious requirement is that the interaction should not be invariant under time reversal, but it is not clear to the present author whether this should be a guiding principle or a goal of the endeavour.

The macroscopic quantum system could well be an LED display that produces a digital readout of, say, the energy of a single spinless particle. How the experimentalist contrives this miracle may be regarded as a black box by the theorist. It should not matter from which angle, and at what distance, one reads the digital display; the latter should be a Euclidean scalar. A direct coupling of the readout to the particle can only be with the Euclidian scalars in its wave function $\psi(x) = \sum_i c_i u_i(x)$ (the u_i are energy eigenfunctions); the only Euclidean scalars here are the expansion coefficients c_i. These observations may be the starting-point for a detailed analysis.

One could also ask a slightly different question. *What are the principles that govern the interaction of a microscopic system with a classical system?*[9] One would like to know whether it is possible to define classical physical systems that *cannot* interact with quantum systems. An affirmative answer to this question will raise new questions about the nature of dark matter and the need for quantizing gravity. In this context, the reader's attention is drawn to the remark by Einstein quoted on page 234.

Superseparability and noninterferometry

The phenomenon of entanglement, which is one of the consequences of the superposition principle, has captivated popular imagination as Einstein's 'spooky action at a distance'. Limitations on the superposition principle in quantum mechanics may allow for possibilities that are, arguably, even spookier.

Suppose that a quantum-mechanical system can exist in vector states ψ_1 and ψ_2 that belong to inequivalent irreducible unitary representations π_1 and π_2, respectively, of the canonical commutation relations. Can any meaning be attached to the expression

$$c_1\psi_1 + c_2\psi_2$$

either mathematically or physically? The mathematical answer is trivially in the affirmative. It is always possible to form the direct sum $\mathfrak{H}$ of the Hilbert spaces $\mathfrak{H}_{1,2}$ that carry the representations π_1, π_2 respectively, and regard $\psi_{1,2}$ as vectors

[8] In the opinion of the present author, this description is roughly equivalent to the description of a single particle, or photon, travelling along two different paths at the same time.

[9] The interaction of quantum and classical systems has been investigated by Blanchard and Jadczyk, but from other points of view; see (Blanchard and Jadczyk, 1993, 1995).

in $\mathfrak{H}$. Then $\mathfrak{H}_{1,2}$ become subspaces of $\mathfrak{H}$, and $c_1\psi_1 + c_2\psi_2$ is formally defined. However, this formal process is physically meaningful only if there is a physically significant observable on $\mathfrak{H}$ such that

$$(\psi_1, O\psi_2) \neq 0$$

for some $\psi_1 \in \mathfrak{H}_1$, $\psi_2 \in \mathfrak{H}_2$. If there is none, then the subspaces $\mathfrak{H}_1$ and $\mathfrak{H}_2$ will be separated by superselection rules, and we shall say that vectors in $\mathfrak{H}_1$ and $\mathfrak{H}_2$ are *superseparated.* If the phenomenon of superseparability exists, then ψ_1 and ψ_2 *may have considerable spatial overlap* at any given time, but they will not be able to interfere with each other.

When inequivalent representations of the CCR are associated with equilibrium states at different temperatures or with different thermodynamic phases, the superposition $c_1\psi_1 + c_2\psi_2$ may not be meaningful, either physically or mathematically. But when a *microscopic* system can be prepared in states that belong to inequivalent representations, as in Reeh's example, the question suddenly takes on a different complexion: the existence or nonexistence of the phenomenon of superseparability may be testable experimentally.

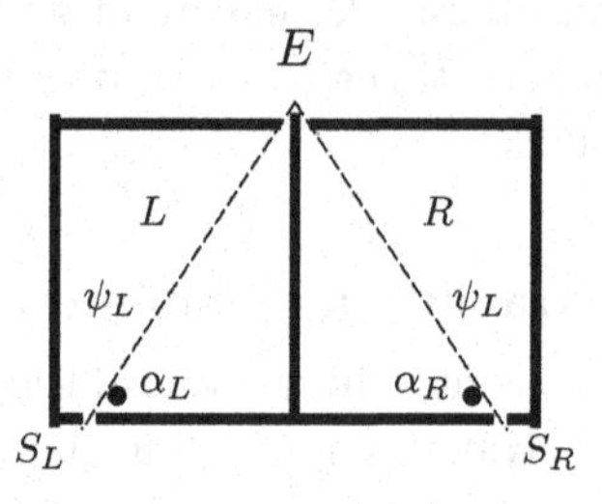

Fig. E.1. Scheme for a noninterferometer

Think of a double-slit interference experiment using a coherent beam of α-particles. If, as is generally assumed, all α-particles belong to the Schrödinger representation, one should expect to observe the familiar double-slit interference pattern. Now consider a modification of the experiment, as shown by the scheme of Fig. E.1. The slits are S_L and S_R, the beams ψ_L and ψ_R, and they meet at E. The difference is that, on the way to E, they pass through two chambers L and R which are isolated, electromagnetically, from each other and which contain adjustable magnetic flux lines of strengths α_1 and α_2, perpendicular to the plane of the paper. These fluxes change the representations of the beams ψ_L and ψ_R, and the geometry precludes the introduction of path differences within the beams. The experiment will consist of looking at interference – or its absence – at E, and how the effect changes as α_1 and α_2 are changed, assuming that the electromagnetic isolation of the chambers L and R is good enough to maintain the beams ψ_1 and ψ_2 in inequivalent representations. As the most startling effect will surely be the total absence of interference, it seems appropriate to describe the scheme of Fig. E.1 as a *noninterferometer.* If the required degree of electromagnetic isolation can indeed be achieved in the laboratory, the nonobservation of superseparability may be less dramatic, but more troubling.

Schmüdgen's operators (7.52) may be written as $x+p_y, p_x$. This suggests that controllable substitutions $p_x \to p_x + \mathrm{i}eA_x$, $p_y \to p_y + \mathrm{i}eA_y$ may result in a greater variety of inequivalent representations. The question seems to be worth investigating, both theoretically and experimentally.

Is nature law abiding?

Born's critique of the notion of precise measurement (page 108) applies equally to classical and quantum mechanics. In quantum mechanics, it applies not only to quantum-mechanical measurement theory as discussed in this book, but also to the measurement of parameters such as charges and masses that appear in current theories. It would make sense to ask how the structure of physical theories would be affected if one left a little wiggle room for physical parameters, meaning that in nature what we call physical parameters are *never* precisely defined, and that theories with precisely defined physical parameters are approximations (to what, one does not know). To keep the discussion from being overly general, let us consider only quantum field theories that have turned out to be useful, despite their well-known difficulties. The field theories that exhibit a sensitive dependence on one of their parameters are theories with zero-mass fields, like the photon or the neutrinos. If there is a *principle of indifference* that asserts that a physical theory constructed with ZFC mathematics should be indifferent to undetectable changes in its physical parameters, then zero-mass theories will fall afoul of this principle. But then an important class of zero-mass theories suffer from the peculiar problem known as the solar neutrino problem. This problem may disappear if one ceases to insist that the neutrinos have rest-masses which are precisely zero. The suggestion is not that the neutrinos have very small but different masses, but that *the notion of precise rest-mass values* is not a sensible one. Nature is not law abiding, in the sense in which we presently understand the phrase 'laws of nature'.

The principle of indifference may be relevant to the problem of the divergent perturbation series of QED, and also to the problem of quantum gravity. The physical argument for quantizing the gravitational field is that interaction with an unquantized gravitational field may lead to violations of the uncertainty principle. But if these violations are undetectable, would one be able to tell whether or not the gravitational field is quantized? Einstein's obstinate refusal to accept quantum mechanics as a satisfactory 'description of reality' is well known, but the following quotation, from a letter to Max Born dated 1 June 1948, seems (at least to the present author) to shed a different light upon it (Einstein, 1971, item 91, p. 178):

> I should just like to add that I am by no means mad about the so-called classical system, but I do consider it necessary to do justice to the principle of general relativity in some way or other, for its heuristic quality is indispensible to real progress.

Perhaps Einstein would have been less opposed to quantum mechanics if he could be persuaded that gravity need not be quantized.

It seems to the present author that something like the principle of indifference is a *sine qua non* for bridging the gap between theoretical and experimental physics, but at the moment one can only speculate on how to express such a principle *mathematically*. One could try a formulation of mathematics that is not based on the ZFC axiomatization of set theory, but has mathematics based on ZFC set theory either as an approximation, or as a singular limit (such as $c \to \infty$ or $\hbar \to 0$). The interested reader may begin by looking at the monograph (MacLane and Moerdijk, 1994) on topos theory, which was initiated by Grothendieck.[10] As a counterpoint, we may mention an article by Cartier on 'the evolution of concepts of space and symmetry', which also starts with work by Grothendieck, but stays within the ZFC axioms (Cartier, 2001).

It will take a mighty heave to cause Dirac's 'mathematical quality in nature' to wobble just a tiny bit.

[10] I seem to recollect that C J Isham and his collaborators have used topos theory in connection with quantum gravity, but I am unable to cite exact references.

Mathematical appendices

A1
Sets and mappings

We begin our discussion of sets with the remark that *we are taking the notion of positive integers as given.* This is probably the approach of the average mathematician; logicians may prefer to consider sets as primitive and the notion of integers as derived from sets. Either way, the subsequent development of mathematics proper will not be affected.

A1.1 Sets

In mathematics, a **set** S is a collection of distinct objects considered as a whole, subject to some restrictions that will be discussed briefly in Section A1.8. The objects constituting a set are called **elements, members** or **points** of the set. Sets are usually denoted by writing their members within curly brackets. This writing can take two forms. If the elements can be enumerated *and* have standard names, one can list them explicitly. Thus, $\{1, 2, 5\}$ is a set that has three elements, the integers $1, 2$, and 5. If the number of elements is large, this may be impractical (although theoretically possible), for example if one wants to define the set of all positive integers less than $N_A = 6.022 \times 10^{23}$ (Avogadro's number). In such cases, one defines the set by the *rule* itself:

$$S = \{n \,|\, n \text{ a positive integer}, n > N_A\}. \tag{A1.1}$$

In the above, it is the right-hand side that defines the set; the set of all n that satisfy the conditions to the right of the vertical bar $|$.[1] S on the left is a name (or symbol) that has been assigned to the set. It would have been better to use the assignment operator := (of some computer languages) instead of the equality sign, but it is the equality sign that is generally used in mathematics, and we shall stay with this practice.

One can extend this method to define sets that *cannot* be enumerated, such as

$$S = \{x \,|\, x \text{ is a real number}, x^2 < 2\}. \tag{A1.2}$$

This is the set of real numbers larger than $-\sqrt{2}$ and smaller than $\sqrt{2}$. The fact that this set cannot be enumerated will be proven in Appendix A2.

[1] The vertical bar $|$ may be replaced by : or ;. All three notations are in current use.

If x is a member of the set S, we write:

$$x \in S. \tag{A1.3}$$

Correspondingly, $x \notin S$ means that x is not a member of S.

The notation $S = T$ means that the sets S and T have the same elements, i.e.,

$$x \in S \Leftrightarrow x \in T.$$

The symbol $\Leftrightarrow$ reads: is equivalent to, or if and only if. The latter was abbreviated by Halmos to **iff**; this abbreviation is now in common use.

It is convenient to define a set that has no elements at all. This set is called, appropriately, the **empty set** and is denoted by $\emptyset$ (a zero with a slash through it).

A1.2 New sets from old

We shall now describe a few procedures for fabricating new sets from old. Other procedures will be described in Sections A1.5 and A1.6.

A1.2.1 Union, intersection and difference

Addition and multiplication are two of the basic rules in algebra for generating new entities from old. Similarly, there are two basic operations for generating new sets from old.[2] They are the **union**

$$A \cup B = \{x \,|\, x \in A \text{ or } x \in B\} \tag{A1.4}$$

and the **intersection**

$$A \cap B = \{x \,|\, x \in A \text{ and } x \in B\}. \tag{A1.5}$$

Thus, if $A = \{1,2,3,4\}$, $B = \{3,4,5,6\}$ and $C = \{5,6\}$, then $A \cup B = \{1,2,3,4,5,6\}$, $A \cap B = \{3,4\}$ and $A \cap C = \emptyset$. Note that $A \cup B = B \cup A$ and $A \cap B = B \cap A$. Two sets X and Y are called **disjoint** if $X \cap Y = \emptyset$.

Union and intersection obey the following **distributive laws**:

$$\begin{aligned} A \cap (B \cup C) &= (A \cap B) \cup (A \cap C); \\ A \cup (B \cap C) &= (A \cup B) \cap (A \cup C). \end{aligned} \tag{A1.6}$$

The proofs consist of verifying – using plain words – that if $x \in L$ then $x \in R$, and vice versa, where L and R are the left and right sides in a given line of (A1.6).

[2] Indeed, in earlier works on set theory, e.g., the book by Hausdorff (English translation, (Hausdorff, 1957)) the union and intersection of A and B were often denoted by $\boldsymbol{A} + \boldsymbol{B}$ and $\boldsymbol{AB}$. The notation we are using is nearly universal by now.

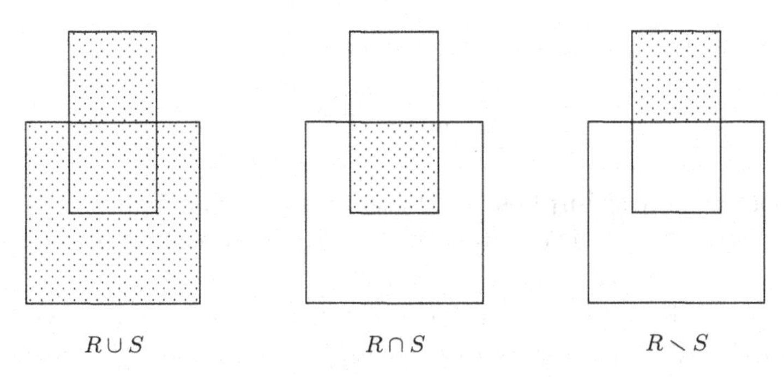

Fig. A1.1. Union, intersection, set-theoretic difference

A further operation that is often used is the **set-theoretic difference**,

$$A \setminus B = \{x \mid a \in A, x \notin B\}. \tag{A1.7}$$

B is not required to be a subset of A. With A and B as in the examples following (A1.5), $A \setminus B = \{1,2\}$ and $B \setminus A = \{5,6\}$. The quantity $A \setminus B$, also called the **complement** of B in A, is written by some authors, including Munkres (Munkres, 1975), as $A - B$, with an ordinary minus sign. The union, intersection and the set-theoretic difference of two sets – a rectangle R and a square S – are shown in Fig. A1.1.

The set-theoretic difference satisfies the following rules, which are special cases of **de Morgan's laws**:

$$\begin{aligned} A \setminus (B \cup C) &= (A \setminus B) \cap (A \setminus C); \\ A \setminus (B \cap C) &= (A \setminus B) \cup (A \setminus C). \end{aligned} \tag{A1.8}$$

The proofs of these are similar to those of the distributive laws.

De Morgan's laws are often verbalized as follows: (a) the complement of a union is the intersection of the complements; b) the complement of an intersection is the union of the complements. In this verbalization the union and intersection can be of more than two sets, but the set A must be understood as given; otherwise the notion of complement itself will not make sense.

A1.2.2 Subsets

From any given set S, one can construct new sets by taking only *some* of the elements of S. Such sets are called **subsets** of S. If T is a subset of S, one writes $T \subset S$ (equivalently, $S \supset T$; the symbols $\subset$ and $\supset$ are modifications of $<$ and $>$).

The formal definition of a subset is as follows:

$$T \subset S \text{ if and only if } x \in T \Longrightarrow x \in S. \tag{A1.9}$$

The symbol $\Rightarrow$ is read **implies**. Note that this definition implies $S \subset S$; every set is a subset of itself. The statement $T \supset S$ is also read as: T is a **superset** of S.

Let us go back, for a moment, to the definition of a subset (A1.9). What happens if $T = \emptyset$? Then the condition $x \in T \Rightarrow x \in S$ is surely satisfied; there is no x for which it is false, because there is no x in T! One says that if $T = \emptyset$, the condition (A1.9) is *vacuously satisfied.* That is, *the empty set is a subset of every set.*

The subsets S and $\emptyset$ of S are called the **improper** subsets of S; every other subset of S is called **proper**. Note that, if $T \subset S$, then $T \setminus S = \emptyset$.

In many contexts, one restricts oneself to subsets of a fixed set X. In these circumstances, the set $A' = X \setminus A$ is known as the **complement**[3] of A in X.

A1.2.3 Boolean algebras

Definition A1.1 A **Boolean algebra** of sets is a pair $(X, \mathcal{B})$ where X is a set and $\mathcal{B}$ a family of subsets of X such that:

(a) $X \in \mathcal{B}$.
(b) If $A, B \in \mathcal{B}$, then $A \cup B \in \mathcal{B}$ and $A \cap B \in \mathcal{B}$.
(c) If $B \in \mathcal{B}$, then $B' \in \mathcal{B}$.

It follows that $\emptyset \in \mathcal{B}$. One says that a Boolean algebra is closed under unions, intersections and complementation. Sets obtained by performing these operations a *finite* number of times upon members of $\mathcal{B}$ will be members of $\mathcal{B}$. The set of all subsets of X is a Boolean algebra.

A1.3 Maps

Let S and T be two sets (the case $S = T$ is included). A **map** (**mapping, function**) f from S to T, written

$$f : S \longrightarrow T, \tag{A1.10}$$

is a rule which assigns, to each point $x \in S$, a *unique* point $y \in T$. One writes $f(x) = y$. The set S is called the **domain** of f, and T the **codomain**. Note the insistence upon the uniqueness of $f(x)$ – functions are single-valued, by

[3] The complement of A is often denoted by A^c in the literature, but we shall adhere to A'.

definition.[4] S is called the **domain** of f. The *subset* $f(S)$ of T defined by[5]

$$f(S) = \{y \mid y \in T, y = f(x) \text{ for some } x \in S\}$$

is called the **range** of f, or the **image** of S under x. For any $V \subset T$, the set

$$f^{-1}[V] = \{x \mid f(x) \in V\} \tag{A1.11}$$

is called the **inverse image** of V. The inverse image is a subset of the domain, and not a function. Clearly, $f(S) \cap W = \emptyset \Leftrightarrow f^{-1}[W] = \emptyset$. A pictorial representation of these definitions is given in Fig. A1.2.

One often wishes to specify the image of a particular point in the domain S under the map (A1.10), and it is useful to have a notation for this. The notation is

$$x \overset{f}{\longmapsto} y, \tag{A1.12}$$

and it is read as follows: f maps x to y.

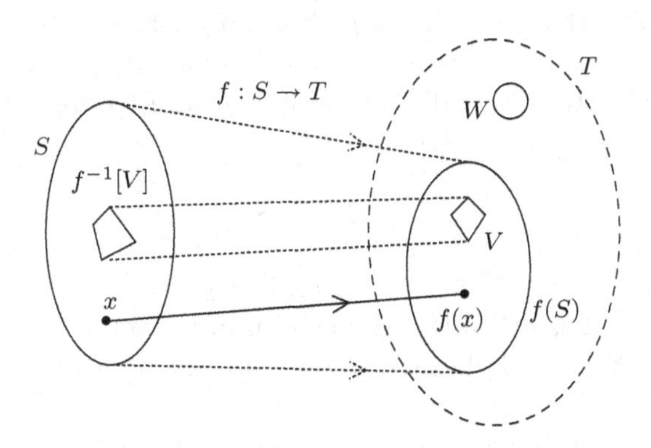

Fig. A1.2. Sets associated with maps

Let X, Y, Z be sets, and $\varphi : X \to Y$, $\psi : Y \to Z$ be maps. One can clearly define a map $\psi \circ \varphi : X \to Z$ (Fig. A1.3)

$$X \overset{\varphi}{\longrightarrow} Y \overset{\psi}{\longrightarrow} Z$$

[4] Till the mid-nineteenth century certain functions – like $\log z$ – were called multiple-valued. However, this underwent a sea-change with the appearance of Riemann's doctoral dissertation in 1851. One no longer talks about multiple-valued functions, and students are introduced quite early to the notion of the Riemann surface of a function. See, for example, the text (Marsden, 1973).

[5] See Notations A1.2 on page 242.

by the formula

$$(\psi \circ \varphi)(x) = \psi(\varphi(x)).$$

This map is called the **composition** of φ and ψ, and is denoted $\psi \circ \varphi$. In Fig. A1.3, $\varphi(X)$ has been denoted by S and $\psi(S) = \psi \circ \varphi(X)$ by T.

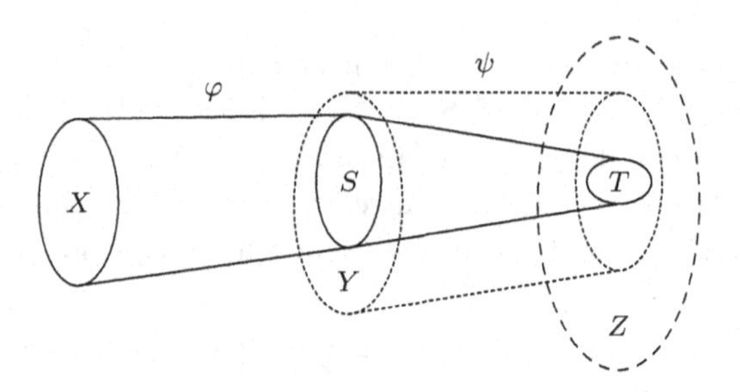

Fig. A1.3. Composition of maps

If $x \neq x'$ implies that $f(x) \neq f(x')$, the map $f : S \to T$ is called **injective**, or **one-to-one**. If $f(S) = T$, then f is called **surjective**, or **onto**. If f is both injective and surjective, it is called **bijective**, or **one-to-one onto**. A bijective map can be inverted. The inverse map

$$g : T \longrightarrow S$$

fulfils the conditions $g(f(x)) = x$ and $f(g(y)) = y$.

A map $\varphi : X \to X$ is called the **identity map** if $\varphi(x) = x$ for all $x \in X$. The identity map of X is often denoted by id_X. In this notation, $g \circ f = \mathrm{id}_S$ and $f \circ g = \mathrm{id}_T$.

We may now give a simple example to illustrate the problems that would arise if the elements of a set are not required to be distinct. Let $A = \{a, b, c\}$, $B = \{1, 2, 3\}$ and $f : A \to B$ the map $f(a) = 1, f(b) = 2, f(c) = 3$. Now set $b = a$. We then obtain $f(a) = 1$ and $f(a) = 2$, i.e., $1 = 2$.

Since electrons are indistinguishable, one cannot talk about a *set* of n electrons in any mathematical context; one can, however, talk about a *system* of n electrons.

Notations A1.2 The image of the point x under the map f is usually written as $f(x)$. The brackets are logically unnecessary, and one could write the same more economically as fx, as is done by quite a few authors. We shall use the notation $f(x)$, which is sanctioned by tradition. The image of a set S under f will be written as $f(S)$. There is no risk of ambiguity here, as sets are denoted by capital letters and elements of sets by small letters. The notation f^{-1} is ambiguous. It is used both to denote inverse images (which are not maps) as in (A1.11), and inverse maps (when the map is invertible). So far we have enclosed the argument of an inverse image in square brackets, as in (A1.11), but we shall revert to the more usual $f^{-1}(V)$ in Appendix A5.

A1.4 Finite and infinite sets

Two sets A and B are called **equivalent**[6] to each other, written $A \sim B$, if they can be mapped bijectively onto each other. If $A \sim B$ and $B \sim C$, then $A \sim C$; sets divide themselves naturally into **equivalence classes**, and every set is equivalent to itself.

A set that contains a finite number of elements has the property that it is not equivalent to any proper subset of itself; the reader is invited to verify this for the three-element set $\{a, b, c\}$. This property can be turned into the formal definition of finite sets:

Definition A1.3 (Finite sets) A set that is equivalent to no proper subset of itself is called a **finite set**.

With each finite set F is associated a unique nonnegative intger n, called the **number of elements** of F. Sets with the same value of n are equivalent to each other; sets with different values of n are not. For example, the proper subsets of $\{a, b, c\}$ are $\{a, b\}, \{b, c\}, \{c, a\}, \{a\}, \{b\}$ and $\{c\}$; none of them is equivalent to $\{a, b, c\}$. These facts may seem to be intuitively obvious, but to the mathematician they need proof. The proofs are simple, but not trivial. The interested reader may consult the book by Munkres (1975, pp. 39–45).

The empty set is finite; as it has no proper subset, it fulfils the finiteness condition trivially. The reader is invited to write this out in full, to appreciate that the word trivial is being used here in a mathematical sense which is different from the dismissive sense in which the word is used in daily life (and, unfortunately, also sometimes in mathematics).

Definition A1.4 (Infinite sets) A set is called **infinite** if it is equivalent to a proper subset of itself.

The set of nonnegative integers is denoted by $\mathbb{N}$, and the set of positive integers by $\mathbf{N}$ (for historical reasons).[7] The set $\mathbb{N}$ is infinite, because it is equivalent to the set $2\mathbb{N} = \{0, 2, 4, \ldots\}$ of nonnegative even integers, which is a proper subset of $\mathbb{N}$. The map $f : \mathbb{N} \to 2\mathbb{N}$, defined by the formula $f(n) = 2n$, proves the assertion.

[6] Other terms in use are **equipollent** and **equipotent**. The term equivalent is used by Fraenkel (Fraenkel, 1953), and we have stayed with it because we shall often refer the reader to this book, which is far more gripping than most textbooks.

[7] The positive integers are called **natural numbers**, after Peano, who also introduced the notation $\mathbf{N}$.

The unit interval $[0, 2]$ is infinite, because it is equivalent to the proper subinterval $[0, 1]$, as shown by the map $f(x) = 2x$.

A set is called **countable**,[8] **countably infinite** or **denumerable** if it is equivalent to $\mathbb{N}$. The set $\mathbb{N}$ itself is countable by definition. The map $f : \mathbb{N} \to \mathbf{N}$ defined by $f(n) = n + 1$ exhibits that $\mathbf{N}$ is countable. The set of integers $\mathbb{Z}$ is countable; the map $f : \mathbb{Z} \to \mathbb{N}$ defined by

$$f(n) = \begin{cases} 2n & \text{if } n \geq 0, \\ -2n - 1 & \text{if } n < 0 \end{cases}$$

is a bijection. An infinite set that is not countable is called **uncountable**.

The following result, which also seems intuitively obvious but requires proof (which we shall omit), is basic.

Theorem A1.5 *A subset of a countable set is either finite or countable.*

Next, recall that a rational number is a number of the form p/q, where p and q are integers, and $q \neq 0$. It is not necessary for p and q to be relatively prime.[9] However, when one talks about the *set* of rational numbers, one has to assume that p and q are relatively prime, to avoid repetitions. The following result may come as a surprise.

Theorem A1.6 *The set of rational numbers is countable.*

Proof First consider the positive numbers of the form p/q, where p and q are positive integers. Set $\Sigma = p + q$. Then $\Sigma \geq 2$. The only rational with $\Sigma = 2$ is the integer 1. There are exactly $n - 1$ rationals with $\Sigma = n$ $(n \geq 2)$, namely

$$\frac{1}{n-1}, \frac{2}{n-2}, \ldots, \frac{n-1}{1},$$

and they are all distinct. They can therefore be collected together in a set. Denote this set by S_n, and its elements by $a_{n,k}$, $k = 1, 2, \ldots, n - 1$. Now consider the set of *symbols* $T = \{a_{n,k} | n = 2, 3, \ldots, k = 1, 2, \ldots, n - 1\}$. This set is arranged in dictionary order, and is clearly countable.

Every number in T appears infinitely often in it; the number j/k reappears as nj/nk in S_{nj+nk} for every integer $n > 1$. *As symbols*, the objects j/k and nj/nk are different; as numbers, they are the same. As an aggregate of symbols, T is a countable set; as an aggregate of numbers, it is not a set. However, one can obtain a set of numbers from T as follows.

[8] Some authors, including Munkres (Munkres, 1975), use the term countable to denote either a finite or a countably infinite set. We shall generally use the term to mean **countably infinite**, but shall be more explicit if there is any risk of ambiguity.

[9] Two integers are said to be **relatively prime** if they have no common factor except unity.

Define recursively[10]

$$S_2^\circ = S_2, \\ S_n^\circ = S_n \smallsetminus S_{n-1}^\circ, \quad n = 3, 4, \ldots \tag{A1.13}$$

The aggregate $T^\circ = \{a_{n,k} | a_{n,k} \in S_n^\circ\}$ is the set of all positive rationals, as well as a set of symbols. As a set of symbols, it is a subset of T. Therefore, by Theorem A1.5, it is either finite or countable. As it contains every nonnegative integer, it cannot be finite, and is therefore countable.

This proves that the set of nonnegative rationals is countable. To include the negative rationals, all one needs is an easy modification of the proof of the result that $\mathbb{Z}$ itself is countable. □

The set of all rationals is denoted by $\mathbb{Q}$.

A1.5 Power sets

It is an elementary combinatorial result that the set of all subsets (including the empty set) of a set of n objects contains exactly 2^n objects; one has the inequality $2^n > n$ for all $n \in \mathbb{N}$.

The set of all subsets of S is called the **power set** of S, and is denoted by $\mathcal{P}(S)$. If $S = \{a_1, a_2, \ldots, a_n\}$, then $\mathcal{P}(S)$ certainly contains the one-point subsets $\{a_1\}, \{a_2\}, \ldots, \{a_n\}$ of S, so that there exists an injective map $a_k \mapsto \{a_k\}$ of S into $\mathcal{P}(S)$. Simple counting shows that there can be no injective map of $\mathcal{P}(S)$ into S.

Except for 'simple counting', the above considerations generalize readily to infinite sets.

The calculus of sets exemplified by the distributive laws (A1.6) and de Morgan's rules (A1.8) may easily be extended to infinite collections of sets. A good grounding in this calculus is essential for every student of mathematics, and therefore every textbook on topology or measure theory begins with this subject.

A1.5.1 Cardinal numbers

The natural numbers can be used for counting, as well as for ordering. When used for counting, they are called **cardinal numbers**: one, two, three, etc. When used for ordering, they are called **ordinal numbers**: first, second, third, etc. The concepts of counting and ordering generalize to infinite sets, but it is not obvious that there is a generalization of the natural numbers that can perform

[10] As the definition of S_n° depends upon that of S_{n-1}°, one cannot define all the S_n° at one stroke, but must proceed one step at a time. Such definitions are called *recursive* or *inductive*.

both tasks. Accordingly, in the transfinite domain, one speaks of **cardinals** and **ordinals**. The cardinal, or **cardinality** (also called **power**, for historical reasons; not to be confused with the *power set*, defined earlier) of the set S is often denoted by $|S|$; we shall use this notation.

The cardinality of a set has the following properties.

Let S and T be two sets. If there exists a bijective map of S into T, the sets S and T have the same cardinality. Cardinality is a total order on the *equivalence classes* (see page 243) of sets. If there exists an injective map of S into T, but *none* of T into S, then the cardinality of T is greater than the cardinality of S. For a finite set, cardinality is just the the number of elements the set has.

The cardinality of the set of nonnegative integers $\mathbb{N}$ is denoted by $\aleph_0$ (pronounced *aleph-null*); $\aleph$, pronounced *aleph*, is the first letter of the Hebrew alphabet. Clearly, $|\mathbb{N}| = |\mathbb{Z}| = |\mathbb{Q}| = \aleph_0$. The cardinality of the continuum $\mathbb{R}$ is denoted by $\aleph$. For any nonempty set S, the cardinality of $\mathcal{P}(S)$ is greater than that of S. (The power set of a set that has only one element is a set with two elements; the other element is the empty set.) One writes $|\mathcal{P}(S)| > |S|$. The fact that a real number admits at least one and at most two decimal representations can be used to show, quite easily, that $|\mathcal{P}(\mathbb{N})| = \aleph$ (see (Hausdorff, 1957, pp. 41–45)). Therefore $\aleph > \aleph_0$. Note that some authors, including Willard, use $\mathfrak{c}$ rather than $\aleph$ for the cardinality of the continuum (Willard, 1970). We are following (Fraenkel, 1953) and (Hausdorff, 1957).

We shall need one more result from the arithmetic of cardinals, which is generally written $\aleph_0 \cdot \aleph_0 = \aleph_0$. It is the multiplicative form of the statement that the union of countably many countable sets is itself countable. The proof of this assertion is rather easy (i.e., it requires no mathematical tools), and the reader is invited to supply it.

Is there an infinite set with cardinality **s** such that $\aleph > s > \aleph_0$? In words, is there a set that is intermediate in size between the rationals and the reals? The hypothesis that *there is no such set* is known as **the continuum hypothesis**. Cantor tried for years to prove it, but failed.[11]

The reader who wishes to supplement our minimal account of this fascinating subject is referred to (Fraenkel, 1953) for a very readable account, with a wealth of historical detail.

The only transfinite ordinal we shall ever need has appeared in disguise, as the *minimal uncountable well-ordered set*. It was defined in Section 4.2, the only place where it is used.

A1.6 Products of sets

Let X, Y be two sets. An **ordered pair** (x, y) consists of two elements, one from X and one from Y, with the one from X being written to the left of the one

[11] Why he failed will be explained at the end of this appendix.

from Y. The **Cartesian product** $X \times Y$ of two sets X and Y is defined to be the set of ordered pairs

$$X \times Y = \{(x, y) | x \in X, y \in Y\}.$$

The ordering requirement allows this definition to be extended to the product of n sets; for example, $X_1 \times X_2 \times X_3$ is defined as $(X_1 \times (X_2 \times X_3))$ or equivalently as $((X_1 \times X_2) \times X_3)$.

To define the product of an arbitrary family of sets, we first reformulate the above procedure. Let $\{1, 2 \ldots, n\}$ be the set of the first n positive integers, and define

$$X = X_1 \cup X_2 \cup \ldots \cup X_n. \tag{A1.14}$$

Then choose a map f

$$f : \{1, 2, \ldots, n\} \to X \tag{A1.15}$$

that satisfies the condition

$$f(k) \in X_k \text{ for } k \in \{1, 2, \ldots, n\}. \tag{A1.16}$$

Set now

$$x_k = f(k). \tag{A1.17}$$

We may then write

$$(x_1, x_2, \ldots, x_n) = f(\{1, 2, \ldots, n\}); \tag{A1.18}$$

the n-tuple $(x_1, x_2, \ldots, x_n)$ is the image of the indexing set $\{1, 2, \ldots, n\}$ under the chosen map f. Clearly, the product $X_1 \times X_2 \times \cdots \times X_n$ is the set of *all* such maps f. The product is the empty set if any of the factor sets is empty.

Note that condition (A1.16) is critical; it ensures that there exist maps

$$\pi_j : X \to X_j, \ j = 1, 2, \ldots, n \tag{A1.19}$$

such that

$$\pi_j(x_1, x_2, \ldots, x_n) = x_j, \quad j = 1, 2, \ldots, n. \tag{A1.20}$$

These maps are called **projections** of X onto the factors X_j.

We are now ready to formulate the general definition. Let A be a set, $\alpha \in A$ and $\{X_\alpha\}_{\alpha \in A}$ an indexed family of sets, indexed by $\alpha \in A$. Define

$$X = \bigcup_{\alpha \in A} X_\alpha, \tag{A1.21}$$

and *assume that there exists a map*

$$f : A \to X \tag{A1.22}$$

that satisfies the conditions

$$f(\alpha) \in X_\alpha \text{ for all } \alpha \in A. \tag{A1.23}$$

The **Cartesian product** $\prod_{\alpha \in A} X_\alpha$ of the X_α is defined to be the set of *all* such maps f. The **projections** π_β are defined by

$$\pi_\beta \left(\prod_{\alpha \in A} X_\alpha \right) = x_\beta, \tag{A1.24}$$

where

$$x_\beta = f(\beta) \in X_\beta. \tag{A1.25}$$

If $X_\alpha = Y$ for all $\alpha \in A$, one writes (A1.21) as

$$X = Y^J, \tag{A1.26}$$

where J is the cardinality of the indexing set A. If A is finite, then J is a nonnegative integer.

A1.7 The axiom of choice and the well-ordering theorem

Suppose that we have n nonempty sets X_k. The existence of n-tuples, i.e., of sets $(x_1, x_2, \dots x_n)$ that contain *exactly one* element from each of the sets X_k, seems obvious; at any rate, it has not been seriously challenged. If, however, the set A is of arbitrarily large cardinality, then the question of existence of maps (A1.22) *satisfying conditions* (A1.23) has, historically, been more divisive.

The assumption that such maps exist is far-reaching, and is known as the **axiom of choice**. It is equivalent to several classical results, such as **Zorn's lemma**, the **well-ordering theorem** and the **Hausdorff maximal principle**, terms that the reader may have come across. We shall define the concept of well-ordering and state the well-ordering theorem.

Definition A1.7 (Well-ordered set) An ordered set is called *well-ordered* if every nonempty subset of it has a first element.

Theorem A1.8 (Well-ordering theorem) *Every set can be well-ordered.*

However, a constructive procedure for well-ordering an arbitrary set has not yet been found.

A1.8 Russell's paradox

Suppose that it is possible to tell uneqivocally whether or not an object is a physicist. Then one may talk about the set of all physicists. This set is clearly not a physicist; therefore it is not a member of itself.

Consider now the set of all nonphysicists. This set is also not a physicist; but, that being the case, it is necessarily a member of itself. That is, a set can (against common sense) be a member of itself.

Finally, consider the set R of *all* sets that are not members of themselves. If R is not a member of itself, then it belongs to the set of all sets that are not members of themselves, i.e., to R; in brief, if $R \notin R$ then $R \in R$. Conversely, if R is a member of itself, then it does not belong to the set of all sets that are not members of themself, i.e., $R \in R$ implies that $R \notin R$. We have an inescapable contradiction, which is known as **Russell's paradox**.

The contradiction arises from the extreme generality of the notion of a set, which finds its expression through the word *all* in the paragraphs above. For application to mathematics, some restrictions would seem to be required. For those who do not wish to delve into these questions, the following restriction would suffice: *A set* cannot *be a member of itself.* This would ensure that the collection of all physicists is a set, but the collection of all nonphysicists is not.

The restrictions generally accepted by mainstream mathematics today are somewhat more subtle, and are known as the **Zermelo–Fraenkel axioms** (ZF, for short) for set theory; see (Fraenkel, Bar-Hillel and Levy, 2001). In this book, ZF is described as 'a modified form of Zermelo's system' (*ibid.*, p. 18).

It has been shown that the axiom of choice is independent of the ZF axioms. The ZF axioms together with the axiom of choice are denoted by ZFC.

We had mentioned the continuum hypothesis earlier. It has turned out that this hypothesis is independent of the ZFC axioms; both the continuum hypothesis and its negation are consistent with ZFC. The consistency (with ZFC) of the continuum hypothesis was proved by Kurt Gödel in 1940; the consistency of its negation was proven by Paul Cohn in 1963.

Terminology A1.9 Quite often one wants to use groupings of objects, but is indifferent to whether or not these groupings form sets in the sense of ZF. In this case one uses the terms **class, collection, aggregate** or **family** to describe them, and the curly brackets notation to write them. This causes no confusion.

A2

The real number system

A2.1 Irrational numbers

Legend has it that a member of the school of Pythagoras proved that $\sqrt{2}$ was irrational, which so upset Pythagoras that he had the discoverer drowned. One wonders how he would have reacted had he been told that square root $\sqrt{m/n}$ is irrational unless m and n are both perfect squares.The proof of this fact is simple. We may assume, without loss of generality, that m and n are relatively prime. Let $\sqrt{m/n} = p/q$, where p and q are also relatively prime. Then $p^2/q^2 = m/n$, or $np^2 = mq^2$. If p and q are relatively prime then so are p^2 and q^2, from which it follows that m must divide p^2, or $m = \lambda p^2$. Substituting, we find that $n = \lambda q^2$. Since m and n are relatively prime, we must have $\lambda = 1$. That is, $m = p^2$ and $n = q^2$.

The number $\sqrt{2}$ is an **algebraic** number; it is a solution of the algebraic equation $x^2 - 2 = 0$. An equation

$$a_0 x^n + a_1 x^{n-1} + a_2 x^{n-2} + \cdots + a_{n-1} x + a_n = 0 \tag{A2.1}$$

with *integer coefficients* a_k, $k = 0, 1, \ldots n$ and $a_0 \neq 0$ is called an **algebraic equation**. The highest power of x on the right-hand side of (A2.1), namely n, is called the **degree** of the algebraic equation.[1] In 1874, Cantor proved the then-surprising result that the set of all algebraic numbers is countable.

According to the fundamental theorem of algebra, a polynomial of degree n with complex coefficients has exactly n roots; these roots may be real or complex, and two or more roots may be equal. It follows that the equation (A2.1) has at most n real solutions. If one could *enumerate* the algebraic equations, Cantor's result would follow. Clearly, there are infinitely many algebraic equations of any given degree n, which makes it impossible to use the degree to enumerate them. Cantor achieved the required enumeration by using the concept of **height** of an algebraic equation, which he defined as follows:

$$h = (n-1) + |a_0| + |a_1| + \cdots + |a_n|. \tag{A2.2}$$

[1] It is also called the degree of the **polynomial** $p_n(x) = b_0 x^n + b_1 x^{n-1} + \cdots + b_n$, where $b_0 \neq 0$ and the b_k are real or complex numbers.

The height is a nonnegative integer, and it does not require much effort to see that there are only a finite number of algebraic equations of height h. It follows that the set of algebraic equations is countable; the countability of the set of algebraic numbers is an immediate consequence of this fact. For details see, for example, (Fraenkel, 1953, pp. 53–57).[2]

Let us temporarily denote the set of real algebraic numbers by $\mathbb{A}$. Then $\mathbb{A} \supset \mathbb{Q}$, where the inclusion is proper, so that the nonempty set $\mathbb{R} \setminus \mathbb{A}$ of non-algebraic numbers consists entirely of irrationals. A real number which is not algebraic is called **transcendental**. The best-known transcendentals[3] are π and e. How numerous are the transcendentals? In the same 1874 paper, Cantor proved the earth-shaking result that the transcendentals 'outnumber' the algebraic numbers.[4]

The proof of theorem A2.1 given below was published by Cantor in 1892. It is based upon the decimal representation of real numbers. We begin by recapitulating the key features of this representation. It will suffice to consider the numbers in the closed interval $[0, 1]$.

Let $b \in [0, 1]$. Then b can be written as

$$b = \frac{b_1}{10} + \frac{b_2}{10^2} + \cdots + \frac{b_n}{10^n} + \cdots, \tag{A2.3}$$

where the b_n are integers from 0 to 9, and n runs over the positive integers. The representation

$$b = 0.b_1 b_2 \ldots b_n \ldots \tag{A2.4}$$

is called a *decimal representation* of the number b.

The decimal representation is not always unique. If $b_k < 9$, then

$$0.b_1 b_2 \ldots b_k \overline{9} \;=\; 0.b_1 b_2 \ldots (b_k + 1)\overline{0}, \tag{A2.5}$$

where the bar over a digit (0 and 9 in the above) or group of digits indicates *recurrence*, i.e., the digit or group of digits is repeated indefinitely to the right. The left-hand side of (A2.5) is called an *infinite* decimal, and the right-hand side a *terminating* decimal.[5] With the single exception of 0, every rational number has a unique infinite decimal representation ($1 = 0.\overline{9}$).

[2] Fraenkel calls h the **amount** of the equation.

[3] The transcendence of e was proven by Hermite in 1873, and that of π by Lindemann in 1884; Lindemann's result established the impossibility of squaring the circle.

[4] This may be seen clearly by using the notion of Lebesgue measure, which is discussed in Appendix A5. The Lebesgue measure of a countable set is zero, and therefore that of the transcendentals in $[0, 1]$ is unity.

[5] Recall that rational numbers p/q such that q has only 2 and 5 as factors is a terminating decimal; all other rationals are represented by recurrent decimals.

In the following, we shall restrict the term *decimal representation* of a number other than zero to mean its *infinite* decimal representation, unless the contrary is explicitly specified.

Theorem A2.1 *The set of real numbers $\mathbb{R}$ is uncountable.*

Proof It will suffice to prove the result for the open interval $(0,1)$. The proof is by contradiction.

Assume that the reals in the interval $(0,1)$ have been enumerated, i.e., arranged in a sequence indexed by $n \in \mathbf{N}$:

$$\{(0,1)\} = \{a_1, a_2, a_3, a_4 \ldots\}$$

We write these numbers in an array

$$\begin{array}{rcl} a_1 & = & 0.a_{11}a_{12}a_{13}a_{14}\ldots \\ a_2 & = & 0.a_{21}a_{22}a_{23}a_{24}\ldots \\ a_3 & = & 0.a_{31}a_{32}a_{33}a_{34}\ldots \\ a_4 & = & 0.a_{41}a_{42}a_{43}a_{44}\ldots \\ \ldots & = & \ldots, \end{array} \tag{A2.6}$$

where a_{ij} is the j-th digit after the decimal point in the decimal expression for a_i. We then form the number

$$d = 0.d_{11}d_{22}d_{33}d_{44}\ldots$$

where

$$d_{nn} = \begin{cases} a_{nn}+1, & \text{if } a_{nn} \neq 9, \\ 0, & \text{if } a_{nn} = 9. \end{cases}$$

By construction, $0 < d < 1$ and d differs from a_k at the k-th decimal place. It therefore differs from every number in the array (A2.6), which contradicts the assumption that (A2.6) contains every number between 0 and 1. This contradiction establishes the desired result. □

The technique by which the above result is established has become known as **Cantor's diagonal method**. We remind the reader that the cardinality of the set $\mathbb{R}$ is denoted by $\aleph$. The fact that $\mathbb{N}$ can be embedded in $\mathbb{R}$, but $\mathbb{R}$ cannot be embedded in $\mathbb{N}$, is expressed as follows:[6] $\aleph_0 < \aleph$.

[6] The word *embedding* is used here in its everyday sense, which translates here to $\mathbb{N} \subset \mathbb{R}$ but $\mathbb{R} \not\subset \mathbb{N}$. In mathematics, the term is used more often in its topological sense, which will be defined on page 262.

A2.2 The real line

Real numbers have a property called **completeness** which is anything but transparent, and which will be of particular interest to us. It is the property that ensures that there exists a real number a to which the series

$$\frac{a_1}{10} + \frac{a_2}{10^2} + \cdots + \frac{a_k}{10^k} + \cdots \tag{A2.7}$$

can converge. This a, and the a_k in (A2.7), are the same as those in (A2.5). It is due to this property that every real number (except zero) has a unique infinite decimal representation.

To explain this property, we shall briefly describe two approaches to the construction of the real numbers from the rationals, one due to Dedekind and the other to Cantor. Both were published in 1872.

A2.2.1 Dedekind's construction

Dedekind's construction was based on the order property of the rationals. The relation $>$ defines a total order on the rationals $\mathbb{Q}$; if r, s are two distinct rationals, then either $r > s$ or $s > r$. (The same total order may also be defined by the relation $<$.)

A partition of $\mathbb{Q}$ into two *nonempty* subsets L (from left) and R (from right) that have the properties

$$\begin{array}{lll} \text{(i)} & L \cap R & = \emptyset, \\ \text{(ii)} & L \cup R & = \mathbb{Q}, \\ \text{(iii)} & \multicolumn{2}{l}{p \in L,\, q \in R \Rightarrow p < q} \end{array} \tag{A2.8}$$

is called a **Dedekind section**, or briefly a **section**, of $\mathbb{Q}$.

A rational number u defines two obvious sections of $\mathbb{Q}$:

$$L_1 = \{q | q < u\}, \quad R_1 = \{q | q \geq u\}$$

and

$$L_2 = \{q | q \leq u\}, \quad R_2 = \{q | q > u\}.$$

In the first, R_1 has a smallest member (namely, u) but L_1 has no largest member; in the second, L_2 has a largest member (again, u), but R_2 has no smallest member. There exist sections that are not defined by rational numbers – which gives the concept its importance. Consider, e.g., the following partition of $\mathbb{Q}$:

$$L = \{r | r < 0, \text{ or } r \geq 0, r^2 < 2\},$$

$$R = \{r | r > 0, r^2 > 2\}.$$

Clearly, L consists of the rationals that are smaller than $\sqrt{2}$, and R of rationals that are greater than $\sqrt{2}$. In this case, *L has no largest member*, and *R has no smallest member.*

The correspondence between a subset of sections of $\mathbb{Q}$ and the rational numbers suggests the possibility of going one step further and *identifying* the rational numbers with these sections. To avoid ambiguity, we choose those sections of $\mathbb{Q}$ for which L has a largest member. For this attempt to be successful, one has to be able to express the laws of arithmetic in terms of sections. This was accomplished by Dedekind, who was then able to take the decisive step and propose the following definition of real numbers:

Definition A2.2 (Real numbers) A **real number** is a section (L, R) of the rationals $\mathbb{Q}$ such that R has no smallest member. If L has a largest member, then the section is a rational number; otherwise it is an irrational number.

It can be shown that the order and arithmetic properties of the rationals continue to hold for the reals. Therefore one can define sections of $\mathbb{R}$ in an obvious manner. However, the properties of sections of $\mathbb{R}$ differ essentially from those of $\mathbb{Q}$. The difference is captured by the following theorem:

Theorem A2.3 *Let (L, R) be a section of the reals. Then either L has a largest member, or R has a smallest number. In other words, the case 'L has no largest number and R has no smallest member' cannot occur.*

Proof Let $L_{\mathbb{Q}} = L \cap \mathbb{Q}$, $R_{\mathbb{Q}} = R \cap \mathbb{Q}$. Then $(L_{\mathbb{Q}}, R_{\mathbb{Q}})$ form a section of $\mathbb{Q}$. There are two possibilities:

(1) $L_{\mathbb{Q}}$ has a largest member, say p.

(2) $L_{\mathbb{Q}}$ has no largest member.

In case (1), p is also the largest member of L. For if there exists $\alpha \in L$, $\alpha > p$, then there exist rational numbers $q \in L$ such that $p < q < \alpha$, which implies that $q \in L_{\mathbb{Q}}$. But this contradicts the assumption that p is the largest member of $\mathbb{Q}$. Finally, if L has a largest member, then R cannot have a smallest member. For suppose that R has a smallest member s. Then $p < s$, and the number $(p+s)/2$ belongs neither to L nor to R, a contradiction.

In case (2), the section $(L_{\mathbb{Q}}, R_{\mathbb{Q}})$ of $\mathbb{Q}$ is a real number β. This number must belong either to L or to R. If it belongs to L, it is the largest member of L, and R does not have a smallest member. If it belongs to R, it is the smallest member of R, and L does not have a largest member. The proofs are exactly the same as in case (1). □

What we have shown above is that, unlike $\mathbb{Q}$, sections of $\mathbb{R}$ give us nothing new. $\mathbb{R}$ is called the **Dedekind completion** of $\mathbb{Q}$, and one says that the set of real numbers is **complete**.

Theorem A2.3 is sometimes called the **least upper bound property** of the real numbers. It could equally well be called the greatest lower bound property, but generally is not.

A2.2.2 The least upper bound property

Let S be a set that is totally ordered by $\prec$, i.e., if $x, y \in S$, $x \neq y$, then either $x \prec y$ (read: x precedes y) or $y \prec x$. A subset T of S is said to be **bounded above** if there exists a point $a \in S$ such that $x \prec a$ for all $x \in T$. The point a is called an **upper bound** for T (in S); it is called the **least upper bound** for T if it is an upper bound, and if any $y \prec a_0$ is not an upper bound. The reader is invited to prove that the least upper bound, if it exists, is unique.

Definition A2.4 (Least upper bound property) A totally ordered set S is said to have the *least upper bound property* if every nonempty subset $T \subset S$ which is bounded above has a least upper bound in S.

The real numbers $\mathbb{R}$ have the least upper bound property; the rationals $\mathbb{Q}$ do not. The set of all rationals less than $\sqrt{2}$ is bounded above in $\mathbb{Q}$, but does not have a least upper bound in $\mathbb{Q}$; it *does* have a least upper bound, namely $\sqrt{2}$ itself, in $\mathbb{R}$. The least upper bound property is equivalent to completeness, but we shall not go into details. The interested reader is referred to the texts (Munkres, 1975; Rudin, 1976).

A2.2.3 Cantor's procedure

Dedekind's completion of $\mathbb{Q}$ was based on the notion of order. However, $\mathbb{Q}$ also has a **metric** (distance between two points) defined on it, namely $d(p, q) = |p-q|$. Cantor's completion of $\mathbb{Q}$ was based on this notion.

Recall that a sequence $\{a_k\}$ of numbers[7] is said to **converge** if there is a number a such that, given $\varepsilon > 0$, there exists a positive integer N such that $|a_n - a| < \varepsilon$ whenever $n > N$. It is called a **Cauchy sequence** if, given $\varepsilon > 0$, there exists a positive integer N such that $|a_m - a_n| < \varepsilon$ for all $m, n > N$. Cauchy sequences do not always converge. Consider the sequence of rationals $1.1, 1.14, 1.141, 1.1415, \ldots$ The nth term a_n of this sequence is obtained by dropping all digits after the nth in the decimal expansion of $\sqrt{2}$. This is a Cauchy sequence. *It does not converge in* $\mathbb{Q}$ because the point it is trying to converge to, namely $\sqrt{2}$, does not belong to $\mathbb{Q}$. However, it *does converge in* $\mathbb{R}$, because $\sqrt{2} \in \mathbb{R}$. A **complete metric space** is one in which every Cauchy sequence converges. $\mathbb{Q}$ is a metric space, but it is not complete; $\mathbb{R}$ is a complete metric space.

[7] Some authors, such as Munkres (Munkres, 1975), reserve the curly brackets notation $\{.\}$ for sets. We shall use it for sets, as well as for sequences, in common with (Friedman, 1970; Apostol, 1974; Rudin, 1976).

But how does one obtain $\mathbb{R}$ from $\mathbb{Q}$ without using Dedekind completion? Cantor approached this problem as follows: if the problem is that some Cauchy sequences have nowhere to converge to, why not define a new space in which the points are the Cauchy sequences themselves? The rational number r would then be the Cauchy sequence $\{a_n = r\}_{n \in \mathbb{N}}$; the number $\sqrt{2}$ will be the Cauchy sequence $1, 1.1, 1.14, 1.141, 1.1415 \ldots$ There are two slight problems with this. One is that convergence or divergence is determined not by the head but by the tail of the sequence, so that two different Cauchy sequences with the same tail will converge to the same point. The other is that even sequences with very different tails may converge to the same point; the sequences $\{-1/n\}$ and $\{+1/n\}$ are both Cauchy, have no point in common, and both converge to 0. However, these problems are easy to dispose of. As the procedure (called **metric completion**) is the same for all incomplete metric spaces[8] (e.g., $\mathbb{Q}^n, n > 1$) – an advantage which is clearly not shared by Dedekind completion – we shall discuss the procedure in some detail. This discussion will be carried out in Appendix A4, as part of a somewhat larger programme.

A2.2.4 Well-ordering and real numbers

On page 248 we had defined well-ordered sets and stated the well-ordering theorem which asserts that every set can be well-ordered. It should therefore be repeated that *no well-ordering of the real numbers is known.*

[8] Metric spaces were defined by Fréchet in 1906, though not under this name; the name is due to Hausdorff, who also described the completion procedure in 1914.

A3

Point-set topology

This section provides a thumbnail sketch of those elements of point-set topology (also called general topology or just plain topology) that are used in this book. The subject grew out of attempts to rid the notion of continuity of its traditional dependence on the notion of distance. It turned out that continuity could be defined without using real numbers at all; the subject could be founded, instead, on the calculus of sets. Unfamiliarity with the latter is perhaps the main source of difficulty for the beginner.

Detailed treatments of the material discussed below may be found in standard textbooks such as (Kelley, 1955, Willard, 1970 and Munkres, 1975). Of these, the one by Munkres will perhaps be the easiest for the physicist.

A3.1 Topological spaces

Point-set topology (usually called topology for short) may be regarded as the study of the notions of convergence of sequences[1] and continuity of maps without using the notion of real numbers. In the theory of functions of a real variable, both of these notions are intimately related to that of neighbourhoods of a point. A **neighbourhood** of a point x on the real line is any subset that contains an open interval $(x-a, x+a)$ around x, where $a > 0$; usually a is a small number, but it does not have to be. In the Euclidean plane $\mathbb{R}^2$, neighbourhoods of a point x are subsets containing **discs** $D_\rho(x) = \{y|0 < d(x,y) < \rho\}$, where $d(x,y)$ is the Euclidean distance between the points x and y. If a, b are two distinct points in the plane, $D_\rho(a)$, $D_\sigma(b)$ neighbourhoods of a and b that have a nonempty intersection, and $c \in D_\rho(a) \cap D_\sigma(b)$, then one can find a real number τ such that

$$D_\tau(c) \subset D_\rho(a) \cap D_\sigma(b), \tag{A3.1}$$

which is easily seen by making a diagram. This observation opens the way to defining neighbourhoods of points in sets on which a notion of distance is not defined. The key notion is that of a topology on a set of points, which is given below; the notion of neighbourhoods comes a little later.

Definition A3.1 (Topology and topological spaces) A topology $\mathcal{T}$ on a point-set X is a collection of subsets U of X that satisfies the following conditions:

[1] A generalization of the notion of sequences is treated in Subsection A3.8.1.

(a) X and $\emptyset$ are members of $\mathcal{T}$.
(b) The union of the members of an arbitrary subfamily of $\mathcal{T}$ is a member of $\mathcal{T}$.
(c) The intersection of a finite number of members of $\mathcal{T}$ is also a member of $\mathcal{T}$.

A pair $(X, \mathcal{T})$ is called a **topological space**. The members U of the family $\mathcal{T}$ are called **open sets**.

This definition is related to our preliminary discussion by the two definitions that follow.

Definition A3.2 (Base for a topology) Let X be a nonempty set. A family $\mathcal{B} = \{B_\alpha | \alpha \in A\}$ of subsets of X is called a **base** (or **basis**) for a topology on X if it satisfies the following conditions:

(a) $\bigcup_{\alpha \in A} B_\alpha = X$.
(b) If $B_1, B_2 \in \mathcal{B}$, $x \in B_1 \cap B_2$, then there exists $B_3 \in \mathcal{B}$ such that $x \in B_3 \subset B_1 \cap B_2$. (*This is the equivalent of* (A3.1).)

Definition A3.3 (Topology generated by a base) The **topology** $\mathcal{T}$ on X generated by the base $\mathcal{B}$ is the family of subsets U of X such that for each $x \in U$, there is a $B \in \mathcal{B}$ such that $x \in B \subset U$. In words, every $U \in \mathcal{T}$ is a union of members of the base.

Every set $B \in \mathcal{B}$ is an open set (briefly: is open). The set X itself is open. Additionally, the empty set $\emptyset$ is open; the condition 'for each $x \in U \ldots$', is trivially satisfied for $U = \emptyset$, because there is no such x.

We have just given two different definitions, Definitions A3.1 and A3.3, of a topology on X. The equivalence of these two definitions is proved in every textbook, and may be summed up as follows: *every topology has a base.* One sometimes says that the family of open sets is closed under arbitrary unions and finite intersections. It is possible to define many distinct topologies on a set X. The two most extreme examples are the **discrete topology**, in which every subset of X is open, and the **indiscrete topology**, in which only X and $\emptyset$ are open. Therefore, to avoid confusion, it may sometimes be necessary to specify the topology explicitly.

We round off the above with the definition of neighbourhoods in topological spaces: a **neighbourhood** of a point x in a topological space X is any subset W of X that contains an open set U containing x: $x \in U \subset W$.

Finally, we give a few examples of topologies:

Examples A3.4

(i) The discrete and the indiscrete topologies on X, defined above.

(ii) A **partial order** $\prec$ on a set X is a binary relation that is **transitive**, i.e., if $x, y, z \in X$, $x \prec y$ and $y \prec x$, then $x \prec z$. If, in addition, the order satisfies the **comparability condition**, i.e., for any two distinct points $a, b \in X$, either $a \prec b$ or $b \prec a$, then the order $\prec$ is said to be **total**, or **linear**. Some authors omit the requirement that a and b be distinct; then one has to require the additional condition that if both $a \prec b$ and $b \prec a$ hold, then $a = b$. (Consider the relations $<$ and $\leq$ on $\mathbb{R}$.)

Let $(X, \prec)$ be a linearly ordered set. The family of subsets $(a, b) = \{x | x \in X, a \prec x \prec b, a \neq b\}$ is a base for a topology on X. (The reader is invited to supply a proof.) This topology is known as the **order topology** on X, and (a, b) is an **open interval** in X. (The condition $a \neq b$ in the definition of (a, b) is unnecessary if $a \prec b \Rightarrow a \neq b$.)

(iii) The relation $<$ is a total order on the real line $\mathbb{R}$. The order topology of this order is known as the **standard** (or **usual**) **topology** on $\mathbb{R}$. (This topology could equally well be defined by the order $>$.)

(iv) Consider the family of all subsets of $\mathbb{R}$ that exclude the origin 0, together with $\mathbb{R}$ itself. This family contains the empty set, and is closed under arbitrary unions and intersections. It therefore defines a topology $\mathcal{T}_{\text{ex}}$ on $\mathbb{R}$, called the **excluded point topology**. (It does not really matter which point is excluded.)

Topologies like the last one are mainly used for providing counter-examples.

A3.1.1 Closed sets, interior, boundary

A **closed set** is defined to be the complement $U' = X \setminus U$ of an open set U in a topological space.[2] The family of closed sets contains X and $\emptyset$, and is closed under finite unions and arbitrary intersections. Let A be an arbitrary subset of X. The union of all open sets *contained in* A is an open set. (Recall that the union of a family of open sets is an open set.) It is called the **interior** of A and denoted A° or $\operatorname{int} A$. The intersection of all closed sets *containing* A is a closed set.[3] It is called the **closure** of A and denoted $\bar{A}$ or $\operatorname{cl} A$. A point b is called a **limit point** of A if $b \in \bar{A}$; b may or may not belong to A. The following is a useful little result:

Theorem A3.5 *A subset A of X is closed if and only if it contains all its limit points.*

The difference $\bar{A} \setminus A^\circ$ is called the **boundary** of A and denoted ∂A or $\operatorname{bd} A$. One has the inclusions $\bar{A} \supset A \supset A^\circ$.

[2] The reader should be warned that the notation $U' = X \setminus U$ is by no means universal. In the literature the superscript $'$ (as in U') is used freely, with many different meanings.

[3] This fact is established using de Morgan's laws, page 239.

Examples A3.6

(i) Let $(X, \mathcal{T})$ be a topological space. Then the set X is both open and closed, irrespective of the topology $\mathcal{T}$. Therefore the boundary of X is empty. The same remarks apply to the empty subset $\emptyset$ of X.

(ii) Let X be endowed with the discrete topology. Then every subset of X is both open and closed, i.e., the boundary of every subset is empty.

(iii) Let $\mathbb{R}$ be the real line with the standard topology. Then the **open intervals** (a, b) are open and **closed intervals** $[a, b]$ are closed. The interior of $[a, b]$ is (a, b), and its boundary is the two-point set $\{a, b\}$ consisting of the end-points. The boundary of a **half-open interval** $(a, b] \subset \mathbb{R}$ is also the two-point set $\{a, b\}$. But – and this is an important part – *if we take* $X = [a, b)$, *then* the boundary of X consists of the one-point set $\{a\}$; the other putative boundary point, b, does not belong to X. Embedding a topological space in a larger topological space can introduce new limit and boundary points.

(iv) Consider the set W of rational numbers that satisfy the inequalities $0 \leq r \leq 1$, i.e., $W = [0, 1] \cap \mathbb{Q}$. We show that W is not closed (in the standard topology on $\mathbb{R}$). The number $\sqrt{2} - 1$ is the limit of the sequence of rational numbers $0.1, 0.14, 0.141, 0.1415$, and so on, all belonging to W. However, it is not a rational number, and therefore does not belong to W, i.e., W does not contain all its limit points.

A3.1.2 Comparison of topologies

Suppose that $\mathcal{T}_1$ and $\mathcal{T}_2$ are two topologies defined on X. Suppose that every open set of $\mathcal{T}_1$ is also an open set of $\mathcal{T}_2$, i.e., $O \in \mathcal{T}_1 \Rightarrow O \in \mathcal{T}_2$, i.e., $\mathcal{T}_1 \subset \mathcal{T}_2$. Then we say that (i) the topologies $\mathcal{T}_1$ and $\mathcal{T}_2$ are **comparable**, and that (ii) $\mathcal{T}_1$ is **coarser** than $\mathcal{T}_2$, or that $\mathcal{T}_2$ is **finer** than $\mathcal{T}_1$; the finer topology includes all the open sets that the coarser one has. The indiscrete topology is the coarsest possible topology on a set, in the sense that it is coarser than any other topology; similarly, the discrete topology is the finest possible topology on a set. The standard topology on $\mathbb{R}$ lies between these two.

The terms **smaller** (for coarser) and **larger** (for finer) are also used for the comparison of topologies. So – and more frequently – are the terms **weaker** and **stronger**, but regrettably the use of the last two is not uniform. If $\mathcal{T}_1 \subset \mathcal{T}_2$, some mathematicians will say that $\mathcal{T}_1$ is stronger than $\mathcal{T}_2$, and some others will say that it is weaker than $\mathcal{T}_2$. The reader who comes across these terms should ascertain what the author means. We shall use only the terms coarser and finer for the comparison of topologies.[4]

[4] Exceptions to this rule will be made when we deal with operators of Hilbert space. The terms *strong* and *weak* are so entrenched there that it would be folly to try to change them.

Two distinct topologies on X are *not* generally comparable. Consider the usual topology $\mathcal{T}$ and the excluded point topology $\mathcal{T}_{\mathrm{ex}}$ on $\mathbb{R}$ (page 259); the interval $(-1, 1)$ is open in $\mathcal{T}$ but not in $\mathcal{T}_{\mathrm{ex}}$; the interval $[1, 2]$ is open in $\mathcal{T}_{\mathrm{ex}}$ but not in $\mathcal{T}$.

A3.2 Continuous functions

Let X, Y be topological spaces. A map

$$f : X \to Y$$

is defined to be **continuous** if the inverse image $f^{-1}[V]$ of an open set $V \subset Y$ is open in X. If $X = Y = \mathbb{R}$, this definition is equivalent to the standard ϵ-δ definition of continuity of functions of a real variable. See, for example (Munkres, 1975, pp. 102–103).

The two examples given below are meant to illustrate the extreme – and not necessarily wanted – generality of the notion of a topological space.

Examples A3.7

(i) Let X be a *discrete* topological space, i.e., let every subset of X be open. Then for any topological space Y and any map $f : X \to Y$, the inverse image $f^{-1}[V] \subset X$ is open for any subset V of Y. In other words, every function from X into another topological space Y is continuous.

(ii) Let now X be an indiscrete topological space, Y a discrete one, and $f : X \to Y$ a continuous map. Then $f^{-1}[V] = \emptyset$ or X for every open set $V \subset Y$. Since one-point sets are open in Y, this would imply that $f^{-1}[\{y\}] = X$ for some $y \in Y$; then this point y has to be unique. This means that f is a constant map; in other words, the only continuous functions are the constants.

As we shall see below, there exists a fertile middle ground between the discrete and the indiscrete topologies.

A3.2.1 Homeomorphisms

Let X, Y be two topological spaces. If there exists a continuous bijection $f : X \to Y$ such that the inverse map $f^{-1} : Y \to X$ is also continuous, then f is said to be a **homeomorphism**[5] between X and Y, and X and Y are said to be **topologically equivalent**, or **homeomorphic**, to each other. Two spaces that are homeomorphic to each other are topologically indistinguishable, although they may be distinguishable in terms of other mathematical structures that are defined on them.

[5] Not to be confused with the algebraic concept of **homomorphism**.

Example A3.8 The real line $\mathbb{R}$ is homeomorphic to the open interval $(0,1)$ if the latter is topologized by the subspace topology[6] inherited from $\mathbb{R}$. The map $f:(0,1)\to\mathbb{R}$ defined by

$$y = f(x) = \frac{x}{1-x^2}$$

is a homeomorphism. We know from elementary calculus that the function f is strictly increasing; it is therefore bijective, the inverse map being

$$x = f^{-1}(y) = \frac{2y}{1+(1+4y^2)^{1/2}}.$$

A strictly increasing map is order-preserving, and therefore maps basis elements on $(0,1)$ (on the X-axis) to basis elements on $\mathbb{R}$ (on the Y-axis). The reader is invited to complete the articulation of the proof.

A property, or quantity, that is invariant under homeomorphisms is called a **topological invariant**. The example just given shows that the distance between two points is most emphatically not a topological invariant.

We end this section with the following definition.

Definition A3.9 (Embeddings) Let X, Y be topological spaces and $f: X\to Y$ a map which is not bijective. Then f is called an **embedding** if it is a homeomorphism $f: X\to f(X)$ onto its range $f(X)$, when $f(X)$ has the subspace topology.

A3.3 New spaces from old

There are three basic constructions for generating new topological spaces out of old: subspaces, products and quotients. We shall consider them in turn.

A3.3.1 Subspaces

Let $(X,\mathcal{U})$ be a topological space and Y a subset of X. It is easy to see that the collection $\{V_\alpha | V_\alpha = U_\alpha \cap Y, U_\alpha \in \mathcal{U}\}$ defines a topology $\mathcal{V}$ on Y. The topology $\mathcal{V}$ is called the **subspace** or the **relative topology**, or the **relativization** of $\mathcal{U}$ to V. The pair $(Y,\mathcal{V})$ is called a **topological subspace** or simply a **subspace** of $(X,\mathcal{U})$.

A3.3.2 Topological products

Let $(X,\mathcal{U})$ and $(Y,\mathcal{V})$ be two topological spaces. It is easily verified that the collection $\mathcal{B} = \{U\times V | U\in\mathcal{U}, V\in\mathcal{V}\}$ is a base for a topology on $X\times Y$. ($\mathcal{B}$ itself is not a topology on $X\times Y$, as the union of two products is not generally a

[6] The subspace topology is defined in Subsection A3.3.1.

product.) This topology is called the **product topology**, and $X \times Y$ furnished with this product is called the **topological product** of X and Y.

Let $(x, y) \in X \times Y$. The maps $\pi_X : X \times Y \to X$ and $\pi_Y : X \times Y \to Y$ defined by $\pi_X(x, y) = x$ and $\pi_Y(x, y) = y$ are called **projections** on the spaces X and Y, respectively. If $X \times Y$ has the product topology, then the projections are continuous.

Let now $\{X_\alpha\}_{\alpha \in A}$ be an arbitrary family of topological spaces. The **product topology** on the Cartesian product $X = \prod_\alpha X_\alpha$ is the topology that has as a base the family of sets of the form

$$B = \prod_{\alpha \in A} U_\alpha,$$

where (i) U_α is open in X_α, and (ii) $U_\alpha = X_\alpha$ for *all but a finite number* of values of α.

Condition (ii) may appear to be counterintuitive, but dropping it would have unfortunate consequences. If it is dropped one would still obtain a topology on X, but this topology will be strictly finer than the product topology; it will be too fine to be of much use. One reason is given below.

Denote by π_α the projection from X to X_α ((A1.24) and (A1.25)). Let $f : W \to X$ be a map from a topological space W to the product space X, and define $f_\alpha = \pi_\alpha \circ f$. Then

Theorem A3.10 *The map $f : W \to X$ is continuous if and only if f_α is continuous for each $\alpha \in A$.*

This result does not hold if the topology on the product $\prod X_\alpha$ is any other than the product topology. For a more detailed discussion, see (Munkres, 1975, pp. 112–115).

A3.3.3 Quotient spaces

In simple cases, the quotient construction may be regarded as a sort of inverse of the product construction; hence the name.

A **relation** R on a set X is a subset[7] of the Cartesian product $X \times X$. The statement $(x, y) \in R$ is abbreviated to xRy, and read as 'x is related to y by R'. A relation R is called an **equivalence relation** if it satisfies the following conditions:

(a) **reflexivity:** aRa,
(b) **symmetry:** $aRb \Rightarrow bRa$, and
(c) **transitivity:** $aRb, bRc \Rightarrow aRc$.

[7] To the nonmathematician, it may seem decidedly odd that the relation $<$ on $\mathbb{R}$ be defined as a subset of $\mathbb{R} \times \mathbb{R}$. The point of the definition given above is that it applies to sets with no mathematical structure on them.

An equivalence relation on a set X partitions X into a set of pairwise-disjoint subsets called **equivalence classes**. The following examples will make this clear.

Examples A3.11

(i) The relation $A \sim B$ between two sets A and B defined on page 243 and called *equivalence* there is an equivalence relation.

(ii) Let $\mathbb{R}^2$ be the Euclidean plane, and x, y the usual Cartesian coordinates of a point on it. Define a relation R on $\mathbb{R}^2$ by $(x, y)R(x', y')$ iff $x = x'$. One sees immediately that R is an equivalence relation; the equivalence classes are lines parallel to the Y-axis.

(iii) Let G be a group, and H a proper subgroup of it. Let $a, b \in G$ and define a relation R on G by aRb if $a^{-1}b \in H$. Since $a^{-1}a = e \in H$, R is reflexive. Since $b^{-1}a = (a^{-1}b)^{-1} \in H$, R is symmetric. If $c \in G$, then the identity $a^{-1}c = a^{-1}b \cdot b^{-1}c$ shows that aRb and bRc together imply aRc, i.e., R is also transitive; it is an equivalence relation. The equivalence classes are left-cosets of H in G.

The set of equivalence classes in X under the relation R is denoted by X/R, or (if only one equivalence relation is being considered) by $X/\!\sim$. Let $\pi : X \to X/\!\sim$ be the map that sends $x \in X$ to its equivalence class in $X/\!\sim$. The **quotient topology** $\mathcal{T}_\pi$ on $X/\!\sim$ is defined as follows:

$$\mathcal{T}_\pi = \{V \subset X/\!\sim | \pi^{-1}[V] \text{ is open in } X\}.$$

The quotient topology is the finest topology on $X/\!\sim$ that makes the **projection** π continuous. The reader is invited to supply a proof.

A3.4 Countability and separation axioms

Unfortunately, the definition of a topological space is a little too general. As we saw in Section A3.2, between some pairs of topological spaces all functions are continuous, while between others only the constants are continuous. Useful (and interesting) topological spaces have to satisfy restrictive conditions that get rid of both of these excesses. The most important of these conditions are given below.

A3.4.1 Second countability and separability

A topological space is said to be **second countable** (or to satisfy the **second axiom of countability**) if its topology has a countable base.[8]

[8] There is also a property called *first countability*. It is considered to be less important than second countability, partly because second countability implies first countability, and partly because mathematicians have overcome most of the problems arising from the failure of first countability. We shall postpone discussion of this concept to Section A3.8.

The real line $\mathbb{R}$ with its usual topology is second countable. Note that the definition requires the space to have *a* countable base; it does not require every base to be countable. The collection of all open intervals is a base, but it is not countable. The collection of all open intervals (r, s), where r and s are rational, is also a base; this base is countable.

A set A is said to be **dense** in a topological space X if X is the closure of A. A topological space is called **separable** if it contains a countable dense subset. The real line $\mathbb{R}$ is separable; the rationals $\mathbb{Q}$ are countable, and are dense in $\mathbb{R}$. It can be proved that a second countable space is separable. However, examples show that not every separable space is second countable.

A3.4.2 Separation properties

The separation conditions discussed below – not to be confused with the separability property defined above – were first studied by Felix Hausdorff; axioms $T_1 - T_4$ that follow are therefore known as the **Hausdorff separation axioms.**[9]

Figure A3.1 provides pictorial representations of the separation axioms. Points are indicated by dots, open sets are bounded by dashed lines, and closed sets by solid lines.

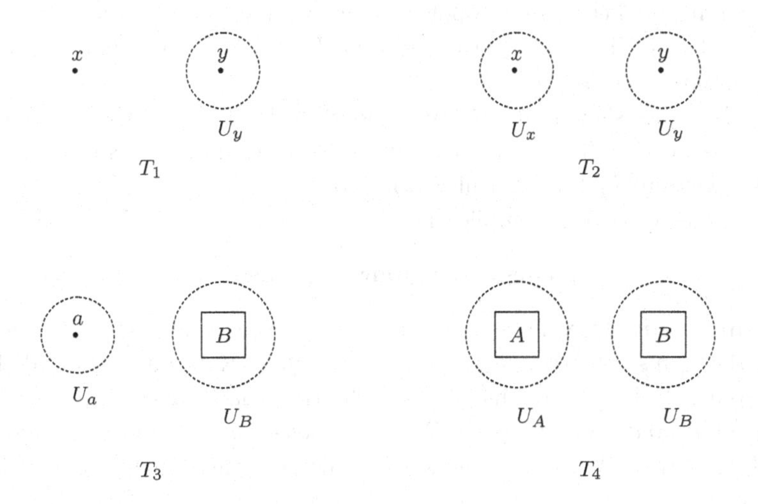

Fig. A3.1. Illustrating the axioms T_1, T_2, T_3 and T_4

[9] The axiom now known as T_1 does not appear in the first edition of Hausdorff's *Mengenlehre* (1914). It seems to be due to Fréchet or to Riesz, and was incorporated by Hausdorff in the third edition (1935) of his book (p. 260 of the English translation (Hausdorff, 1957)). The term separation axiom (*Trennungsaxiom* in German), the T_n-notation and the T_4-axiom were introduced by Tietze in 1923. T_1-spaces are sometimes known as **Fréchet spaces**.

(i) **Axiom T_1:** Let X be a topological space, and x, y any two distinct points in it. Then there exists an open set $U_y \subset X$ such that $y \in U_y$ but $x \notin U_y$.

One-point sets are closed in T_1-spaces. Let $a \in X$. For each $y \in X$, $y \neq a$, choose an open set U_y as above. The union

$$W = \bigcup_{\substack{y \in X \\ y \neq a}} U_y$$

is an open set. Therefore its complement $X \setminus W$ is closed. But this complement is exactly the one-point set $\{a\}$.

(ii) **Axiom T_2:** Let X be a topological space, and x, y any two distinct points in it. Then there exist open sets $U_x,\ U_y \subset X$ such that $x \in U_x, y \in U_y$ and $U_x \cap U_y = \emptyset$.

T_2-spaces are known as **Hausdorff spaces**. One says that, in a Hausdorff space, two distinct points can be separated by open sets. A Hausdorff space is clearly a T_1-space.

(iii) **Axiom T_3:** Let X be a topological space, $a \in X$ and B any closed set disjoint from a: $a \notin B$. Then there exist open sets $U_a,\ U_B \subset X$ such that $a \in U_a,\ B \subset U_B$ and $U_a \cap U_B = \emptyset$.

Examples show that a T_3-space does not have to be T_1; but if it is T_1 (written: $T_3 + T_1$), then it is Hausdorff (take B to be a one-point set). A T_3-space that is also T_1 is called **regular**.

(iv) **Axiom T_4:** Let X be a topological space, and $A, B \subset X$ any two disjoint closed sets. Then there exist open sets $U_A,\ U_B \subset X$ such that $A \subset U_A, B \subset U_B$ and $U_A \cap U_B = \emptyset$.

Examples show that a T_4-space does not have to be either T_3 or T_1. But if it is T_1, then it is clearly T_3 and therefore regular. A T_4-space that is also T_1 (written: $T_4 + T_1$) is called **normal**.

Thus we have the implications

$$\textbf{normal} \Rightarrow \textbf{regular} \Rightarrow \textbf{Hausdorff} \Rightarrow T_1.$$

Note on terminology There is no consistency, even at the textbook level, in the use of the terms *regular* and *normal*. For example, Kelley and Willard (Kelley, 1955; Willard, 1970) interchange the definitions of T_3 and regular spaces, as well as of T_4 and normal spaces. We have chosen our definitions to agree with (Munkres, 1975). However, Munkres does not use the designations T_2, T_3 and T_4 at all.

A3.4.2.1 The T_0-separation property

A separation axiom which is weaker than T_1 was introduced by Kolmogoroff. It is known as the **Kolmogoroff axiom** or, rather more frequently, as the **T_0-axiom**.The precise statement is as follows:

A topological space X is said to satisfy the T_0-separation axiom if, for $x, y \in X, x \neq y$, there exists either an open set U_x such that $x \in U_x, y \notin U_x$ or an open set U_y such that $y \in U_y, x \notin U_y$.

The family of intervals $\{(a, \infty) | a \in \mathbb{R}\}$, together with $\mathbb{R}$ itself and the empty set, defines a topology $\mathcal{T}$ on the set $X = \mathbb{R}$. This topology is not T_1; if $a < b$, then there is no interval (x, ∞) which contains a but not b. However, there are intervals (x, ∞) that contain b but not a. That is, the topology $\mathcal{T}$ is T_0, but not T_1.

T_0-spaces are seldom encountered, except in the theory of topological groups. As topological groups form the substratum for much of theoretical physics, the subject is briefly discussed in Section A3.9. That is the only reason why the T_0-separation axiom is being mentioned here.

A3.4.3 Separation by continuous functions

When does a topological space have a sufficiently large supply of continuous real-valued function to be really useful? The definitive result in this direction was proved by Urysohn, and is known as **Urysohn's lemma.**

Theorem A3.12 (Urysohn's lemma) *Let X be a normal space, and A, B closed disjoint subsets of it. Then there exists a continuous function $f : X \to \mathbb{R}$ such that $f(A) = 0$ and $f(B) = 1$.*

The proof is a work of genius, and rewarding to go through. A very readable account may be found in (Munkres, 1975, pp. 207–211).

The conclusion of Urysohn's lemma is often described as **separation by continuous functions.** The result holds under slightly weaker conditions than assumed in Theorem A3.12; it is not necessary for one-point sets to be closed. T_4 separation is sufficient but not necessary, but the weaker T_3 condition is necessary but not sufficient. T_3-spaces which have the Urysohn separation property were once jokingly called $T_{3\frac{1}{2}}$, but the term has stuck. The bourgeois term, **completely regular**, has survived. A **Tychonoff space** is a completely regular T_1-space. We have the implications

$$\textbf{normal} \Rightarrow \textbf{Tychonoff} \Rightarrow \textbf{regular} \Rightarrow \textbf{Hausdorff} \Rightarrow T_1;$$

$$T_4 \Rightarrow T_{3\frac{1}{2}} = \textbf{completely regular} \Rightarrow T_3.$$

Note on terminology Tychonoff spaces are called completely regular by Munkres (Munkres, 1975), who does not use the term Tychonoff space at all.

These terminological disparities do not cause trouble because one-point sets are generally closed in the topological spaces used in most areas of mathematics.

A3.5 Metric spaces and the metric topology

Metric spaces are much more structured than topological spaces. 'When is a topological space metrizable?' is an important question in point-set topology. In this section we shall define metric spaces, describe the metric topology, and state a metrization theorem due to Urysohn.

Definition A3.13 (Metric) A **metric** on a point-set X is a map $d : X \times X \to \mathbb{R}$ that satisfies the conditions:

(a) $d(x, y) = d(y, x)$ for all $x, y \in X$,

(b) $d(x, y) \geq 0$ for all $x, y \in X$,

(c) $d(x, z) \leq d(x, y) + d(y, z)$ for all $x, y, z \in X$,

(d) $d(x, y) = 0$ iff $x = y$.

Condition (c) is known as the **triangle inequality**. If one drops condition (d), the structure obtained is called a **semimetric**. We shall have occasion to use this concept in Appendix A5.

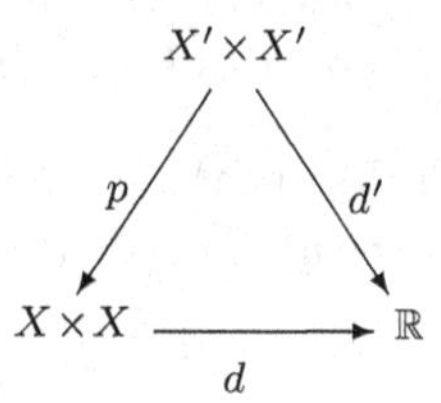

Fig. A3.2. From semimetric to metric

Let d' be a semimetric on X'. Define on X' a relation R as follows: $x'Ry'$ if $d'(x', y') = 0$. The reader is invited to verify that R is an equivalence relation on X'. Denote by x the equivalence class of $x' \in X'$, by X the set of these equivalence classes, and by $\pi : X' \to X$ the projection map which sends $x' \in X'$ to its equivalence class $x \in X$. Finally, let $p = \pi \times \pi : X' \times X' \to X \times X$ be the map which sends the ordered pair $(x', y') \in X' \times Y'$ to the ordered pair $(x, y) \in X \times Y$, where $x = \pi(x')$ and $y = \pi(y')$. Then there is a unique map $d : X \times X \to \mathbb{R}$ such that

$$d' = d \circ p. \tag{A3.2}$$

Equation (A3.2) means that the two 'routes' from $X' \times X'$ to $\mathbb{R}$ in Fig. A3.2 yield the same map. Such a diagram is called a **commutative diagram**. The reader is invited to verify that d is a metric on X.

Let $x, y \in \mathbb{R}^n$. The function $d(x, y)$ defined by

$$d(x, y) = [(x_1 - y_1)^2 + (x_2 - y_2)^2 + \cdots + (x_n - y_n)^2]^{\frac{1}{2}} \tag{A3.3}$$

is a metric on $\mathbb{R}^n$. Proof of the triangle inequality requires a little work, but we shall content ourselves with assuming the result. This metric is known as the **Euclidean metric** on $\mathbb{R}^n$. The spaces $\mathbb{R}^n$ furnished with the metric (A3.3) are known as **Euclidean spaces**.

Note that, according to Definition A3.13, the Minkowski 'metric'

$$s^2(x, y) = (x_0 - y_0)^2 - (x_1 - y_1)^2 - (x_2 - y_2)^2 - \cdots - (x_{n-1} - y_{n-1})^2 \quad \text{(A3.4)}$$

on $\mathbb{R}^n$ is not a metric at all; it violates every condition except the first of Definition A3.13. In mathematics, it would generally be called an indefinite quadratic form. We shall call it the **Minkowski form**.

In the literature, the term **pseudometric** is used both for the Minkowski form and for the semimetric. To avoid confusion, we shall not use this term at all.

A metric on X defines a topology on X, called the **metric topology**. Let d be a metric on X, and define the **open ball** with centre a and radius $r > 0$ as follows:

$$B_r(a) = \{x | x \in X, d(a, x) < r\}.$$

The family of open balls is a base for a topology on X; this topology is the metric topology.

Different metrics may induce the same topology on a set X. For example, the function

$$d(a, b) = \max\{|a_1 - b_1|, |a_2 - b_2|\}$$

(where (x_1, x_2) are the Cartesian coordinates of the point x) defines a metric on $\mathbb{R}^2$ (it requires some work to show this) which is manifestly different from the Euclidean metric $[(x_1 - y_1)^2 + (x_2 - y_2)^2]^{1/2}$; nevertheless, the topology induced by this metric on $\mathbb{R}^2$ is the same as that induced by the Euclidean metric.

The reader is invited to show that the metric topology on $\mathbb{R}$ is the same as the order topology on it.

Let (X, d) and (Y, δ) be metric spaces. A map $f : X \to Y$ is called an **isometry** if it preserves distances, i.e., if

$$d(x, y) = \delta(f(x), f(y)) \text{ for all } x, y \in X.$$

A3.5.1 Metrizability

A topological space $(X, \mathcal{T})$ is said to be **metrizable** if it admits a metric d such that the metric topology $\mathcal{T}_d$ induced by it on X coincides with the topology $\mathcal{T}$.

Theorem A3.14 (Urysohn's metrization theorem) *A regular second countable space is metrizable.*

One can show that regularity is necessary for metrizability, but that the second countability condition can be weakened. Details may be found in (Munkres, 1975, Chapter 6).

The *metric completion* of the metric space $\mathbb{Q}$ was touched upon in Subsection A2.2.3. We need to discuss the completion of an arbitrary metric space (X, d) in some detail, but shall postpone this discussion to Appendix A4.

A3.6 Compactness

A **cover** of a set X is a collection of subsets $\{V_\alpha\}$ such that $X = \cup_\alpha V_\alpha$. The straight line can be covered in many ways by open intervals of unit length. None of these covers has a finite subcover, for the simple reason that every finite collection of open unit intervals is bound to have a finite total length. Now consider the unit circle, i.e., the circle of unit radius. Its circumference is 2π. Seven open arcs of unit length will therefore suffice to cover the unit circle. These simple examples exhibit the difference between compact and noncompact topological spaces.

A cover of a topological space by open sets is called an **open cover**. A topological space is called **compact** if every open cover has a finite subcover. The emphasis is on the word *every*, for since X is open in itself, every topological space has at least one finite open subcover. A space that is not compact is called **noncompact**. The reader is invited to prove the following result: *compactness is a topological invariant.* That is, if X and Y are homeomorphic and one of them is compact, then so is the other.

A3.6.1 Some properties of compact spaces

Compact spaces have many useful properties (which is the mathematician's way of saying that compactness is the source of quite a bit of mathematical structure). We begin with some that do not involve sets of real numbers:

(i) A closed subset of a compact space is compact.
(ii) If X is compact and $f : X \to Y$ is continuous, then $f(X)$ is a compact subspace of Y. This result is generally stated as follows: the continuous image of a compact space is compact.

Next, compactness combines nicely with the Hausdorff property to yield strong results:

(i) A compact subset of a Hausdorff space is closed.
(ii) A compact Hausdorff space is normal.

Let us now turn our attention to the Euclidean spaces $\mathbb{R}^n$. These are metric spaces, but we are considering only their topological aspects, the topologies being defined by the Euclidean metric. Recall that for $n = 1$, the metric topology coincides with the order topology, and is called the usual topology.

Since $\mathbb{R}$ is noncompact, and the open interval $(0, 1)$ is homeomorphic to it, it follows that $(0, 1)$ is noncompact. However, the *closed* interval $[0, 1]$ is compact (in the subspace topology inherited from $\mathbb{R}$). To prove it, one needs the least upper bound property of $\mathbb{R}$, in one form or another. Details may be found in (Munkres, 1975, pp. 173–174).

Theorem A3.15 *Every closed interval of the real line is compact.*

The reader is invited to make the connections necessary to prove that

Theorem A3.16 *A continuous real-valued function on a compact space has a maximum and a minimum value.*

We are now ready to enunciate the **Tychonoff theorem.**

Theorem A3.17 (Tychonoff's theorem) *A product of arbitrarily many compact spaces is compact.*

Using the definition of the product topology, it is quite easy to prove this result for the product of two compact spaces. Using that proof as a model, one can prove it for a product of n compact spaces. However, the proof of the general theorem requires the axiom of choice. Tychonoff's theorem is one of the most important in general topology, and a detailed proof will be found in every textbook on the subject.

Tychonoff announced the proof of his theorem, which used the axiom of choice, in 1929. The importance of his theorem was realized immediately, but at that time mathematicians were still not happy with the axiom of choice. Many attempts were made to prove Tychonoff's theorem without using the axiom of choice, but they were unsuccessful. This led Kakutani to conjecture that the Tychonoff theorem was *equivalent* to the axiom of choice. In 1950, Kelly proved Kakutani's conjecture, which meant that rejecting the axiom of choice would imply rejecting Tychonoff's theorem. Most mathematicians were unwilling to pay this price.

Since the closed unit interval $[0, 1]$ is compact, it follows from Tychonoff's theorem that the cubes $[0, 1]^J$ are compact. Here J is the cardinality of the indexing set that defines the product (see A1.26).

A subset A of a metric space (X, d) is said to be **bounded** if there exists a real number K such that $d(a, b) \leq K$ for all $a, b \in A$.

Theorem A3.18 *A subset A of $\mathbb{R}^n$ is compact (in the metric topology) iff it is closed and bounded.*

A3.6.2 Local compactness

A topological space X is said to be **locally compact** if, for each $x \in X$, there exist subsets C_x, U_x of X such that C_x is compact, U_x is open, and $x \in U_x \subset C_x$; in other words, if every point has a compact neighbourhood.

The spaces $\mathbb{R}^n$ are locally compact for all (finite) n. Infinite-dimensional Hilbert space is not locally compact.

A3.6.3 One-point compactification

Several procedures have been devised for turning a locally compact space X into a compact space. We shall describe one of these.

Definition A3.19 (One-point compactification) Let X be a locally compact Hausdorff space, and take an object (often denoted by the symbol ∞) that is not in X. Let $Y = X \cup \{\infty\}$. The collection of subsets (of Y)

(a) U, where U is open in X, and

(b) $Y \setminus C$, where C is a compact subset of X,

defines a topology on Y. This topology makes Y into a compact space, which is called the **one-point compactification** of X.

It takes a little work to show that the above defines a topology on Y, and that Y is compact. The sets C are closed in X, and therefore the subspace topology that X inherits from Y is the same as its original topology. (Had it been otherwise, the term 'compactification' would have been ill advised.)

The one-point compactification of the real line is the circle S_1. The one-point compactification of $\mathbb{R}^n, n \geq 1$, is called the *n-sphere* S_n. The n-sphere is homeomorphic to the surface of the closed unit ball in Euclidean $\mathbb{R}^{n+1}$. The proofs have required the development of the subjects called combinatorial, algebraic and differential topology, and are the work of many mathematicians spread over a hundred years.

A3.7 Connectedness and path-connectedness

Let X be a topological space. If there exist two nonempty open subsets U, V of X such that $U \cap V = \emptyset$ and $U \cup V = X$, then the pair U, V is said to form a **separation** (or **disconnection**) of X, and U, V are called **components** of X. If X has no disconnection, i.e., it has only one component, then X is said to be **connected**. The real line $\mathbb{R}$ is connected, and so is every interval, finite or semi-infinite, of it; $\mathbb{R} \setminus \{O\}$ (the real line with the origin removed) is disconnected; its components are $(-\infty, 0)$ and $(0, \infty)$. All indiscrete spaces are connected, and all discrete spaces containing two or more points are disconnected. A space is called **totally disconnected** if its connected subsets are exclusively one-point sets. A discrete space is totally disconnected.

The following is a useful result. Let X be a connected topological space, Y a topological space and $f : X \to Y$ a continuous map. Then $f(X)$ is connected. *The continuous image of a connected space is connected.* It follows that if Y is disconnected, $f(X)$ must lie entirely in one of its components.

The *intermediate value theorem* of calculus is a consequence of the connectedness of $\mathbb{R}$.

Theorem A3.20 (Intermediate value theorem) *Let X be a connected space and $f : X \to \mathbb{R}$ a continuous map. If $a, c \in X$, $f(a) < f(c)$, then for any r such that $f(a) < r < f(c)$, there exists $b \in X$ such that $f(b) = r$.*

Let X be a topological space. A **path** from a to b in X is a continuous map $f : [0, 1] \to X$ such that $f(0) = a$ and $f(1) = b$. The space X is said to be **path-connected** if every pair of points $x, y \in X$ can be joined by a path.

It is rather easy to see that a path-connected space is connected. Examples show that the converse is not necessarily true.

A3.8 The theory of convergence

Recall that on the real line a sequence $\{x_n\}$ is said to **converge** to a point x if, given $\varepsilon > 0$, there exists an integer N such that $|x_k - x| < \varepsilon$ for all $k > N$. **Cauchy's criterion of convergence** states that the sequence $\{a_n\}$ is convergent if, given $\varepsilon > 0$, there exists an integer N such that $|a_k - a_l| < \varepsilon$ for all $k, l > N$. If $\{a_n\}$ is convergent, then the limit $\lim_{n\to\infty} a_n$ exists, and is unique.

Both the definition of convergence and Cauchy's criterion generalize to arbitrary metric spaces (X, d); all one has to do is to replace $|a_k - a_l|$ by $d(a_k, a_l)$. A sequence $\{a_n\}$ that has the property that $d(a_k, a_l) < \varepsilon$ whenever $k, l > N$ is called a **Cauchy sequence**. If a Cauchy sequence converges, then it converges to a unique point. A metric space in which every Cauchy sequence converges is called **complete**. The real line is complete; the open interval $(0, 1)$ is not; the sequences $\{a_n = 1/n\}$ and $\{b_n = 1 - 1/n\}$, $n = 2, 3 \ldots$ do not converge, because the points 0 and 1 do not belong to the interval $(0, 1)$.

On a topological space X, a sequence $\{x_n\}$ is said to converge to x $(x \in X)$ if every neighbourhood of x contains points of the sequence other than x. If U is a neighbourhood of x, one says that $\{x_n\}$ is **ultimately in** U if $x_k \in U$ for all $k > N$, where N is some positive integer. On an indiscrete topological space, every sequence converges to every point of the space. On a discrete topological space, the only sequences which converge are the constant sequences: $x_n = x$ for all $n \in \mathbb{N}$. However, in a Hausdorff space, a sequence can converge to only one point. For suppose that $\{x_n\}$ converges to x (one writes $x_n \to x$) and $y \in X$, $y \neq x$. Then there exist open sets U_x, U_y such that $x \in U_x$ $y \in U_y$ and $U_x \cap U_y = \emptyset$. Since $x_n \to x$, the sequence $\{x_n\}$ is ultimately in U_x, and therefore it cannot be ultimately in U_y.

A3.8.1 Filters

A topological space X is said to have a **countable base at** x if there exists a countable set of open sets $\{U_n | x \in U_n \text{ for all } n\}$ such that if V is an open set containing the point x, then $U_k \subset V$ for some k. If X has a countable base at each of its points, then it is said to be **first countable**, or to satisfy the **first axiom of countability**. (It is obvious that a second countable space is first countable.) It turns out that in first countable spaces all limit points can be detected by means of convergent sequences, and therefore the topology of the space can be expressed entirely in terms of convergent sequences.[10] However, there are spaces that are not first countable, and one has to generalize the notion of sequences if one wants to express the topology in similar terms. Two essentially equivalent generalizations have been proposed: **nets**, which involve a generalization of the notion of order, and the rather different objects called **filters**. We shall encounter spaces that are not first countable only in the context of uniformities and uniform completion. We shall treat these subjects in Appendix A4, basing ourselves, in the main, on the text by James (James, 1999). As this text uses filters, a brief discussion of filters is given below.

Definition A3.21 (Filter on a set) A **filter** $\mathcal{F}$ on a set X is a collection of *nonempty* subsets of X such that

(a) If $F \in \mathcal{F}$ and $G \supset F$, then $G \in \mathcal{F}$.

(b) If $F_1, F_2 \in \mathcal{F}$ then $F_1 \cap F_2 \in \mathcal{F}$.

The definition implies that the intersection of a finite number of members of a filter is nonempty; however, this does not necessarily apply to intersections of infinitely many elements of a filter.

On any set X, the subsets containing the point x form a filter, called the **principal filter** generated by x. We shall denote it by $\mathcal{P}_x$. In the same way, any nonempty subset W of X generates a filter on X. If $\{x_n\}$ is a sequence of points on X, then the family of subsets M of X such that $\{x_n\}$ is ultimately in M is a filter on X. It is called the **elementary filter** associated with the sequence $\{x_n\}$. If $\{y_n\}$ is another sequence such that $x_k = y_k$ for $k > N$, then the elementary filters associated with $\{x_n\}$ and $\{y_n\}$ are the same.

Definition A3.22 (Refinement of a filter) If $\mathcal{F}$ and $\mathcal{G}$ are two filters on X such that $\mathcal{F} \supset \mathcal{G}$ (i.e., $G \in \mathcal{G} \Rightarrow G \in \mathcal{F}$), then $\mathcal{F}$ is said to **refine** $\mathcal{G}$, or to be a **refinement of** $\mathcal{G}$.

A3.8.1.1 Filters on topological spaces

In a topological space X, the family of neighbourhoods of a point x is clearly a filter on X; we denote it by $\mathcal{N}_x$, and call it the **neighbourhood filter** at x.

[10] For details, see (Munkres, 1975, p. 190).

The intersection of all members of $\mathcal{N}_x$ is nonempty; it contains at least the point x.

The principal filter at x is clearly a refinement of the neighbourhood filter at x.

Definition A3.23 (Limit points and convergence in terms of filters) Let X be a topological space and $\mathcal{F}$ a filter on it. If $\mathcal{F}$ is a refinement of the neighbourhood filter $\mathcal{N}_x$ of $x \in X$, then x is said to be a **limit point** of $\mathcal{F}$. If x is a limit point of $\mathcal{F}$, then $\mathcal{F}$ is said to **converge** to x, and written $\mathcal{F} \to x$.

Clearly, if $\mathcal{F}$ converges to x then so does any refinement of $\mathcal{F}$. For any $x \in X$, the principal filter at x refines the neighbourhood filter at x, and therefore converges to x. However, a filter on X does not have to converge; if it does, it may converge to more than one point. If, for example, every neighbourhood of x is also a neighbourhood of y, $y \neq x$ (X is not Hausdorff), then $\mathcal{F} \to x$ implies that $\mathcal{F} \to y$. However, on a Hausdorff space a filter can converge to at most one point:

Theorem A3.24 *Let X be a Hausdorff space, $x, y \in X$, $x \neq y$. If $\mathcal{F}$ is a filter on X that converges to X, then $\mathcal{F}$ cannot converge to y.*

Proof Since X is Hausdorff, x and y are separated by disjoint neighbourhoods U_x and U_y. Being disjoint, they cannot both belong to the same filter. That is, if $\mathcal{F} \to x$, then there exists a neighbourhood U_y of y such that $U_y \notin \mathcal{F}$, which is the same as saying that $\mathcal{F}$ does not converge to y. □

A3.9 Topological groups

A **topological group** is a group G which is also a topological space such that the group operations (the multiplication map $G \times G \to G$ and the inverse map $G \to G$, defined by $g \mapsto g^{-1}$) are continuous. The topology on $G \times G$ is the product topology. The basic results on topological groups are easy to establish, and may therefore be found in the exercise sections of most books on general topology, as well as in the very readable book by Pontrjagin (Pontrjagin, 1939).

If a topological group is a T_0-space, then the group structure forces it to be a Tychonoff space. For this reason, works on abstract harmonic analysis[11] generally assume that the topological groups under consideration[12] are T_0. However, a T_0-topological group may fail to be normal, and therefore need not be metrizable.[13]

[11] We quote from (Hewitt and Ross, 1963), p. 1, paragraph 2: 'Using a fundamental construction published in 1933 by A. Haar, A. Weil in 1940 showed that Fourier series and integrals are but special cases of a construct which can be produced on a very wide class of topological groups.' *Abstract harmonic analysis* is the subject that grew out of Weil's observation.

[12] See (Hewitt and Ross, 1963, p. 83 and p. 19).

[13] See (Hewitt and Ross, 1963, pp. 70–75).

Continuous groups of interest to physicists, such as the Poincaré or Galilei groups and their subgroups, the group $SU(3)$, etc., usually carry 'natural' topologies. The Poincaré group P is the semidirect product[14] of the group T of space-time translations and the homogeneous Lorentz group L: $P = T \wedge L$. The Lorentz group L is a *closed* subgroup of P, meaning that it is a closed set in the topological space P. It is also a normal subgroup of P, so that the quotient space P/L is a group; it is the translation group T. Quotient spaces will be important to us only in the context of Lie groups. A **Lie group** is a topological group in which the multiplication and inverse maps are **analytic**, i.e., functions that have convergent Taylor series expansions.

When is a topological group a Lie group? This is Hilbert's fifth problem, which was solved in 1952 by Gleason, and independently by Montgomery and Zippin. The answer can be formulated quite simply. A topological group G is said to have *no small subgroups* (NSS) if there is a neighbourhood of the identity e in which the only subgroup of G is $\{e\}$. The answer is: *A topological group is a Lie group if and only if it is locally compact and has no small subgroups.* The abbreviation NSS, understood as an adjective, is sometimes used to describe such topological groups.

[14] There seems to be no standard notation for semidirect products.

A4
Completions

There are many differences between the spaces $\mathbb{R}^n$ and $\mathbb{Q}^n$, but the one we shall single out is that differentiable functions[1] can be defined on $\mathbb{R}^n$, but not on $\mathbb{Q}^n$. It is this fact that invests the process of *completion* – i.e., passage from $\mathbb{Q}^n$ to $\mathbb{R}^n$ – with so much interest.

A completion process requires more structure than topology. We have already discussed the Dedekind completion of the rationals, which is based on the concept of order and cannot be extended to sets that are not totally ordered. The most important class of spaces that can be completed are the metric spaces; a metric, as we have already noted, imposes more structure than a topology. Finally, there are the structures called *uniformities*, weaker than metrics but stronger than topologies, that can also be completed. Remarkably, the completion of uniform spaces, unlike that of metric spaces, does not require the explicit use of real numbers.

We shall discuss metric completion, uniformities and uniform completion in this appendix. A metric space can be completed in at least two different ways (with the same result); one can be generalized to uniform spaces, and the other cannot. We shall discuss only the former. Similarly, uniformities can be defined in at least three different but equivalent ways; we shall choose the one which is best adapted to generalizing the procedure of metric completion. The summaries given below will not provide balanced pictures of their subjects.

A4.1 Metric completion

Recall that a metric space is called **complete** if every Cauchy sequence in it converges. A metric d on X is called a **complete metric** if the space (X, d) is a complete metric space. A topological space is called **completely metrizable** if it admits a complete metric. A completely metrizable space is also called **topologically complete**; the property of topological completeness, unlike that of metric completeness, is preserved under homeomorphisms. One should note that a completely metrizable topological space X may be metrizable with two different metrics, one of which is complete and the other is not. A counterintuitive example is provided by the space $\mathbb{P} = \mathbb{R} \setminus \mathbb{Q}$ of the irrationals, which is known to be completely metrizable; it is clearly not complete in the usual metric (inherited from $\mathbb{R}$).

[1] In the usual sense; variants of the differential calculus have been devised on discontinua, but these have remained more exotic than useful.

A topological space may be metrizable without being completely metrizable; the standard example is that of the rationals[2] $\mathbb{Q}$. By contrast, a *metric* space (X, d) can always be completed, *by adding more points to* X. To see how this is to be achieved, let (X, d) be a metric space and consider the set of Cauchy sequences on it. Define two Cauchy sequences $\{x_n\}$ and $\{y_n\}$ to be equivalent, written $\{x_n\} \sim \{y_n\}$, if

$$\lim_{n\to\infty} d(x_n, y_n) = 0. \tag{A4.1}$$

The reflexivity and symmetry of $\sim$ are obvious; transitivity, i.e.,

$$\{a_n\} \sim \{b_n\}, \ \ \{b_n\} \sim \{c_n\} \Rightarrow \{a_n\} \sim \{c_n\},$$

follows from the triangle inequality

$$d(a_n, c_n) \leq d(a_n, b_n) + d(b_n, c_n).$$

Two sequences that share the same tail are clearly equivalent, but there is more; an equivalence class consists of all Cauchy sequences on X that 'try to converge' to the same point. If the sequences in an equivalence class converge to a point $a \in X$, then *this equivalence class contains the constant sequence* $\{a_n = a\}$. Since a metric space is Hausdorff, a convergent sequence cannot converge to more than one point; an equivalence class cannot contain two distinct constant sequences.

The metric space (X, d) is complete if every equivalence class contains a constant sequence.

Definition A4.1 (Metric completion) Let (X, d) be a metric space. Its **metric completion** is a metric space (X^*, d^*) such that:

(a) There is an injective map $\imath : X \to X^*$ such that $\imath(X)$ is dense in X^*. (The topology of X^* is the topology induced by the metric d^*.)
(b) The bijective map $\imath : X \to \imath(X)$ so defined is an isometry.

The map $\imath$ is called an **isometric embedding** of X into X^*.

One has to prove that the definition is not empty of content, i.e., the object defined actually exists. The proof is constructive, and straightforward. One would also like to know whether or not the object constructed is unique. The affirmative answer will be provided after the construction.

The construction of X^* from X is as follows. The space X^* is just the set of equivalence classes of Cauchy sequences on X as defined by (A4.1). We denote the equivalence class of the sequence $\{x_n\}$ in X by $[x_n]$. If $[x_n]$ contains the

[2] The proofs of the assertions $\mathbb{P}$ *is completely metrizable but* $\mathbb{Q}$ *is not* require much more machinery than we are able to develop here. The interested reader is referred to (Willard, 1970).

constant sequence $a_n = a$, we write it as $[a]$ for brevity. When $[a]$ is considered as a point in X^*, we denote it by a. If $[x_n]$ does not contain any constant sequence, then it determines a point in X^* which is not in X.

Next, define a function d^* on X^* by

$$d^*([x_n], [y_n]) = \lim_{n\to\infty} d(x_n, y_n). \tag{A4.2}$$

Clearly, $d^*([a], [b]) = d(a, b)$. The function d^* is nonnegative, and is symmetric in its arguments. The reader is invited to prove that it satisfies the triangle inequality. Furthermore, if $[x_n] \neq [y_n]$, then $d^*([x_n], [y_n]) \neq 0$. Thus d^* is a metric on X^*. The map $\imath : X \to X^*$ defined by $\imath(x) = [x]$ is an isometric embedding. Any point in X^* that is not in $\imath(X)$ is a limit point of $\imath(X)$ in the topology of the metric d^*, and therefore $\imath(X)$ is dense in X^*. Finally, every equivalence class of Cauchy sequences in (X^*, d^*) contains a constant sequence; X^* is complete, and *its* completion would add no new points.

We have proved that:

(i) d^* is a metric on X^*.
(ii) (X^*, d^*) is a complete metric space.
(iii) $\imath : X \to X^*$ defined by $\imath(x) = [x]$ embeds X densely and isometrically in X^*.

There remains the question of uniqueness. This is easily settled. If (Y, δ) is another completion of (X, d), then X is densely embedded in Y. Let $\jmath : X \to Y$ be this embedding. We shall take the liberty of writing $\jmath(x) = x$ and describing it as 'a point in X' for every $x \in X$. Then the points of Y are limits of Cauchy sequences in X, and the metric δ on Y is defined by the same formula (A4.2) as d^*. Define now a map $\varphi : Y \to X^*$ by $\varphi(y) = [x_n]$, where $x_n \to y \in Y$. This map is an isometry, and, if $y \in X$, then $\varphi(y) = [y]$, i.e. φ leaves X pointwise fixed. That is, the completion is unique up to an isometry that leaves X pointwise fixed.

We conclude this section with defining the concept of total boundedness for metric spaces, which establishes the connection between completeness and compactness. This concept will later be generalized to uniform spaces. Recall that, in a metric space X, the open ball of radius $\varepsilon > 0$ and centre a is the set $B_\varepsilon(a) = \{x | d(x, a) < \varepsilon\}$. Here $x, a \in X$ and d is the metric on X.

Definition A4.2 A metric space is called **totally bounded** if it can be covered by a finite number of ε-balls for every $\varepsilon > 0$.

The significance of the concept of total boundedness lies in the following results.

Proposition A4.3 *A metric space is totally bounded if and only if each sequence has a Cauchy subsequence.*

Proposition A4.4 *A metric space is compact if and only if it is totally bounded and complete.*

Examples A4.5 Consider the intervals $(0,1)$ and $[0,1]$ of $\mathbb{R}$, and the set $W = \mathbb{Q} \cap [0,1]$. All are totally bounded. However, neither $(0,1)$ nor W is compact, whereas $[0,1]$ is. W is not compact because it is not closed (page 260).

Because of this connection with compactness, a totally bounded metric space is also known as a **precompact** space.

A4.2 Uniformities

If we look upon topology as growing out of the attempt to free the notion of continuity from its dependence on real numbers, we may equally well look upon uniformities as growing out of the attempt to free the notion of completeness from *its* dependence on real numbers. However, this is only a part of the story; one has to look at history to provide a wider perspective, and an explanation for the use of the term.

Consider, therefore, the theory of functions of several real variables, and let $f : D \to \mathbb{R}$, where $D \subset \mathbb{R}^n$. The function f is said to be continuous at the point $x_0 \in S$ if, given $\varepsilon > 0$, there exists $\delta > 0$ such that $|f(x) - f(x_0)| < \varepsilon$ whenever $d(x, x_0) < \delta$. The number δ generally depends both on x_0 and on ε; $\delta = \delta(x_0, \varepsilon)$. A function f defined on a domain D is said to be continuous if it is continuous at every point in D.

Contrast this with the definition of uniform continuity. The function f is said to be uniformly continuous if, given $\varepsilon > 0$, there exists $\delta > 0$ such that $|f(x) - f(y)| < \varepsilon$ whenever $d(x, y) < \delta$. There is no reference to any specific point x_0; the definition relates, instead, to *pairs of points* $x, y \in D$ which are close to each other (in some sense), but could be located anywhere in the set D. It is this last feature which is reflected in the term *uniform*.

A metric obviously defines a notion of closeness between two points, but the notion itself may be definable without recourse to a metric. Let us go back to Cauchy's criterion of convergence for a sequence $\{s_n\}$ of real or complex numbers, which says that the sequence converges if, given $\varepsilon > 0$, there is a positive integer N such that $|s_k - s_l| < \varepsilon$ whenever $k, l > N$. In a metric space which is not necessarily complete, a sequence $\{s_n\}$ is called a Cauchy sequence if it satisfies the above condition.

We now attempt to exploit the following fact: both $\mathbb{R}$ and $\mathbb{C}$ are Abelian topological groups, with 0 and $0 + \mathrm{i}0$ as their respective identities.[3] This enables Cauchy's criterion to be paraphrased in terms of neighbourhoods of the identity: the sequence $\{s_n\}$ is Cauchy if, given any neighbourhood of the identity V, one has $s_k - s_l \in V$ whenever $k, l > N$. If now G is a non-Abelian (meaning not

[3] The reader is invited to verify that $\mathbb{R}$ and $\mathbb{C}$ are indeed topological groups under addition.

necessarily Abelian) topological group, then the sequence $\{g_n\}$ may reasonably be called Cauchy if, given any neighbourhood V of the identity e, there exists a positive integer N such that $g_l^{-1}g_k \in V$ whenever $k, l > N$. We have got rid of the metric by using the notion of group multiplication. However, one can do better; one does not need any predefined mathematical structure on X.

One does need some definitions from the calculus of sets. Recall that a **relation** on a set X was defined (on page 263) to be a subset of the Cartesian product $X \times X$. The following operations are defined on the family of relations on X:

(i) **Inverse** If E is a relation, then the inverse relation E^{-1} is defined to be

$$E^{-1} = \{(x, y) | (y, x) \in E\}.$$

(ii) **Composition** If E and F are relations on X, then their composition $E \circ F$ is defined to be

$$E \circ F = \{(x, z) | (x, y) \in E \text{ and } (y, z) \in F \text{ for some } y \in X\}.$$

A relation E is called **symmetric** if $E = E^{-1}$. Finally, the **diagonal** Δ of a set $X \times X$ is the subset $\Delta = \{(x, x) | x \in X\}$.

A4.2.1 Definition and general properties

We are now ready to define uniformities.

Definition A4.6 (Uniformities) A *uniformity* on a set X is a filter $\mathcal{E}$ on the Cartesian product $X \times X$, consisting of subsets called **entourages** or **surroundings**, such that:

(a) If $E \in \mathcal{E}$, then $\Delta \subset E$.

(b) If $E \in \mathcal{E}$, then there exists an entourage F such that $F \subset E^{-1}$.

(c) If $E \in \mathcal{E}$, then there exists an entourage F such that $F \circ F \subset E$.

The pair $(X, \mathcal{E})$ is called a **uniform space**.

If $E \in \mathcal{E}$, then $E^{-1} \in \mathcal{E}$; condition (b) is satisfied by taking $F = E^{-1}$. Consequently, some authors, e.g., (Kelley, 1955) replace condition (b) by

(b′) If $E \in \mathcal{E}$ then $E^{-1} \in \mathcal{E}$.

A uniformity $\mathcal{E}$ on X is called **Hausdorff** (or **separated**, or **separating**) if

$$\bigcap_{E \in \mathcal{E}} E = \Delta.$$

Definition A4.6(c) has been described by Kelley as 'a vestigial form of the triangle inequality'. The composition operation $\circ$ is also reminiscent of the multiplication operation on topological groups. Definition A4.6 does not involve any number system, nor, for that matter, any topology on X. However, we are still some distance from our stated goal, which is to define a notion of completeness without using the real number system.

It should be noted that there are two other (but equivalent) definitions of uniformities; by means of *uniform covers* (roughly speaking, covers by sets of the same 'size') and via a family of *semimetrics*.[4] These definitions are generally more suitable for applications. We refer the reader to the texts (Kelley, 1955; Willard, 1970) for details, or to the monograph (Borchers and Sen, 2006) for a summary, and for further references.

Examples A4.7 (Examples of uniformities)

(i) For any set X, the family of all supersets of the diagonal Δ defines a uniformity called the **discrete uniformity**, and the family consisting of the single set $X \times X$ a uniformity called the **trivial uniformity**.

These examples establish that uniformities exist; otherwise, they are of no interest.

(ii) Let (X, d) be a metric space, $\epsilon > 0$, and define

$$D_\epsilon = \{(x,y)|(x,y) \in X \times X,\, d(x,y) < \epsilon\}.$$

The family of all supersets of the D_ϵ for all $\epsilon > 0$ defines a uniformity on X called the **metric uniformity** of the metric d.

A uniformity $\mathcal{E}$ on X defines a topology on X. Let $E \in \mathcal{E}$ and define subsets $E[x]$ of X by

$$E[x] = \{y|(x,y) \in E, x \in X\}.$$

The **topology of the uniformity** $\mathcal{E}$, or the **uniform topology**, is the family $\mathcal{T}$ of all subsets U of X such that for each $x \in X$ there is an $E \in \mathcal{E}$ such that $E[x] \subset U$. It is clear from the definition that the family $\mathcal{T}$ covers X, and is closed under arbitrary unions. It is not difficult to show that it is closed under finite intersections (Kelley, 1955). It should come as no surprise that the uniform topology of the discrete uniformity is discrete, while that of the indiscrete uniformity is indiscrete.

While the topology of a uniformity is uniquely determined by the uniformity, examples show that different uniformities may generate the same topology. Thus a uniformity has more structure than a topology. Now let X be a point-set and d, d' two different metrics on it. As we saw above, each metric determines a unique

[4] In the theory of uniformities, the preferred term seems to be *pseudometric*.

uniformity on X. However, different metrics may generate the same uniformity on X (for example the metrics d and $2d$). Thus a metric has more structure than a uniformity.

One should note the following result, which is important.

Theorem A4.8 *The topology $\mathcal{T}$ of a uniformity $\mathcal{E}$ on X is Hausdorff if and only if $\mathcal{E}$ is Hausdorff.*

A4.2.2 Uniformizability of a topological space

A topological space $(X, \mathcal{T})$ is called **uniformizable** if there exists a uniformity $\mathcal{E}$ on X such that the uniform topology it induces is $\mathcal{T}$.

When is a topological space uniformizable? This important question is answered by the following result.

Theorem A4.9 (Uniformizability of a topological space) *A topological space is uniformizable if and only if it is completely regular.*

Recall that we do not require one-point sets to be closed in a completely regular space (page 267).

A4.2.3 Metrizability of a uniform space

A uniform space $(X, \mathcal{E})$ is called **metrizable** if there exists a metric d on X such that the uniformity it induces on X is $\mathcal{E}$.

When is a uniform space metrizable? This question is answered by the following result.

Theorem A4.10 (Metrizability of a uniform space) *A uniform space is metrizable if and only if it is Hausdorff and has a countable base.*

We are now within sight of our goal, which is to establish that the notion of completeness can be extended to uniform spaces without the use of real numbers.

A4.3 Complete uniform spaces

We begin with some basic definitions and results.

Definition A4.11 (Uniform continuity) Let $(X, \mathcal{E})$ and $(Y, \mathcal{K})$ be uniform spaces. A map $f : X \to Y$ is said to be **uniformly continuous** if for each $K \in \mathcal{K}$ there is some $E \in \mathcal{E}$ such that $(x_1, x_2) \in E \Rightarrow (f(x_1), f(x_2)) \in K$.

The reader is invited to prove that a uniformly continuous function is continuous (in the uniform topologies).

Next, we define the notion of uniform equivalence of two uniform spaces. This notion is analogous to the notion of homeomorphism of topological spaces, but there appears to be no shorter term for it.

Definition A4.12 (Uniform equivalence) The uniform spaces X and Y are said to be **uniformly equivalent** if there exists a bijection $\phi : X \to Y$ such that both ϕ and ϕ^{-1} are uniformly continuous.

Finally, we come to the definition of Cauchy sequences in uniform spaces.

Definition A4.13 (Cauchy sequences in uniform spaces) A sequence of points $\{x_n\}$ in a uniform space X is called a **Cauchy sequence** if, for each entourage E in $X \times X$, there is an integer N such that $(x_k, x_l) \in E$ whenever $k, l > N$.

This definition does not involve the real numbers. But what justifies use of the term Cauchy sequence? This very reasonable question is answered by the following result.

Proposition A4.14 *Let $\{x_n\}_{n \in \mathbb{N}}$ be a sequence of points in the uniform space X. If this sequence converges in the uniform topology, then it is a Cauchy sequence.*

A uniformly continuous function maps Cauchy sequences to Cauchy sequences.

Proposition A4.15 *Let X and Y be uniform spaces and $f : X \to Y$ a uniformly continuous function. If $\{x_n\}_{n \in \mathbb{N}}$ is a Cauchy sequence in X, then $\{f(x_n)\}_{n \in \mathbb{N}}$ is a Cauchy sequence in Y.*

A4.3.1 Uniform completion

A metric space is necessarily first countable (in the metric topology), but a uniform topology need not be first countable; i.e., its topology may not be describable by sequences. It will surely be describable by filters, and therefore it is pleasing that Cauchy filters are definable as easily as Cauchy sequences:

Definition A4.16 (Cauchy filters in uniform spaces) A filter $\mathcal{F}$ in the uniform space $(X, \mathcal{E})$ is called a *Cauchy filter* if, for each entourage E of the uniformity, there exists a member F of $\mathcal{F}$ such that $F \times F \subset E$.

Observe that this definition, like Definition A4.13, has no dependence on the notion of real numbers.

Proposition A4.14 has an analogue for filters. It is:

Proposition A4.17 *Let $\mathcal{F}$ be a filter on the uniform space X. If $\mathcal{F}$ converges in the uniform topology, then it is a Cauchy filter.*

We now come to the key definition of uniform completeness:

Definition A4.18 (Uniformly complete spaces) A uniform space X is called (**uniformly**) **complete** if every Cauchy filter on X converges to a point in X.

It is of course not necessary for a uniform space to be uniformly complete. However, a uniform space can be completed. The uniform completion of a uniform space is defined as follows:

Definition A4.19 (Uniform completion of a uniform space) A **uniform completion** of a uniform space $(X, \mathcal{E})$ is a pair $(\imath, (X^*, \mathcal{E}^*))$ where $(X^*, \mathcal{E}^*)$ is a complete uniform space and $\imath : X \to X^*$ is a uniform embedding of X as a dense subspace of X^*.

It is always possible to complete a uniform space:

Theorem A4.20 (Existence of uniform completions) *Let $(X, \mathcal{E})$ be a uniform space. There exists a complete uniform space $(X^*, \mathcal{E}^*)$ such that X can be densely and uniformly embedded in X^*. The space X^* is unique in the sense that, if Y is any uniformly complete space that has X as a dense subspace, then X^* and Y are uniformly isomorphic under an isomorphism that leaves X pointwise fixed. Furthermore, X^* is Hausdorff if X is Hausdorff.*

The proof of this theorem requires a little more machinery than we have developed so far, and therefore we shall content ourselves with referring the reader to (James, 1999). It is modelled after the existence proof of the metric completion given earlier. One starts by defining the space $\hat{X}$ of Cauchy filters on X; if X is Hausdorff, then it turns out that $X^* = \hat{X}$.

A4.3.2 Miscellaneous results

We list below a definition, and a number of results that we shall use.

Definition A4.21 A uniform space X is said to be **sequentially complete** if every Cauchy sequence in X converges.

Theorem A4.22 *A complete uniform space is sequentially complete.*

Theorem A4.23 *A closed subset of a complete uniform space is complete.*

Theorem A4.24 *A complete subspace of a Hausdorff uniform space is closed.*

Theorem A4.25 *If X is a compact Hausdorff space, then there exists a unique uniformity on X that generates its topology.*

A4.3.3 Complete uniformizability

In Section A4.2, we tried consciously to avoid using real numbers. Despite our efforts – the theoretical physicist will be pleased to learn – we have not succeeded in avoiding them. What follows is a summary of the relevant definitions and results.

There is an analogue of the notion of complete metrizability in the theory of uniform spaces, which is the following:

Definition A4.26 (Complete uniformizability) A topological space X is said to be **completely uniformizable** if there exists a uniformity in which X is complete, and which induces the topology of X.

Every complete uniform space is completely uniformizable in its uniform topology. Now:

Theorem A4.27 (Shirota's theorem) *A topological space is completely uniformizable if and only if*

(1) *X is Tychonoff;*

(2) *every closed discrete subspace of X has nonmeasurable cardinal; and*

(3) *X can be embedded as a closed subspace in $\mathbb{R}^J$ for some J.*

The cardinal of a set S is called **nonmeasurable** if every countably additive two-valued measure taking the values 0 and 1 (see Subsection A5.4.2, page 301) defined on all subsets of S and vanishing on one-point sets vanishes. All known cardinals are nonmeasurable. Whether or not measurable cardinals exist is an undecidable problem in ZFC.

We shall therefore assume that a uniformly complete Tychonoff space is a closed subspace of some $\mathbb{R}^J$. Mathematically, this will not lead us to inconsistencies. Physically, we are unlikely, at least in the foreseeable future, to call upon large cardinals the nonmeasurability of which is open to doubt. We refer the interested reader to the monograph (Fraenkel, Bar-Hillel and Levy, 2001).

A4.3.4 Total boundedness

There exists a relation between completeness and compactness in uniform spaces, as there is in metric spaces. Once again, it is expressed through the concept of total boundedness.

Definition A4.28 A uniformity $\mathcal{E}$ is called **totally bounded** if, for each $E \in \mathcal{E}$, there exists a finite cover $\{U_1, \ldots, U_m\}$ such that $U_k \times U_k \in E$ for $k = 1, \ldots, n$.

The analogues of Propositions A4.3 and A4.4 for uniform spaces are

Proposition A4.29 *A uniform space is totally bounded if and only if each filter has a Cauchy subfilter.*

Proposition A4.30 *A uniform space is compact if and only if it is totally bounded and complete.*

Examples A4.5 illustrate these propositions as well, as in these cases uniform completion is the same as metric completion.

It should be remarked that equivalent definitions of total boundedness or **precompactness** can be given using uniform covers (Willard, 1970) or pseudometrics (Kelley, 1955). It will not be necessary for us to go into details. We need the concept of total boundedness only for Proposition A4.30, which plays a key role in Chapter 4.

A5
Measure and integral

In quantum mechanics one requires the spaces of square-integrable wave functions to be complete; this cannot be achieved with Riemann-integrable functions. One also needs to determine the structure of self-adjoint operators (observables), the paradigm for which is the diagonalization of $n \times n$ Hermitian matrices. This is, however, a vastly more complex enterprise, and requires a deep understanding of the nature of these operators. Among the tools required for this endeavour, part of which is sketched in Appendix A6, are *measures and integrals.* This appendix will provide an introduction to these subjects tailored to the specific needs of this book.

Historically, the integral now known by his name was announced by Lebesgue in 1902, four years before metric spaces were defined by Fréchet, and more than a decade before the completion process for metric spaces was devised by Hausdorff. Lebesgue's theory was based on the notion of *measure*, which is a generalization of geometrical concepts such as length, area and volume, and of physical concepts such as mass and charge distributions, both discrete and continuous. In the 1920s (possibly earlier), it was noticed that an integral defined a metric[1] on the space of integrable functions, and that the metric space of absolutely Riemann-integrable functions was incomplete. Its completion turned out to be the space of Lebesgue-integrable functions with the metric defined by the Lebesgue integral. This made it possible to develop the 'theory of functions' using only the notion of *sets of measure zero.* However, this simplification is no longer available when one tries to understand, say, the spectrum of a Hamiltonian operator.

For practical purposes such as the numerical solution of differential equations, it is enough to know that the space of Riemann-integrable functions is a metric space which can be completed by the procedure described in detail in Appendix A4; from here, approximation theorems take over. The completion process is abstract, but the elements it adjoins to function spaces are not structureless (as opposed, say, to points in Euclidean geometry). Some aspects of

[1] Actually, a semimetric (see page 268), rather than a metric; we can make this semimetric into a metric by passing to the space of *equivalence classes* of Riemann-integrable functions. Two functions f and g belong to the same equivalence class if the Riemann integral $\int |f(x) - g(x)| \mathrm{d}x$ vanishes. This definition is incomplete as long as we do not have a characterization of the class of Riemann-integrable functions. Fortunately, this lacuna does not affect the discussion in Subsection A5.2.2, which is restricted to a space of *continuous* functions.

their structure cannot be discerned by the tools we have developed so far, which are chiefly topological. For example, the necessary and sufficient conditions for a function of a real variable to be Riemann-integrable are not purely topological. Certain problems, in mathematics as well as in physics, can only be addressed by representing elements of the completed spaces concretely (as functions or their equivalence classes) with sufficient resolution, and we have to develop the tools that make such representations possible.

This appendix is organized as follows. We begin with the general notion of normed spaces because, since the publication of Banach's classic book Théorie des opérations linéaires in 1932, the emphasis in the study of linear spaces has shifted from the metric to the norm. We continue with a discussion of function spaces as normed spaces, with the focus on the problem of norm-completeness. This will naturally lead us to the Lebesgue integral, but we shall first give an account of the Riemann–Stieltjes integral; for us, it will represent the utilitarian aspect of the theory. This being done, we shall move on to σ-algebras, measures and integrals in the abstract setting. As in physics one is working mainly with spaces of functions defined on topological groups, this will be followed by a brief survey of Haar measure. Next, we shall turn to the Lebesgue integral itself, define the L^p-spaces, and state the Riesz–Fischer theorem in its general form. We shall then use the insights gained from the Riemann–Stieltjes integral to introduce noninvariant measures on $\mathbb{R}$, concluding with the Lebesgue decomposition theorem. In the final section, we shall touch upon the subject of differentiation, and indicate 'what would have happened' had we concentrated, from the outset, not on the property of completeness but on the property of differentiability.

A5.1 Normed spaces

Given any set X, the set of real-valued functions on it is a linear space; if f, g are two such functions and a, b two real numbers, then $af + bg$ is again a real-valued function on X. The same holds if the word real is replaced by the word complex in the last sentence. On any linear space, one can define a quantity analogous to the length of a vector in an Euclidean space.[2] The precise definition is given below.

Definition A5.1 (Norm) Let $\mathfrak{F}$ be a linear space over $\mathbb{R}$ (or $\mathbb{C}$). A **norm** on $\mathfrak{F}$, denoted $||\cdot||$, is a map $\mathfrak{F} \to \mathbb{R}$ that satisfies

(a) $||af|| = |a| \cdot ||f||$ for all $a \in \mathbb{R}$ (or $\mathbb{C}$) and $f \in \mathfrak{F}$.

(b) $||f|| \geq 0$ for all $f \in \mathfrak{F}$.

(c) $||f|| = 0$ iff $f = 0$.

(d) $||f + g|| \leq ||f|| + ||g||$ for all $f \in \mathfrak{F}$.

[2] It is important to point out that this notion of length is *not* accompanied by a sense of direction, or of the angle between two directions. This last notion requires an additional mathematical structure, which can be defined in only one special case. This case will be studied in Subsection A5.2.3.

Condition (d) will be recognized as the **triangle inequality**. If condition (c) is dropped, the structure obtained is called a **seminorm**. Owing to the triangle inequality, a norm induces a metric (likewise, a seminorm a semimetric) ϱ on $\mathfrak{F}$ in the following obvious manner:

$$\varrho(f,g) = ||f-g||. \tag{A5.1}$$

However, metrics induced by norms have the property of **translation invariance**:

$$\varrho(f+h, g+h) = \varrho(f,g). \tag{A5.2}$$

Conversely, a translation invariant metric ϱ on a linear space induces a norm by the obvious formula $||f|| = \varrho(f,0)$.

A linear space equipped with a norm is called a **normed space**. The notions of convergence, Cauchy sequence and completeness, defined on metric spaces in Section A3.5, extend to normed spaces in an obvious manner. If one restricts oneself to translation invariant metrics on linear spaces, and to norms and metrics induced by each other (as we shall do in this appendix), *convergence and completeness in the norm become identical with convergence and completeness in the metric*, and the choice of which word to use is no longer restricted by mathematical considerations.

A normed space which is complete in its norm is called a **Banach space**.

A5.2 Function spaces

One can define inequivalent norms on a linear space, just as one can define inequivalent metrics on a set. We shall define many such norms below on function spaces, but shall exploit only three of them. One of these will be used only to provide an example,[3] but the other two will be bread-and-butter norms in quantum mechanics.

A5.2.1 The sup norm

Let C be the set of continuous real-valued functions on the closed interval[4] $[a,b]$. The function $||\cdot||_\infty : C \times C \to [0,\infty)$ defined by

$$||f||_\infty = \sup_{x\in[a,b]} |f(x)| = \varrho(f,0), \quad f \in C \tag{A5.3}$$

[3] To avoid creating a false impression, one should add that this norm, the sup norm, is of considerable importance in mathematical analysis.

[4] We remind the reader that, in the language of the theory of functions, continuity at an end-point of a closed or half-open interval means one-sided continuity. The distinction is unnecessary in the language of topology.

is a norm on C, called the **sup norm.**[5] It is clear that $||f||_\infty \geq 0$, equality holding iff $f = 0$. The reader is invited to prove, using the fact that a continuous real-valued function on a closed interval is bounded, that the triangle inequality for the sup norm

$$||f + g||_\infty \leq ||f||_\infty + ||g||_\infty$$

follows from the triangle inequality for real-valued functions

$$|f(x) + g(x)| \leq |f(x)| + |g(x)| \tag{A5.4}$$

which holds pointwise at each $x \in [a, b]$. The normed space $(C, ||\cdot||_\infty)$ and the equivalent metric space (C, ϱ) are denoted by $\mathcal{C}[a, b]$.

The space $\mathcal{C}[a, b]$ is complete, as we prove below. Let $\{f_n\}$ be a Cauchy sequence in it. For each $\gamma \in [a, b]$, the sequence of real numbers $\{f_n(\gamma)\}$ is a Cauchy sequence. Therefore, by the completeness of real numbers, it converges to a limit $f(\gamma)$. The set of these limits for all $\gamma \in [a, b]$ defines a real-valued function f on $[a, b]$; one says that the sequence $\{f_n(x)\}$ converges **pointwise** to the function $f(x)$. In fact, the convergence is uniform, as the following argument shows. Since $\{f_n\}$ is a Cauchy sequence, given $\epsilon > 0$ there exists a positive integer N such that $||f_n - f_{n+j}||_\infty < \epsilon/2$ for all $n > N$ and $j > 0$. Then for all $x \in [a, b]$

$$|f_n(x) - f(x)| = \lim_{j\to\infty} |f_n(x) - f_{n+j}(x)| \leq \frac{\epsilon}{2},$$

so that $|f_n(x) - f(x)| < \epsilon$ for $n > N$. It is a key result in the theory of uniform convergence that the limit of a uniformly convergent sequence of continuous functions is continuous (see, e.g., the text (Apostol, 1974)). Therefore f is continuous, $f \in C$ and $\mathcal{C}[a, b]$ is complete.

The reader is invited to ascertain that the definition of the norm (A5.3) makes little use of the structure of the underlying space $[a, b]$ on which the functions are defined, in the sense that it remains valid if the interval $[a, b]$ is replaced by any compact space X. What is essential is that a continuous real-valued function on X have a maximum and a minimum value, for which the compactness of X is sufficient.

The famous *Weierstrass approximation theorem* states that a function $f \in \mathcal{C}[a,b]$ can be approximated uniformly as well as one likes by polynomials on $[a, b]$. That is, for any $f \in \mathcal{C}[a, b]$ and any $\varepsilon > 0$, there exists a sequence $\{f_n(x)\}$ of polynomials and a positive integer N such that $||f_n - f||_\infty < \varepsilon$ for $n > N$.

[5] The terms **supremum** and **infimum** (abbreviated **sup** and **inf** respectively) denote the least upper bound and the greatest lower bound of a set of real numbers. They were introduced into analysis some decades before their existence was proved by Dedekind and Cantor, and continue to be used in analysis.

In topological language, we would say that the polynomials are dense in $f \in \mathcal{C}[a,b]$ in the topology induced by the norm $||\cdot||_\infty$. It is fairly obvious that a polynomial in $\mathcal{C}[a,b]$ can be uniformly approximated by polynomials with *rational* coefficients, and the latter form a countable set (page 251). It follows from these that $\mathcal{C}[a,b]$ is separable (page 265).

A5.2.2 The 1*-norm. Incompleteness of the Riemann integral*

The functional[6]

$$||f||_1 = \int_a^b |f(x)|\mathrm{d}x, \tag{A5.5}$$

where the integral is a Riemann integral, also defines a norm on C, as is easily verified. It should be noted that the critical triangle inequality follows once again from the triangle inequality for functions (A5.4) and does not depend much on the integral itself. However, the normed space $(C, ||\cdot||_1)$ is not complete, as we shall now show by constructing a simple counterexample.

Fix a point $\gamma \in (a,b)$, and let N be an integer such that $N > 1-(\gamma - a)$. Then $\gamma - 1/n > a$ for $n > N$. For such n, define the functions $f_n(x)$ on $[a,b]$ as follows:

$$f_n(x) = \begin{cases} 0, & \text{if} \quad a \le x < \gamma - \dfrac{1}{n}, \\ 1 - n(\gamma - x), & \text{if} \quad \gamma - \dfrac{1}{n} \le x < \gamma, \\ 1, & \text{if} \quad \gamma \le x \le b. \end{cases} \tag{A5.6}$$

Then, for $\gamma - 1/n \le x \le \gamma$, one has $0 \le f_n(x) \le 1$, and $f_n(x)$ is continuous. The graph of $f_n(x)$ is shown in Fig. A5.1. The sloping part of the graph of $f_m(x)$, where $n > m > N$, is also shown in the figure, by a dashed line. The norm $||f_m - f_n||_1$, which is the Riemann integral of $|f_m(x) - f_n(x)|$ from a to b, is the area of the triangle with vertices $(\gamma - 1/m, 0), (\gamma - 1/n, 0)$ and $(\gamma, 1)$ on the figure. This area is

$$\frac{1}{2}\left(\frac{1}{m} - \frac{1}{n}\right) < \frac{1}{2m},$$

[6] In current mathematical usage, the term **functional** is used to denote a map, generally linear, from a linear space into the real or complex numbers. The idea is to distinguish between maps such as $f : \mathbb{R} \to \mathbb{C}$ and $f : \mathcal{C}[a,b] \to \mathbb{C}$, because points of $\mathbb{R}$ are structureless, while points of $\mathcal{C}[a,b]$ are not. The importance of the distinction was first appreciated by Volterra, who defined a concept he called *functions of lines*, around 1913. The name was later changed to functionals by Hadamard, and it has stuck.

which shows that the sequence $\{f_n(x)\}_{n>N}$ is a Cauchy sequence in the norm $||\cdot||_1$. However, the function $f(x)$ to which it converges is

$$f(x) = \begin{cases} 0, & \text{for} \quad a \leq x < \gamma, \\ 1, & \text{for} \quad \gamma \leq x \leq b, \end{cases}$$

which is discontinuous; it does not belong to $\mathcal{C}[a,b]$. This is clear from Fig. A5.1, and may also be verified analytically with ease.

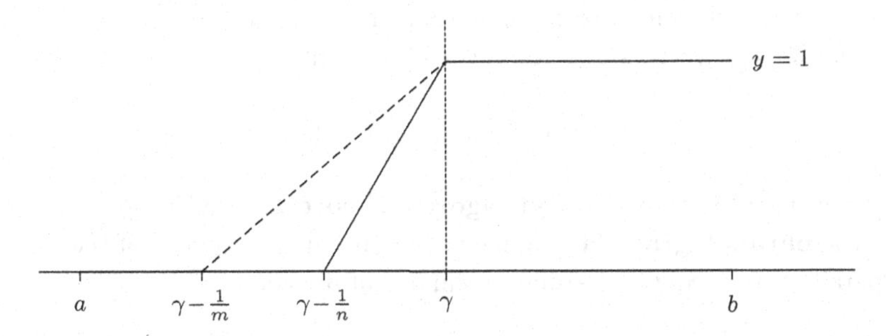

Fig. A5.1. Illustrating the incompleteness of the Riemann integral

It may be argued at this point that, instead of considering C, one ought to consider the space of all Riemann-integrable functions on $[a,b]$, which is larger than C. This is a valid point (although we do not yet have the tools to give a characterization of this space), but it turns out that this space, when normed with $||\cdot||_1$ as defined by (A5.5), is also incomplete.

A5.2.3 The p-norms; the 2-norm

Using the Riemann integral, one can define an infinity of norms on the space $C[a,b]$. Let $p \in [1,\infty)$ and $f \in C[a,b]$. Then

$$||f||_p = \left[\int_a^b |f(x)|^p \mathrm{d}x\right]^{1/p} \tag{A5.7}$$

defines a norm on $C[a,b]$, sometimes called the ***p*-norm**. The definition can be extended to: (i) spaces of real- or complex-valued functions on $\mathbb{R}^n$ or I^n, where I^n is a closed cube in n dimensions; (ii) linear spaces of *sequences* of real or complex numbers, subject to the appropriate convergence conditions. The convergence condition in the p-norm on the space of sequences of complex numbers $z = \{z_n\}, n \in \mathbb{N}$ is

$$||z||_p = \left[\sum_0^\infty |z_n|^p\right]^{1/p} < \infty. \tag{A5.8}$$

The essential part of establishing that the quantities $||f||_p$ and $||z||_p$ are indeed norms is proving the triangle inequality. This is accomplished by using the classical inequalities of Hölder and Minkowski. We omit the lengthy details, which may be found in any textbook on analysis, e.g., (Friedman, 1970).

Of particular interest is the 2-norm, both for function spaces and for sequence spaces, because it generalizes the notion of Euclidean space. Recall that in Euclidean $\mathbb{R}^n$ two vectors u, v are perpendicular iff

$$||u+v||^2 = ||u||^2 + ||v||^2,$$

which is just a restatement of **Pythagoras' theorem**. The Euclidean norm also obeys the **parallelogram law**, namely that the sum of squares of the diagonals of a parallelogram equals the sum of squares of its sides,

$$||u+v||^2 + ||u-v||^2 = 2||u||^2 + 2||v||^2, \tag{A5.9}$$

which is proven by direct computation. As a consequence of the parallelogram law, one finds that the **inner product** on a Euclidean space (which is also called the **scalar product**) is related to the norm by the following identity, known as the **polarization identity**:

$$(u, v) = \tfrac{1}{4}\left[||u+v||^2 - ||u-v||^2\right]. \tag{A5.10}$$

Note that (A5.10) is valid only for linear spaces over $\mathbb{R}$. For linear spaces over $\mathbb{C}$, it has to be replaced by

$$(u, v) = \tfrac{1}{4}\left[(||u+v||^2 - ||u-v||^2) + \mathrm{i}(||u+\mathrm{i}v||^2 - ||u-\mathrm{i}v||^2)\right]. \tag{A5.11}$$

Again, one verifies (A5.11) by direct computation.

A Banach space in which the norm satisfies the parallelogram law (A5.9) is a **Hilbert space**, with the inner product *defined* by (A5.10) or (A5.11).

It can be shown (we shall not stop to prove it) that of all the p-norms, the only one that satisfies the parallellogram law is the 2-norm. It is therefore the only one in which one can define an inner product, i.e., the notion of an angle.

It follows from the completeness of real and complex numbers that sequence spaces are complete in the p-norms; the reader would undoubtedly have surmised that the function spaces $\mathcal{C}[a, b]$ are not, as long as the p-norms are defined by the Riemann integral.

A5.3 The Riemann–Stieltjes integral

We begin with a quotation from Apostol (Apostol, 1974, pp. 140–141):

> The *Riemann-Stieltjes integral* [written $\int_a^b f(x)\mathrm{d}\alpha(x)$]... involves two functions f and α... When α has a continuous derivative... the Stieltjes integral $\int_a^b f(x)\mathrm{d}\alpha(x)$ becomes the Riemann integral $\int_a^b f(x)\alpha'(x)\mathrm{d}x$. However, the Stieltjes integral still makes sense when α is not differentiable or even when α is discontinuous. In fact, it is in dealing with *discontinuous* α that the importance of the Stieltjes integral becomes apparent. By a suitable choice of a discontinuous α, any finite or infinite sum can be expressed as a Stieltjes integral, and summation and ordinary Riemann integration become special cases of this more general process. Problems in physics which involve mass distributions that are partly discrete and partly continuous can also be treated by using Stieltjes integrals.

In this section, we shall work with a fixed real interval $[a, b]$, denoted by I, and the set of *bounded* real-valued functions[7] on I. This set will be denoted by F.

A **partition** of I is a finite set of points $P = \{a = x_0, x_1, \ldots, x_n = b\}$ such that $x_k < x_{k+1}$ for $k = 0, 1, \ldots, n$. If Q is another partition of I and $P \subset Q$, we say that Q **refines** P, or Q is **finer** than P. The **norm** of the partition P is defined to be the length of the largest subinterval in P, and is denoted by $||P||$. $Q \supset P$ implies that $||Q|| \leq ||P||$; refinements do not have to decrease the norm of a partition, but one is mostly interested in those that do.

We shall now define the Riemann–Stieltjes integral.[8] The reader will notice that the definition is modelled after a standard definition of the Riemann integral, and, like the latter, is not an effective computational tool.

We begin by selecting, arbitrarily, a particular function in F. This function will be denoted by α. We define

$$\Delta\alpha_k = \alpha(x_k) - \alpha(x_{k-1}). \tag{A5.12}$$

Then:

Definition A5.2 (Riemann–Stieltjes integral) Let f be any function in F, and α a fixed function in F. Let $P = \{x_0, \ldots, x_n\}$ be a partition of I, and t_k

[7] Note that we do not require continuity. The reason for emphasizing the boundedness will become clear at the beginning of Subsection A5.3.1.

[8] There are several inequivalent definitions of the Riemann–Stieltjes integral in the literature. The differences between them are not significant for our purposes. The Riemann–Stieltjes integral is also known simply as the Stieltjes integral. However, as there is also a Lebesgue–Stieltjes integral, which we shall not define, we have chosen to stay with the name *Riemann–Stieltjes integral* to avoid any possibility of confusion.

any point in $[x_{k-1}, x_k]$. A sum

$$S(P, f, \alpha) = \sum_{k=1}^{n} f(t_k)\Delta\alpha_k$$

is called a **Riemann–Stieltjes sum** of f with respect to α.

If there exists a number A with the property that for any $\varepsilon > 0$, there is a partition P_ε of I such that for every $P \supset P_\varepsilon$,

$$|S(P, f, \alpha) - A| < \varepsilon$$

for every choice of points $t_k \in [x_{k-1}, x_k]$, we say that f is **Riemann–Stieltjes integrable with respect to** α on I (written $f \in \boldsymbol{R}(\alpha)$ on I), and write

$$\int_a^b f(x)\mathrm{d}\alpha(x) = \int_a^b f\mathrm{d}\alpha = A.$$

The number A, if it exists, is unique. The functions f and α are called the **integrand** and the **integrator** respectively.

The integrator does not have to be monotonic, but for us it will be enough to consider those that are *monotonic nondecreasing*. Note that if $\alpha(x) = x$, then the integral becomes the Riemann integral, i.e., $\boldsymbol{R}(x)$ consists of the Riemann integrable functions on $[a, b]$. In this case we write $\mathrm{d}x$ for $\mathrm{d}\alpha(x)$. If $\alpha(x)$ is differentiable on (a, b) and $\alpha'(x)$ is continuous, then we have the pleasing result

$$\int_a^b f(x)\mathrm{d}\alpha(x) = \int_a^b f(x)\alpha'(x)\mathrm{d}x. \tag{A5.13}$$

The Riemann–Stieltjes integral is linear in both f and α. More precisely:

Theorem A5.3 (First linearity property) *If $f, g \in \boldsymbol{R}(\alpha)$, then for any two $c_1, c_2 \in \mathbb{R}$, $c_1 f + c_2 g \in \boldsymbol{R}(\alpha)$.*

Theorem A5.4 (Second linearity property) *If $f \in \boldsymbol{R}(\alpha)$ and $f \in \boldsymbol{R}(\beta)$, then $f \in \boldsymbol{R}(c_1\alpha + c_2\beta)$ for any two $c_1, c_2 \in \mathbb{R}$, and*

$$\int_a^b f\mathrm{d}(c_1\alpha + c_2\beta) = c_1 \int_a^b f\mathrm{d}\alpha + c_2 \int_a^b f\mathrm{d}\beta.$$

We also have the familiar result for integration by parts, which holds for Riemann–Stieltjes integrals as well:

Theorem A5.5 (Integration by parts) *If $f \in \boldsymbol{R}(\alpha)$ on I, then $\alpha \in \boldsymbol{R}(f)$ on I, and*

$$\int_a^b f(x)\mathrm{d}\alpha(x) + \int_a^b \alpha(x)\mathrm{d}f(x) = f(b)\alpha(b) - f(a)\alpha(a).$$

Finally, we have

Theorem A5.6 *If $c \in [a, b]$, then*

$$\int_a^c f\mathrm{d}\alpha + \int_c^b f\mathrm{d}\alpha = \int_a^b f\mathrm{d}\alpha.$$

The straightforward proofs of these theorems may be found in (Apostol, 1974).

A5.3.1 Step functions as integrators

If the integrator is once-differentiable, then we see from (A5.13) that the Riemann–Stieltjes integral reduces to a Riemann integral. We now come to the heart of the matter, i.e., to integrators that are not continuous.

First, some notations and terminology. Let $\kappa(x)$ be bounded on I, and assume that the limits (recall that $\varepsilon > 0$ by convention)

$$\lim_{\varepsilon \to 0}(\kappa(x) + \varepsilon) \text{ and } \lim_{\varepsilon \to 0}(\kappa(x) - \varepsilon)$$

exist (i) at all $x \in (a, b)$, (ii) from the right at a, and (iii) from the left at b. We write these limits as $\kappa(x+)$ and $\kappa(x-)$ respectively. If $\kappa(x) = \kappa(x\pm)$, it is called **continuous from the right** (or **right-continuous**, plus sign), and **continuous from the left** (or **left-continuous**, minus sign), respectively, at x.

Let $\tau \in (a, b)$, and define a function α as follows. The values $\alpha(a)$, $\alpha(\tau)$ and $\alpha(b)$ are assigned arbitrarily, and for other values of $x \in I$, $\alpha(x)$ is defined by

$$\alpha(x) = \begin{cases} \alpha(a), & a \leq x < \tau, \\ \alpha(b), & \tau < x \leq b. \end{cases} \tag{A5.14}$$

The function α is an example of a *step function*; step functions are defined in their generality in Definition A5.8.

Theorem A5.7 *Let f be continuous on I, and α be as in (A5.14). Then*

$$\int_a^b f \mathrm{d}\alpha = f(\tau)[\alpha(\tau+) - \alpha(\tau-)].$$

Proof Since any partition of I that does not contain the point τ is refined by the one obtained by adjoining τ to it, there is no loss of generality in considering only those partitions that contain the point τ. Let P be such a partition, with $x_k = \tau$. Owing to the form of α, every term but two in the Riemann–Stieltjes sum $S(P, f, \alpha)$ vanishes identically, and we have:

$$S(P, f, \alpha) = f(t_k)[\alpha(\tau) - \alpha(\tau-)] + f(t_{k+1})(\alpha(\tau+) - \alpha(\tau)].$$

Therefore we may write

$$\begin{aligned}\Delta &= S(P, f, \alpha) - f(\tau)[\alpha(\tau+) - \alpha(\tau-)] \\ &= (f(t_k) - f(\tau))[\alpha(\tau) - \alpha(\tau-)] + (f(t_{k+1} - f(\tau))[\alpha(\tau+) - \alpha(\tau)] \\ &= (f(t_k) - f(\tau))[\alpha(\tau) - \alpha(a)] + (f(t_{k+1} - f(\tau))[\alpha(b) - \alpha(\tau)].\end{aligned}$$

Therefore

$$|\Delta| \leq |(f(t_k) - f(\tau))| \cdot |\alpha(\tau) - \alpha(a)| + |(f(t_{k+1} - f(\tau))| \cdot |\alpha(b) - \alpha(\tau)|.$$

Now, since f is continuous, for any $\varepsilon > 0$ we can find $\delta > 0$ such that $|f(x) - f(\tau)| < \varepsilon$ whenever $|x - \tau| < \delta$. Therefore, for any partition P_ε with $||P_\varepsilon|| < \delta$, we have

$$|\Delta| \leq \varepsilon|\alpha(\tau) - \alpha(a)| + \varepsilon|\alpha(b) - \alpha(\tau)|,$$

which proves the theorem. □

If $\alpha(a) = \alpha(\tau-) = 0$ and $\alpha(b) = \alpha(\tau+) = 1$, the result of the theorem becomes

$$\int_a^b f \mathrm{d}\alpha = f(\tau).$$

Physicists are familiar with this result in the form

$$\int_a^b f(x)\delta(x - \tau)\mathrm{d}x = f(\tau), \tag{A5.15}$$

where $\delta(x - \tau)$ is Dirac's delta-function. Integrals with the delta-function in the integrand may thus be interpreted as Riemann–Stieltjes integrals with step functions as integrators. Observe that the value of $\alpha(x)$ at $x = \tau$ does not play a role; all that is necessary is for $\alpha(x)$ to be defined at $x = \tau$. Indeed, **step functions** are defined as follows:

Definition A5.8 (Step functions) Let $P = \{a = x_1, \ldots, x_n = b\}$ be a partition of $[a, b]$. A function $\alpha : [a, b] \to \mathbb{R}$ is called a **step function** if $\alpha(x) = \text{const}$ on all open intervals (x_k, x_{k+1}). For $1 \leq k < n$, the number $h_k = \alpha(x_k+) - \alpha(x_k-)$ is called the **jump**[9] of α at x_k. The jump at x_1 is $\alpha(x_1+) - \alpha(x_1)$, and the jump at x_n is $\alpha(x_n) - \alpha(x_n-)$.

Using Theorem A5.7, the reader is invited to establish the following:

Theorem A5.9 *Let α be a step function on I, with jumps h_k at x_k. Then, if f is continuous on I, $\int_a^b f d\alpha$ exists and*

$$\int_{-n}^{b} f(x) d\alpha(x) = \sum_{k=1}^{n} f(x_k) h_k.$$

Step functions, as defined above, are not required to be continuous either from the left or from the right at their discontinuities. If one tightens this condition when using them as integrators, e.g., by requiring α to be right-continuous, then one can relax the continuity condition on f; it would be enough for f to be left-continuous at the discontinuities of α. For details, see (Apostol, 1974).

As a last example, let $a = -n$ and define $\alpha(x)$ as follows:

$$\alpha(x) = \begin{cases} -k, & x \in [-k, -k+1), \; k = n, n-1, \ldots, 1; \\ x, & x \in [0, b]. \end{cases} \tag{A5.16}$$

Then, under suitable continuity conditions on f,

$$\int_{-n}^{b} f(x) d\alpha(x) = \sum_{k=1}^{n} f(-k) + \int_{0}^{b} f(x) dx,$$

where the integral on the right is an ordinary Riemann integral.

A5.4 σ-algebras, measures and integrals

If the notions of Euclidean geometry coexist in harmony with the notion of a set, the following question ought to make sense: *Does every subset of three-dimensional Euclidean space have a volume?* Banach and Tarski proved the following result (which has since become known as the **Banach–Tarski paradox**) in 1924. It is possible to take a ball of unit radius in three-space, divide

[9] Apostol (Apostol, 1974) uses the notation α_k for the jump of the step function α at x_k.

it into a finite number of pieces, and then, *using only rotations and translations*, assemble the pieces to form two balls, each of unit radius. The construction made very explicit use of the axiom of choice. However, the pieces into which Banach and Tarski had divided the unit ball were subsets for which *the concept of volume was not well defined.* That the re-assembly would double the volume was no more and no less surprising than the fact that a ball with a well-defined volume could be decomposed into a finite number of pieces, none of which could be assigned a volume!

If it is not possible to assign a measure[10] to every subset of $\mathbb{R}^n$ (which seems to be the case if one wants to stay with set theory and the axiom of choice), there are two questions to be answered:

(i) What are the measurable sets?
(ii) How does one assign a measure to a measurable set?

A5.4.1 Measurable sets

Although we were led to the above questions by a paradox in Euclidean space, they have been answered in a much more general setting. The class of measurable sets is defined collectively on an arbitrary set of points, by a process which is broadly similar to the definition of topologies.

Definition A5.10 (σ-rings and σ-algebras) Let X be a set. A family $\mathcal{R}$ of subsets of X is called a **σ-ring** if:

(a) $\emptyset \in \mathcal{R}$.
(b) If $A, B \in \mathcal{R}$, then $A \smallsetminus B \in \mathcal{R}$.
(c) If $A_n \in \mathcal{R}$ for $n \in \mathbf{N}$, then

$$\bigcup_{n=1}^{\infty} A_n \in \mathcal{R}.$$

If, in addition, $X \in \mathcal{R}$, then $\mathcal{R}$ is called a **σ-algebra**.

Properties of σ-rings and σ-algebras are studied in detail in books on measure theory; see, for example, the short but lucid text by Friedman (Friedman, 1970).[11] Our next step will be based on the following straightforward result, which we shall state without proof:

Theorem A5.11 (The σ-algebra generated by a family of subsets) *Let $\mathcal{Q}$ be a family of subsets of X which includes X itself. Then there exists a unique*

[10] For the moment, we are using the term *measure* in a loose sense; the precise definition will be given presently.

[11] We shall not need the extra generality afforded by σ-rings over σ-algebras. We have defined the concept purely to make it easier for the reader to consult standard texts.

smallest σ-algebra $\mathcal{R}_{\mathcal{Q}}$ which contains every member of $\mathcal{Q}$. Smallest *means here that if $\mathcal{R}_1$ is any σ-algebra such that $\mathcal{R}_1 \supset \mathcal{Q}$, then $\mathcal{R}_1 \supset \mathcal{R}_{\mathcal{Q}} \supset \mathcal{Q}$.*

The σ-algebra $\mathcal{R}_{\mathcal{Q}}$ is said to be **generated** by $\mathcal{Q}$.

The set X of Definition A5.10 was not a topological space. We shall, however, be chiefly concerned with σ-algebras on topological spaces, which are defined as follows, and have a special name:

Definition A5.12 (Borel structures) Let X be a topological space. The σ-algebra generated by the open sets of X is called a **Borel structure** on X, and will be denoted by $\mathcal{B}$. Elements of $\mathcal{B}$ are called **Borel sets.**

Theorem A5.11 ensures that the Borel structure on X is unique. ($\mathcal{B}$ is a σ-algebra and not only a σ-ring because X itself is an open set.) Observe that if X is a T_1-space, then one-point subsets are Borel. We now define:

Definition A5.13 (Measurable sets) Let $\mathcal{R}$ be a σ-algebra on a set X. The elements of $\mathcal{R}$ are called **measurable subsets** of X.

It follows trivially that the measurability or nonmeasurability of a subset of X depends on the σ-algebra being used. If one is working with a T_1 topological space and its Borel structure, then one-point sets are measurable, as are countable sets and their complements. Therefore a nonmeasurable set must necessarily be uncountable, and have an uncountable complement (which will also be nonmeasurable). Of course, it remains to be demonstrated that nonmeasurable sets indeed exist.

Such demonstrations are generally provided by constructing examples. (Recall Weierstrass' famous proof of existence of an everywhere continuous but nowhere differentiable function.) Nonmeasurable sets are not easy to construct, which may be why they are so seldom encountered.

Let us restate the above in different words. Note, first of all, that – except in Section A7.2 on the general theory of probability – we shall be working almost exclusively with measures on topological spaces (indeed, on topological groups), so that the relevant σ-algebra will be the Borel structure $\mathcal{B}$ on it. The latter, by Theorem A5.11, is unique. It is clearly a subset of the power set $\mathcal{P}(X)$ of X (see page 245). What we still do not know is whether or not $\mathcal{B}$ is a *proper* subset of $\mathcal{P}(X)$.

A5.4.2 Measures

We now turn to the definition of measures. For this we need to augment the set of real numbers. The set of **extended real numbers** is defined to be the union $\mathbb{R} \cup \{-\infty, \infty\}$, with the assumption that $-\infty < x < \infty$ for any $x \in \mathbb{R}$; $-\infty$ is to be included among the negatives, and $+\infty$ among the positives. This set is sometimes denoted by $\mathbb{R}^\star$. It is topologized in the same manner as a closed interval of the reals. We also need to augment the definition of a function:

Definition A5.14 (Set functions) A **set function** is a function the domain of definition of which is a class, or family, of sets.

These definitions are put to use in the definition of measures.

Definition A5.15 (Measures on X) A **measure** μ on $(X, \mathcal{R})$ is a set function $\mu : \mathcal{R} \to [0, \infty]$ from a σ-algebra $\mathcal{R}$ on X into the extended nonnegative real numbers that has the following properties:

(a) μ is **countably additive**;[12] i.e., if $\{B_k\}_{k=1}^{\infty}$ is any pairwise-disjoint countable subcollection of $\mathcal{R}$, then

$$\mu\left(\bigcup_{k=1}^{\infty} B_k\right) = \sum_{k=1}^{\infty} \mu(B_k);$$

(b) $\mu(\emptyset) = 0$.

It follows from these conditions that $\mu(A) \leq \mu(B)$ for $A \subset B$, $A, B \in \mathcal{R}$

An immediate consequence of the above definition is that *measures can be multiplied by positive numbers, and added*; if μ_1, μ_2 are measures on $(X, \mathcal{R})$ and $a, b \in [0, \infty)$, then $a\mu_1 + b\mu_2$ is again a measure on $(X, \mathcal{R})$.

A measure μ on $(X, \mathcal{R})$ is called σ-**finite** if there exists a countable subcollection $\{B_k\}$ of $\mathcal{R}$ which covers X and is such that $\mu(B_k) < \infty$ for all k. It is called **finite** if $\mu(X) < \infty$. For example, the usual measure of length on the real line (as we shall see) is σ-finite but not finite. Counting measures (i.e., measures that count the number of points in a subset) on the power sets of uncountable sets are neither finite nor σ-finite.

A triple $(X, \mathcal{R}, \mu)$ is called a **measure space**. Similarly, a triple $(X, \mathcal{B}, \mu)$, with X a topological space, is called a **Borel space**. A measure μ on a Borel space is sometimes called a **Borel measure**.

It may seem almost too trivial to state that, given a measure μ on X, a subset $E \in \mathcal{R}$ is said to be of **measure zero** if $\mu(E) = 0$; however, it brings us to one of the most important concepts in measure theory:

Definition A5.16 (Almost everywhere) A property P is said to hold **almost everywhere** (abbreviated a.e.) if it holds everywhere except on a set of measure zero.

Important examples are functions that are equal a.e., continuous a.e., bounded a.e., differentiable a.e., and sequences of functions that converge a.e.

We end this section by stating the following results:

(i) Let $(X, \mathcal{R}, \mu)$ be a measure space and S a measurable subset of X. Using the fact that the intersection of two members of $\mathcal{R}$ belongs to $\mathcal{R}$, one finds

[12] Friedman uses the term **completely additive** (Friedman, 1970).

that the collection

$$\mathcal{R}_S = \{A \cap S | S \in \mathcal{R}\}$$

is a σ-algebra on S, and that the restriction of μ to $\mathcal{R}_S$ is a measure on $(S, \mathcal{R}_S)$.

(ii) Let $(X, \mathcal{B}_X, \mu)$ and $(Y, \mathcal{B}_Y, \nu)$ be two Borel spaces. If S_X and S_Y are Borel sets in X and Y, then $S_X \times S_Y$ is a Borel set in $X \times Y$, and $\mu \times \nu$, defined by

$$(\mu \times \nu)(S_X \times S_Y) = \mu(S_X) \cdot \nu(S_Y)$$

is a measure on $X \times Y$.

The first of these is easy to prove, but the second is more involved. See (Friedman, 1970, Sec. 2.15).

A5.4.3 Integrals

The definition of the Riemann integral may be viewed as a two-step process. The first step consists of defining a suitable family of approximants. In the traditional picture of the Riemann integral as the area under a curve, the approximants are represented by histograms; the step-function that is the upper boundary of the histogram is an approximation to the curve itself. The second step consists of choosing a suitable sequence of approximants, and of proving, or testing, their convergence. The definition of the integral in the general setting of measure spaces follows the same paradigm, with histograms being replaced by objects that are more subtle, and the notion of convergence amended to reflect this subtlety.

We begin with a few definitions.

Definition A5.17 (Measurable function) An extended real-valued function f on $(X, \mathcal{R})$ is said to be **measurable** if

(a) $f^{-1}(U)$ is a measurable subset of X (i.e., $f^{-1}(U) \in \mathcal{R}$) for any open set U in $\mathbb{R}$.

(b) $f^{-1}(-\infty)$ and $f^{-1}(\infty)$ are measurable sets.

Note that f need not be continuous; indeed, X need not be a topological space. But if it is, and if $\mathcal{R}$ is replaced by $\mathcal{B}$ in the above, f is said to be **Borel-measurable**, or simply **Borel**. A continuous function is necessarily Borel. Measurable functions on $(X, \mathcal{R})$ form a linear space over $\mathbb{R}$.

Recall now that the **characteristic function** χ^S of a subset $S \subset X$ is defined as follows:[13]

$$\chi^S(x) = \begin{cases} 1, & \text{if} \quad x \in E, \\ 0, & \text{if} \quad x \in X \setminus E. \end{cases}$$

The part of histograms in the Riemann integral is played by *simple functions*, which are defined as follows, in terms of characteristic functions:

Definition A5.18 (Simple functions) A function $f : X \to \mathbb{R}$ is called a **simple function** if it assumes only a finite number of distinct values. Formally, if f is a simple function, then there exist distinct nonzero real numbers α_k, $k = 1, \ldots, n$ and pairwise-disjoint measurable subsets S_k, $k = 1, \ldots, n$ of X such that

$$f(x) = \sum_{k=1}^{n} \alpha_k \chi^{S_k}(x).$$

It will be convenient, for later use, to define the set

$$S_0 = X \setminus \bigcup_{k=1}^{n} S_k. \tag{A5.17}$$

Theorem A5.19 (Approximation by simple functions) *Let f be a nonnegative measurable function. Then there exists a sequence $\{f_n\}$ of nonnegative simple functions such that $\{f_n(x)\}$ converges pointwise to $f(x)$.*

We shall not prove this theorem (for a proof, see (Friedman, 1970)), but shall illustrate it by an example which will also illustrate the difference between approximation by a simple function and by a Riemann sum. For this we need the auxiliary result – which the reader is invited to prove – that if f is a measurable extended real-valued function on X, then the inverse images of closed and half-open intervals on $\mathbb{R}$ are measurable.

Let f be a bounded measurable function on X that takes its values in $[a, b]$. Set

$$\alpha_k = a + \frac{b-a}{n} k, \quad 0 \leq k \leq n,$$

and

$$S_k = f^{-1}([\alpha_k, \alpha_{k+1}]).$$

[13] The characteristic function is usually written as χ_S. We have chosen the superscript notation because we find TEX's χ^S and χ^{S_k} to be more pleasing, visually, than its χ_S and χ_{S_k}.

Then

$$f(x) \approx \sum_{k=1}^{n} \alpha_k \chi^{S_k} \tag{A5.18}$$

is an approximation to f by a simple function; the larger n is, the better the approximation. The following, in an obvious notation,

$$f(x) \approx \sum_{j=1}^{m} m_j \Delta_j \tag{A5.19}$$

is an approximation to f by a Riemann sum. The essential difference between (A5.18) and (A5.19) is that the first is obtained by subdividing the *range* of f, whereas the second is obtained by subdividing its *domain*.

Definition A5.20 (Integrable simple functions) A simple function $f(x) = \sum_0^n \alpha_k \chi^{S_k}$, where $\alpha_0 = 0$, is said to be **integrable** with respect to the measure μ if $\mu(S_k) < \infty$ for $1 \le k \le n$. If $\mu(S_0) = \infty$, we agree to set $0 \cdot \mu(S_0) = 0$. The integral of f is defined as follows:

$$\int_X f \mathrm{d}\mu = \sum_{k=0}^{n} \alpha_k \mu(S_k).$$

It is easy to see that the integrable simple functions on $(X, \mathcal{R}, \mu)$ form a linear space over $\mathbb{R}$.

Definition A5.21 (Cauchy sequence in the mean) A sequence $\{f_n\}$ of simple functions is said to be a **Cauchy sequence in the mean** if

$$\int |f_m - f_n| \mathrm{d}\mu \to 0 \quad \text{as } m, n \to \infty.$$

The reader will surely have surmised what will come next:

Definition A5.22 (Integrable functions and integrals) Let f be an extended real-valued function of the measure space $(X, \mathcal{R}, \mu)$. The function f is said to be **integrable** if there exists a sequence $\{f_n\}$ of *integrable* simple functions such that

(a) $\{f_n\}$ is a Cauchy sequence in the mean.

(b) $\{f_n\}$ converges to f a.e.

If f is an integrable function, then its **integral** (with respect to the measure μ) is defined to be

$$\int_X f \mathrm{d}\mu = \lim_{n \to \infty} \int_X f_n \mathrm{d}\mu.$$

Of course it has to be proved that the limit on the right exists, but that is straightforward. The $d\mu$ on the right is sometimes written as $d\mu(x)$.

A5.5 Haar measure

Let G be a locally compact σ-compact[14] topological group (Section A3.9, page 275), $\mathcal{B}$ the Borel structure on it and μ a measure on $(G, \mathcal{B})$. The measure μ is said to be **left-invariant** if, for any $B \in \mathcal{B}$ and any $g \in G$, $\mu(gB) = \mu(B)$; it is said to be **right-invariant** if $\mu(Bg) = \mu(B)$; and **bi-invariant**, or simply **invariant**, if $\mu(gB) = \mu(Bg) = \mu(B)$. Since the modern study of these measures was initiated by Haar in 1933, they are known as **Haar measures.**[15]

The fundamental theorem on Haar measures is:

Theorem A5.23 (Haar measures) *A locally compact topological group admits a left-invariant and a right-invariant measure. These measures are unique up to normalization.*

If a topological group admits an invariant measure (which will then be unique up to normalization), it is called **unimodular**.

It follows trivially from Theorem A5.23 that an Abelian topological group is unimodular. Therefore the spaces $\mathbb{R}^n$, considered as topological translation groups, are unimodular. The unique translation-invariant measures on them are the **Lebesgue measures** on $\mathbb{R}^n$; therefore, if one manages to construct a translation-invariant measure on $\mathbb{R}$, that measure will have to be the Lebesgue measure on it.

Every compact topological group is unimodular. Fortunately, most of the important noncompact groups in physics, such as the Lorentz groups, are unimodular. It follows easily that the Euclidean and inhomogeneous Lorentz groups are unimodular. It is *not* true that a closed subgroup of a unimodular group is necessarily unimodular; for an example, see (Hewitt and Ross, 1963, p. 201).

Let G be a unimodular topological group, and D a unitary representation of it on a Hilbert space $\mathfrak{H}$. The representation D is called **weakly measurable** if its matrix elements $D_{\varphi,\psi}(g) = (\varphi, D(g)\psi)$ are measurable functions of g for all $\varphi, \psi \in \mathfrak{H}$. The following theorem of von Neumann (Hewitt and Ross, 1963; Mackey, 1976) essentially settles all continuity problems for unitary group representations in physics.

[14] A topological space is said to be **σ-compact** if it is a countable union of compact subsets. Every connected locally compact topological group is σ-compact. All topological groups we shall consider in this section will be assumed to be locally compact and σ-compact.

[15] Invariant integrals on groups were studied as early as 1895 by A. Hurwicz. They were used by Schur to develop the theory of group characters in 1924. Schur's results were used a little later by Peter and Weyl to prove the famous Peter–Weyl theorem (the full reducibility of representations of compact groups). The Hurwicz invariant integral was used by Wigner in 1931 to calculate the Clebsch–Gordan coefficients of the rotation group in three dimensions.

Theorem A5.24 *A Hilbert space representation of a group with a Haar measure is strongly continuous iff it is weakly measurable.*

The notion of strong continuity has been defined in the section on Stone's theorem, page 133.

A5.6 The Lebesgue integral and the L^p-spaces

In this section we shall define the Lebesgue measure on $\mathbb{R}$ (and on $[a, b] \subset \mathbb{R}$) and then the function spaces L^p, which are complete in the p-norm determined by the Lebesgue integral. We begin with a general construction which will be exploited again in Section A5.7.

A5.6.1 Construction of measures

Let $\alpha(x)$ be a real-valued, monotonic, nondecreasing function on $\mathbb{R}$ which is continuous from the right. Then the limits $\alpha(x+)$ and $\alpha(x-)$ exist at all x. For any open interval (a, b), define

$$v(a, b) = \alpha(b-) - \alpha(a+).$$

Then $v(a, b)$ is nonnegative for any open interval (a, b).

Let $\{I_n\}$ be any finite or countable collection of *pairwise-disjoint* open intervals, and set $O = \cup_n I_n$. Denote the family of all such O by $\mathcal{J}$.

For $B \in \mathcal{B}$, let $O_B \in \mathcal{J}$ be such that $B \subset O_B$ (i.e., O_B covers B by a finite or countable union of pairwise-disjoint open intervals). Set

$$v(O_B) = \sum_{n=1}^{\infty} v(I_n).$$

Then $v(O_B)$ is nonnegative for every O_B. Finally, define

$$\mu_\alpha(B) = \inf_{\text{all } O_B} v(O_B). \tag{A5.20}$$

It can be shown that μ_α is a measure on $\mathbb{R}$. The proof is an adaptation of the standard textbook definition of the Lebesgue measure on $\mathbb{R}$.

A5.6.2 The Lebesgue measure on $\mathbb{R}$

If $\alpha(x) = x$, then $v(a, b) = b - a$. In this case the measure μ_α, denoted μ, is called the **Lebesgue measure** on $\mathbb{R}$. The Lebesgue measure of a one-point set is zero, and that of an interval – open, closed or half-open – is its length. It is clearly translation-invariant, $\mu(B) = \mu(B + x)$, and is therefore the unique Haar measure on $\mathbb{R}$. The Lebesgue integral is usually denoted by $\int f \mathrm{d}x$, instead of $\int f \mathrm{d}\mu$ or $\int f \mathrm{d}\mu(x)$, with appropriate limits.

Every countable subset of $\mathbb{R}$ has zero Lebesgue measure. Somewhat surprisingly, there exist *uncountable* subsets of measure zero, and we shall soon come to them.

Mutatis mutandis, one can define a measure on closed intervals $[a, b]$ on $\mathbb{R}$ by the same procedure. This measure is also called the Lebesgue measure, but on $[a, b]$; the notation for the integral is unchanged, but with limits of integration a and b.

For any $\varepsilon > 0$, a set of Lebesgue measure zero can be covered by a countable collection of open intervals of total length $< \varepsilon$. This result[16] can be turned into the definition of sets of measure zero long before the general concept of measure is defined. This makes it possible to define the Lebesgue integral before defining the Lebesgue measure. We refer the interested reader to Part I of the monograph (Riesz and Sz-Nagy, 1955) for a systematic development, and to the text (Apostol, 1974) for a shorter account.

One of Lebesgue's deepest theorems is:

Theorem A5.25 (Lebesgue) *A monotonic function is differentiable a.e.*

The following characterization of the class of Riemann-integrable functions is also due to Lebesgue:

Theorem A5.26 (Riemann-integrable functions) *A bounded function on $[a, b]$ is Riemann-integrable if and only if it is continuous a.e.*

We see that one needs a concept from the Lebesgue theory to characterize the class of Riemann-integrable functions, which is why this characterization could not be given earlier. We shall have no occasion to use this result, but Theorem A5.25 will be used to construct noninvariant measures in Section A5.7.

A5.6.3 The space $L^1([a, b], \mathrm{d}x)$

If we try to define a norm on the space of measurable functions on $[a, b]$ via the Lebesgue integral, i.e., by defining

$$||f - g||_1 = \int_a^b |f(x) - g(x)| \mathrm{d}x, \tag{A5.21}$$

we run into the problem that $||f - g|| = 0 \not\Rightarrow f = g$, for, if $f = g$ a.e., the integral (A5.21) will vanish; the expression $|| \cdot ||_1$ defined by (A5.21) is a seminorm, and not a norm. However, this difficulty disappears if we pass to the *equivalence*

[16] This is a far-from-trivial result – and one which does not have a constructive proof – as one can see immediately by trying to cover the rationals $\mathbb{Q}$ on $\mathbb{R}$ by a countable collection of intervals of total length $< \varepsilon$. If one uses this as a *definition* of sets of measure zero, then, when one eventually defines a measure, one has to prove that the measure of a set of measure zero is zero!

classes of functions that are equal a.e. The space of these equivalence classes, with the norm defined by the Lebesgue integral, is denoted by $L^1([a,b], dx)$. The completeness of $L^1([a,b], dx)$ is the content of Lebesgue's monotone convergence and dominated convergence theorems, which actually antedate the general notion of completion. They may be found in every text on the subject, for instance (Friedman, 1970).

Strictly speaking, one ought not to speak of *functions* in $L^1([a,b], dx)$, but one does! The reason why one does not come to grief is the following: if the equivalence class of a function contains a continuous function, then it contains no other continuous function. The same statement holds if one replaces 'continuous' by 'differentiable'. Moreover, left- and right-hand limits at a point are unaffected by changing the value of the function at that point, so that if a function has a jump every member of its equivalence class has that jump.

Mutatis mutandis, the same considerations apply to the space $L^1(\mathbb{R}, dx)$. It should be mentioned that the spaces $L^1(\mathbb{R}, dx)$ and $L^1([a,b], dx)$ are often denoted simply as $L^1(\mathbb{R})$ and $L^1([a,b])$.

A5.6.4 The spaces L^p

Earlier we defined the p-norms $||\cdot||_p$, $p \in [1,\infty)$ on $C[a,b]$ by (A5.7), using the Riemann integral, and pointed out that $C[a,b]$ is not complete under the 1-norm. We saw above that these difficulties disappear if (i) one defines the 1-norm via the Lebesgue integral, and (ii) one enlarges the space $C[a,b]$ to the space of equivalence classes of measurable functions that are equal a.e. The space of these equivalence classes, it turns out, is complete in every p-norm, when the p-norm is defined via the Lebesgue integral. The spaces themselves are denoted $L^p([a,b], dx)$, and their completeness is nowadays known as the **Riesz–Fischer theorem.**

Mutatis mutandis, the same considerations apply to the spaces[17] $L^p(\mathbb{R}, dx)$ and $L^p(\mathbb{R}^n, d\mu)$, where $d\mu = dx_1 \ldots dx_n = d^n x$. By using the Weierstrass approximation theorem in several variables, it can be shown that the L^p-spaces, for $1 \leq p < \infty$, are separable. The reader is invited to ascertain that the proof requires only a slight extension of the proof of separability of $\mathcal{C}[a,b]$.

Proofs of the above assertions, including the Riesz–Fischer theorem, may be found in every introductory text on functional analysis, such as (Friedman, 1970). It should be mentioned that a few authors use the notation L_p rather than L^p.

A5.6.5 The space L^∞

The definition A5.3 of $||\cdot||_\infty$ is no longer useful in the context of Lebesgue integration. The necessary changes are almost self-evident, but they do require a change of terminology.

[17] These spaces are often denoted simply by $L^p(\mathbb{R}^n)$, $1 \leq n < \infty$.

A real-valued measurable function f on X, where $X = [a, b]^n$ or $X = \mathbb{R}^n$, $n \in \mathbf{N}$ is said to be **essentially bounded** if there exists a positive number K such that

$$|f(x)| \leq K \text{ a.e.}$$

The greatest lower bound of all such K is called the **essential supremum** of f, and written

$$\text{ess} \sup_{x \in X} |f(x)| \text{ or ess} \sup_X |f|. \tag{A5.22}$$

The class of all measurable and essentially bounded functions on X is denoted by $L^\infty(X, \mathrm{d}\mu)$. It is a complete normed space under the norm

$$||f||_\infty = \text{ess} \sup_X |f|. \tag{A5.23}$$

Unlike the spaces $L^p(X, \mathrm{d}\mu)$ for $p < \infty$, the space L^∞ is not separable. The proof of this fact is rather simple, but it requires some basic notions of linear analysis which we have not provided.

A5.7 Noninvariant measures on $\mathbb{R}$

We shall now construct some noninvariant measures on $\mathbb{R}$. We begin with two definitions, which apply to any measure space:

Definition A5.27 Let $\mathcal{R}$ be a σ-algebra on X, and μ, ν two measures on $(X, \mathcal{R})$. The measure ν is said to be **absolutely continuous** with respect to μ, written $\nu \ll \mu$, if $\mu(S) = 0 \Rightarrow \nu(S) = 0$, for $S \in \mathcal{R}$.

Absolute continuity is not symmetric ($\nu \ll \mu \not\Rightarrow \mu \ll \nu$), but it is transitive; $\nu \ll \mu, \mu \ll \lambda$ together imply $\nu \ll \lambda$.

The following situation is in stark contrast to absolute continuity:

Definition A5.28 Let $\mathcal{R}$ be a σ-algebra on X, and μ, ν two measures on $(X, \mathcal{R})$. The measures μ, ν are said to be **mutually singular** (written $\mu \perp \nu$) if there exist two subsets $V, W \in \mathbb{R}$ such that $V \cup W = X$, $V \cap W = \emptyset$ and $\mu(V) = \nu(W) = 0$.

Examples A5.29 (Noninvariant measures on $\mathbb{R}$)

(i) Let the function $\alpha(x)$ which was used to define the measure μ_α in (A5.20) be a step function with a single jump at $x = c$. This case has been discussed in Subsection A5.3.1. This function defines a measure which is said to be **concentrated at** c. Formally, this measure is the Dirac measure δ_c,

multiplied by a constant.[18] Since countable subsets of $\mathbb{R}$ are measurable, a slight generalization of the above procedure allows one to define measures that are concentrated on a countable set of distinct points. Denote such a measure by μ_1. For any subset A of $\mathbb{R}$, $\mu_1(A \setminus \{x_n\}) = 0$.
If X is a countable subset of $\mathbb{R}$, then a measure concentrated on X is called a **pure point** measure, and denoted by μ_{pp}. The measure μ_1 constructed above is a pure point measure. One sees immediately that a pure point measure and a Lebesgue measure on $\mathbb{R}$ are mutually singular.

(ii) The function $\alpha(x)$ introduced at the beginning of Subsection A5.6.1 is monotonic nondecreasing. By Lebesgue's theorem, $\alpha'(x)$ exists a.e., and therefore $\alpha'(x) \geq 0$ a.e. Let now $\alpha'(x) > 0$ a.e., and denote the measure defined by it by μ_2. Then if a subset of $\mathbb{R}$ is of Lebesgue measure zero, it is of μ_2-measure zero. μ_2 is absolutely continuous with respect to the Lebesgue measure; it is not an invariant measure.
A measure which is absolutely continuous with respect to the Lebesgue measure is denoted by μ_{ac}. Note that μ is absolutely continuous with respect to itself. Probability measures[19] on $\mathbb{R}^n$ (like $\bar{\psi}(x)\psi(x)\mathrm{d}^n x$, where ψ is a Schrödinger wave function) are noninvariant measures which are absolutely continuous with respect to Lebesgue measure.

(iii) Begin by defining the **Cantor set C**. Remove the *open middle third* (the interval $(\frac{1}{3}, \frac{2}{3})$) from $[0, 1]$. At the second step, remove the open middle third from each of the remaining two subintervals $[0, \frac{1}{3}]$ and $[\frac{2}{3}, 1]$; at the third step, remove the open middle thirds of each of the four remaining subintervals, and so *ad infinitum.* The set **C** that remains is called the Cantor set.
At each step, the set that remains is a closed set. An easy calculation shows that the total length of the intervals removed is 1, so that **C** has Lebesgue measure zero. By looking at the infinite ternary representations[20] of numbers between 0 and 1, one finds that the ternary representations of the points (numbers) that are not removed do not contain any 1s. It is then trivial to establish a $(1, 1)$ correspondence between this set, and the set of infinite binary representations of numbers in $[0, 1]$. This means that **C** is uncountable.
Let $x \in [0, 1]$. Then

$$x = \sum_{n=1}^{\infty} \frac{a_n}{3^n}, \quad \text{where} \quad a_n = 0, 1 \text{ or } 2,$$

[18] Dirac's δ-function behaves like a measure under integration and like a distribution (Section A5.8) under differentiation. When it is used as a distribution, it is written $\delta(x - c)$, but the same object, used as a measure, is written, in the mathematical literature, as δ_c, where $\delta_c(A) = 1$ if $c \in A$ and $\delta_c(A) = 0$ if $c \notin A$.

[19] A measure μ on X is called a **probability measure** if $\mu(X) = 1$. See Section A7.2.

[20] See the discussion surrounding (A2.5) on page 251.

is its infinite ternary expansion. Let $n = n(x)$ be the first index for which $a_n = 1$. If there is no such index (i.e., if $x \in \mathbf{C}$), then write $n(x) = \infty$. Define the function ψ by

$$\psi(x) = \sum_{1 \le j \le n} \frac{a_j}{2^{j+1}} + \frac{1}{2^{n(x)}}.$$

Then ψ, called the **Cantor function**, is a continuous, monotonic function which increases from 0 at 0 to 1 at 1, is constant on the intervals that have been removed, and is differentiable a.e., with $\psi' = 0$ wherever it is defined. It is a generalization of the notion of a step function, and its graph is known as the **devil's staircase.**[21] As ψ is monotonic nondecreasing, we may define a measure $\mu_3 = \mu_\psi$. As ψ is continuous, $\mu_3(x) = 0$ for every x. Since $\psi' = 0$ on the intervals that have been removed, $\mu_3([0,1] \setminus \mathbf{C}) = 0$. Therefore μ_3 differs from zero on an uncountable set of Lebesgue measure zero.
A measure like μ_3 which is singular both with respect to pure point measures on $[0,1] \setminus C$ and the Lebesgue measure is called **singular continuous**, and written μ_{sc}.

The following fundamental result was established by Lebesgue:

Theorem A5.30 (The Lebesgue decomposition theorem) *Let μ be a Borel measure* (page 302) *on $\mathbb{R}$. Then μ can be decomposed uniquely as follows:*

$$\mu = \mu_{\mathsf{pp}} + \mu_{\mathsf{ac}} + \mu_{\mathsf{sc}}.$$

A5.7.1 Signed measures; the Radon–Nikodym theorem

We now consider a simple generalization of the notion of measure. The aim of this generalization is to be able to state a result known as the Radon–Nikodym theorem in the form in which it will be needed later.

Definition A5.31 A **signed measure** on a measure space $(X, \mathcal{R})$ is a set function $\mu : \mathcal{R} \to [-\infty, \infty]$ which satisfies the following conditions:

(a) μ is countably additive.

(b) $\mu(\emptyset) = 0$.

(c) μ takes on at most one of the values $-\infty, +\infty$.

A signed measure can be decomposed into a positive and a negative part. This decomposition is effected in two steps:

[21] The Cantor function is defined, and some of its properties given as exercises in (Halmos, 1950, p. 83). Its graph, the devil's staircase, is pictured in almost every book on fractals. See, for instance, (Schroeder, 1991, p. 168).

Proposition A5.32 (Hahn decomposition) *Let μ be a signed measure on $(X, \mathcal{R})$. Then there exists a partition of X into two measurable sets A and B (i.e., $A \cup B = X$, $A \cap B = \emptyset$) such that*

$$\mu(A \cap Z) \geq 0 \text{ and } \mu(B \cap Z) \leq 0$$

for any measurable set $Z \subset X$.

Clearly, the sets A and B are not unique; they can be altered by adding or subtracting subsets of X of measure zero. Now define

$$\mu^+(Z) = \mu(A \cap Z) \text{ and } \mu^-(Z) = -\mu(B \cap Z). \tag{A5.24}$$

Then $\mu^+(Z)$ and $\mu^-(Z)$ are measures on X, and at least one of them is finite. It can be shown that they are independent of the particular Hahn decomposition. Now:

Proposition A5.33 (Jordan decomposition) *Let μ be a signed measure on $(X, \mathcal{R})$. Then*

$$\mu = \mu^+ - \mu^-,$$

where μ^+ and μ^- are given by (A5.24).

We now define a quantity $|\mu|$ by

$$|\mu|(Z) = \mu^+(Z) + \mu^-(Z). \tag{A5.25}$$

It is clear that $|\mu|$ is a measure on $(X, \mathcal{R})$. Therefore the notion of absolute continuity is defined with respect to $|\mu|$. We extend this to the definition of absolute continuity with respect to signed measures:

Definition A5.34 Let μ, ν be signed measures on $(X, \mathcal{R})$. We shall say that ν is **absolutely continuous** with respect to μ, and write $\nu \ll \mu$, if $\nu(A) = 0$ for any measurable set A for which $|\mu|(Z) = 0$.

We are now in a position to state the Radon–Nikodym theorem in the required form.

Theorem A5.35 (Radon–Nikodym theorem) *Let $(X, \mathcal{R}, \mu)$ be a σ-finite measure space and ν a σ-finite signed measure on $(X, \mathcal{R})$ which is absolutely continuous with respect to μ. Then there exists a nonnegative measurable function f on X such that for each set S in $\mathcal{R}$,*

$$\nu(S) = \int_S f \mathrm{d}\mu.$$

If there exists another measurable function g such that $\nu(S) = \int_S g \, \mathrm{d}\mu$, then $f = g$ a.e.

The function f of the above theorem is called the **Radon–Nikodym derivative** of ν with respect to μ, and is written

$$f = \frac{\mathrm{d}\nu}{\mathrm{d}\mu}. \tag{A5.26}$$

Since $\mathbb{R}^n$ is second countable, its Borel structure satisfies the countability condition of Theorem A5.35 for every n.

The Radon–Nikodym theorem will be used to prove the existence of conditional expectations in Section A7.3.

A5.8 Differentiation

In the theory of the Riemann integral, the relation

$$\frac{\mathrm{d}}{\mathrm{d}x} \int_a^x f(t)\mathrm{d}t = f(x)$$

holds if f is continuous at x. In the theory of the Lebesgue integral, one can show that this relation is replaced by

$$\frac{\mathrm{d}}{\mathrm{d}x} \int_a^x f(t)\mathrm{d}t = f(x) \text{ a.e.} \tag{A5.27}$$

Denoting, temporarily, the equivalence class of the function $f(x)$ by $[f(x)]$, we may rewrite (A5.27) as

$$\frac{\mathrm{d}}{\mathrm{d}x} \int_a^x [f(t)]\mathrm{d}t = [f(x)],$$

which shows that differentiation is the inverse of integration in the theory of the Lebesgue integral as well.

Similarly, if $f(x)$ is a differentiable function, then

$$\frac{\mathrm{d}}{\mathrm{d}x}[f(x)] = [f'(x)].$$

However, differentiable functions are rare even among continuous functions, to say nothing of L^1. It has been shown that, in a measure-theoretic sense, almost every continuous function is nowhere-differentiable.

In view of this fact – which has wide-ranging repercussions for mathematicians and mathematical physicists, as we shall glimpse in Appendix A6 – the reader may justly point out that we have got ourselves into this situation by according primacy to the notion of integration, and requiring the norm defined by the integral to be complete. What would happen if we were to give primacy, instead, to the notion of differentiation and tried to build a space that consists only of differentiable functions and is complete?

The answer to this question was provided by Laurent Schwartz in his *Théorie des distributions* (Schwartz, 1957, 1959).[22] It is indeed possible to define families of objects – called **distributions** by Schwartz – that are infinitely differentiable. They are not functions, but linear functionals on suitable classes of **test-functions**; in this they resemble quantized, rather than classical fields.[23] Spaces of test-functions are linear spaces over $\mathbb{R}$ with topologies defined by a family of seminorms; they are called **topological vector spaces**. The family of seminorms actually defines a uniformity on the space, which is complete in this uniformity (Appendix A4). The topology on them is just the topology of the uniformity. It would take us too far afield to go into further details. The interested reader is referred to the summary in Chapter 2 of (Streater and Wightman, 1964), the detailed treatment in Chapter 6 of (Rudin, 1974), and to the references cited therein. However, the notion of uniformities is not used in the texts cited above.

[22] Distributions are called **generalized functions** by Gelfand and the Soviet school, and **ideal functions** by Courant in the English translation of Part II (Courant and Hilbert, 1962).

[23] The physical antecedents of the theory of distributions, in particular the insights of Niels Bohr in 1931, have been discussed in Section 6.1.

A6

Hilbert space, operators and spectral theory

In this appendix we shall make precise the notions of Hilbert space and operators on Hilbert space, and state and explain the various cases of the spectral theorem for self-adjoint operators on Hilbert space. The integral used for defining Hilbert spaces of square-integrable functions will be the Lebesgue integral; the one used for stating the spectral theorem will be, in the first instance, the Riemann–Stieltjes integral.

A6.1 Hilbert space

The term *Hilbert space* was introduced in the late 1920s, by von Neumann who defined an abstract Hilbert space $\mathfrak{H}$ by the axioms that are summarized below (in his own enumeration).

A *Linearity* $\mathfrak{H}$ is a linear space over $\mathbb{C}$.

B *Inner product* A Hermitian **inner product** $(f, g), f, g \in \mathfrak{H}$ is defined on $\mathfrak{H}$. Von Neumann defined the inner product to be linear in the first argument and *antilinear* in the second, i.e., $(af, g) = a(f, g)$ and $(f, ag) = \bar{a}(f, g)$. The bar above a letter denotes its complex conjugate. This is the convention used by mathematicians, but physicists, following Dirac, use the opposite convention: an inner product which is antilinear in the first argument and linear in the second:[1]

$$(af, g) = \bar{a}(f, g) \text{ and } (f, ag) = a(f, g).$$

We shall use the physicists' convention; the reader who consults a mathematics text should bear this in mind. The hermiticity condition

$$(f, g) = \overline{(g, f)}$$

remains unaffected. Note that the inner product induces a metric

$$d(f, g) = |(f - g, f - g)| = \sqrt{(f - g, f - g)^2}.$$

[1] Note that the inner product defined by (A5.11) is that of the mathematician.

The quantity $||f||$ defined by $||f||^2 = (f, f)$ is called the **norm** of f. It satisfies the conditions

(i) $||cf|| = |c| \cdot ||f||$,
(ii) $|(f, g)| \leq ||f|| \, ||g||$ (the **Schwarz inequality**), and
(iii) $||f + g|| \leq ||f|| + ||g||$ (the **triangle inequality**)

for all $c \in \mathbb{C}$ and $f, g \in \mathfrak{H}$. The Schwarz and triangle inequalities are fundamental.

C *Dimension* Either there are exactly n linearly independent vectors in $\mathfrak{H}$, in which case axiom **C** is denoted by $\mathbf{C}^{(n)}$, or else there is no maximum number of linearly independent vectors in $\mathfrak{H}$, in which case axiom **C** is denoted by $\mathbf{C}^{(\infty)}$.

Von Neumann points out that **C** is essentially not a new axiom; if **A** and **B** hold, then either $\mathbf{C}^{(n)}$ or $\mathbf{C}^{(\infty)}$ must hold.
The axioms **D** and **E** that follow are needed only if $\mathbf{C}^{(\infty)}$ holds; they are theorems (from **A** and **B**) if $\mathbf{C}^{(n)}$ holds.

D *Completeness* $\mathfrak{H}$ is complete.
E *Separability* $\mathfrak{H}$ is separable (in the topological sense; the topology of Hilbert space is the metric topology).[2]

Von Neumann then defines **orthonormal sets** of vectors in $\mathfrak{H}$. An orthonormal set $\mathcal{O} \subset \mathfrak{H}$ is called **complete** if it is not a proper subset of any other orthonormal set. He then proves the following results (we confine ourselves to infinite-dimensional Hilbert spaces; in this book the term *Hilbert space* generally means infinite-dimensional Hilbert space over the complex numbers; exceptions, used mainly in Chapter 10, are identified explicitly):

(i) A complete orthonormal set is countably infinite.
(ii) Let $\{e_k\}$ be a complete orthonormal set in $\mathfrak{H}$. Then any vector $f \in \mathfrak{H}$ has a unique expansion

$$f = \sum_{k=1}^{\infty} f_k e_k,$$

which means that

$$||f - \sum_{k=1}^{n} f_k e_k|| \to 0 \text{ as } n \to \infty,$$

[2] Separability is no longer assumed in the mathematical literature (see (Friedman, 1970, pp. 201, 216)); it continues to be fundamental in quantum mechanics.

and the $f_k \in \mathbb{C}$ are defined by

$$f_k = (e_k, f)$$

with

$$\sum_{k=1}^{\infty} |f_k|^2 < \infty.$$

The notation $< \infty$ means 'is finite'. We remind the reader that the inner product used for defining the expansion coefficients f_k is that of the physicist.

It should be noted that the separability condition is essential for the above results; in fact, one can prove that separability is equivalent to the existence of a *countable* orthonormal set which is complete, a fact that makes it possible for the working physicist to ignore this condition. A complete orthonormal set is often called an **orthonormal basis**. Detailed accounts of the above may nowadays be found in many textbooks, but it is still rewarding to read the original (von Neumann, 1955). A more recent treatment, aimed specifically at mathematical physicists, will be found in (Reed and Simon, 1972).

Examples A6.1 (Examples of Hilbert spaces)

(i) The space l_2 of all sequences $\{z_n\}$ of complex numbers such that

$$\sum_{k=1}^{\infty} |z_n|^2 < \infty,$$

with multiplication by scalars and addition of vectors being defined in the obvious manner: $\alpha\{z_n\} = \{\alpha z_n\}$ and $\{x_n\} + \{y_n\} = \{x_n + y_n\}$. The inner product of $u = \{x_k\}$ and $v = \{y_k\}$ is defined to be

$$(u, v) = \sum_{k=1}^{\infty} \bar{x}_k y_k.$$

It is obvious that (u, v) defined above is antilinear in the first argument and linear in the second. We omit the proofs of the Schwarz and triangle inequalities, which may be found in almost every text on analysis, e.g., (Rudin, 1976, pp. 15–17). It is easily seen that the set of vectors $e_1 = (1, 0, \ldots), e_2 = (0, 1, 0, \ldots)$, etc. forms a complete orthonormal set in l_2, which assures separability. The vector $\{z_n\}$ may be written as

$$\{z_n\} = \sum_{n=1}^{\infty} z_n e_n.$$

(ii) Denote by $L^2(\mathbb{R}^n)$ the space of complex-valued functions on $\mathbb{R}^n$ that are square-integrable with respect to the Lebesgue measure $d\mu = d^n x$, i.e., of functions $f = u + iv$ such that $\int |f|^2 d\mu = \int (u^2 + v^2) d\mu < \infty$. This space[3] is a Hilbert space under the inner product

$$(f, g) = \int_{\mathbb{R}^n} \bar{f}(x) g(x) d\mu.$$

The completeness of this space is the content of the celebrated **Riesz–Fischer theorem**. The complex case subsumes the real case. Starting from any vector in $L^2(\mathbb{R}^n)$, a complete orthonormal set of vectors may be constructed by the **Schmidt orthogonalization process.**

The spaces l_2 and $L^2(\mathbb{R}^n)$ are isomorphic. Let $\{e_k\}$ be any complete orthonormal set in $L^2(\mathbb{R}^n)$ and define, for $f \in L^2(\mathbb{R}^n)$,

$$f_k = (e_k, f).$$

The f_k are called the **Fourier coefficients** of f relative to the $\{e_k\}$. The sequence $\{f_k\}$ belongs to l_2, and the formula

$$f \longrightarrow \{f_k\}$$

is an isometry of l_2 and $L^2(\mathbb{R}^n)$ (the result is independent of n):

$$\sum_{n=1}^{\infty} |f_n|^2 = \int_{\mathbb{R}^n} |f(x)|^2 d\mu. \tag{A6.1}$$

This was the form in which the Riesz–Fischer theorem was proven by Riesz in 1907. Observe that the same result holds if $L^2(\mathbb{R}^n)$ is replaced by $L^2(I^n)$, where $I^n = [a, b]^n$ and the measure on I^n is the Lebesgue measure.

A6.1.1 Direct sums

We assume that the reader is familiar with the notion of **direct sum** of finite-dimensional vector spaces from linear algebra. This notion generalizes to the direct sum of countably many Hilbert spaces as follows.

Definition A6.2 (Direct sum of countably many Hilbert spaces) Let $\mathfrak{H}^{(n)}, n \in \mathbf{N}$ be a countable set of Hilbert spaces, and let $\phi^{(n)} \in \mathfrak{H}^{(n)}$. The

[3] In the expressions $L^2(\mathbb{R}^n)$, in which no measure is specified, it is understood that the measure being used is the Lebesgue measure.

(infinite) direct sum $\mathfrak{H}$ of the Hilbert spaces $\mathfrak{H}^{(n)}$, written

$$\mathfrak{H} = \bigoplus_{n=1}^{\infty} \mathfrak{H}^{(n)}$$

is defined as follows:

(a) The vectors $\phi \in \mathfrak{H}$ are the sequences $(\phi^{(1)}, \ldots \phi^{(n)}, \ldots)$, with addition and multiplication by scalars defined componentwise, which satisfy the convergence condition

$$\sum_{n=1}^{\infty} ||\phi^{(n)}||^2 < \infty.$$

Here $||\phi^{(n)}||$ is the norm of $\phi^{(n)}$ in $\mathfrak{H}^{(n)}$.

(b) The inner product (ϕ, ψ) in $\mathfrak{H}$ is defined as

$$(\phi, \psi) = \sum_{n=1}^{\infty} (\phi^{(n)}, \psi^{(n)}), \tag{A6.2}$$

where $(\phi^{(n)}, \psi^{(n)})$ is the inner product in $\mathfrak{H}^{(n)}$.

It is easily seen that (A6.2) indeed defines an inner product on $\mathfrak{H}$, and that

$$||\phi||^2 = (\phi, \phi) = \sum_{n=1}^{\infty} ||\phi^{(n)}||^2,$$

as it should be. The completeness and separability of $\mathfrak{H}$ are also easy to verify (using the fact that the union of countably many countable sets is countable; see page 246).

Mutatis mutandis, the above remarks apply to (i) finite direct sums of Hilbert spaces, and (ii) infinite direct sums of finite-dimensional vector spaces. Note that the subspace of $\mathfrak{H}$ defined by $\mathfrak{H}_k = \{\phi \in \mathfrak{H} | \phi^{(j)} = 0 \, \text{for} \, j \neq k\}$ is identifiable with its kth component $\mathfrak{H}^{(k)}$; the map $\phi^{(k)} \leftrightarrow (0, \ldots, \phi^{(k)}, 0, \ldots)$ provides the identification.

Let $\mathfrak{H} = \mathfrak{H}^{(1)} \oplus \mathfrak{H}^{(2)}$, and denote by $\mathfrak{H}_1$ and $\mathfrak{H}_2$ the corresponding subspaces of $\mathfrak{H}$. If $\phi \in \mathfrak{H}_1, \psi \in \mathfrak{H}_2$, then $(\phi, \psi) = 0$. This may be expressed as follows: in words, that $\mathfrak{H}_1$ and $\mathfrak{H}_2$ are **orthogonal complements** of each other; in symbols, $\mathfrak{H}_1^{\perp} = \mathfrak{H}_2$, and $\mathfrak{H}_2^{\perp} = \mathfrak{H}_1$.

A6.1.2 Tensor products

Next, we shall define the **tensor product** of two Hilbert spaces.[4] The process is formally identical with constructing the two-particle Hilbert space out of the one-particle Hilbert spaces of two *nonidentical* particles, which is the picture we may initially keep in mind. We shall give two equivalent definitions; the concrete one used by von Neumann, and the slightly more abstract one in current use. We have used both definitions in Chapters 8–11.

Definition A6.3 (Tensor product of Hilbert spaces, concrete) Let $\mathfrak{H}^{\mathrm{I}} = L^2(\mathbb{R}^M, \mathrm{d}q)$ and $\mathfrak{H}^{\mathrm{II}} = L^2(\mathbb{R}^N, \mathrm{d}r)$, where M, N are fixed positive integers, $q \in \mathbb{R}^M$, $r \in \mathbb{R}^N$ and $\mathrm{d}q$, $\mathrm{d}r$ the respective Lebesgue measures. Let $\{\varphi_m^{\mathrm{I}}(q)\}$ and $\{\varphi_n^{\mathrm{II}}(r)\}$ be orthonormal bases in $\mathfrak{H}^{\mathrm{I}}$ and $\mathfrak{H}^{\mathrm{II}}$, and define

$$\phi_{mn}(q,r) = \varphi_m^{\mathrm{I}}(q)\varphi_n^{\mathrm{II}}(r).$$

The $\phi_{mn}(q,r)$ can obviously be multiplied by complex numbers, and added together. They form a linear space $\mathfrak{H}$ over $\mathbb{C}$, consisting of vectors of the form

$$\Phi(q,r) = \sum_{m,n=1}^{\infty} f_{mn}\phi_{mn}(q,r),$$

where the coefficients f_{mn} are restricted by

$$\sum_{m,n=1}^{\infty} |f_{mn}|^2 < \infty.$$

Define $(\phi_{mn}, \phi_{m',n'})$ by

$$(\phi_{mn}, \phi_{m',n'}) = (\varphi_m^{\mathrm{I}}, \varphi_{m'}^{\mathrm{I}}) \cdot (\varphi_n^{\mathrm{II}}, \varphi_{n'}^{\mathrm{II}}) = \delta_{mm'}\delta_{nn'}. \tag{A6.3}$$

Formula (A6.3) defines an inner product on $\mathfrak{H}$. Let $\Psi(q,r) = \sum_{ij} g_{ij}\phi_{ij}$. Then

$$(\Phi, \Psi) = \sum_{m,n=1}^{\infty} \bar{f}_{mn} g_{mn} \tag{A6.4}$$

is an inner product if the left-hand side is antilinear in the first argument and linear in the second. This is the inner product on $\mathfrak{H}$, and it induces a norm in the usual manner. One proves, as in the case of l_2, that the norm satisfies the Schwarz and triangle inequalities, and that $\mathfrak{H}$ is complete in the norm. Separability is assured by the very definition. We have made use of the isometry between $\mathfrak{H}$ and l_2 to frame the definition.

[4] In the theory of group representations (which, historically, came earlier) the tensor product was called the **direct product**. This term is also used in Wigner's lecture notes (Wigner, 1983).

We shall now give the abstract definition, which does not depend on the fact that the vectors in $\mathfrak{H}^{\mathrm{I}}$ and $\mathfrak{H}^{\mathrm{II}}$ can be multiplied together as complex-valued functions.

Definition A6.4 (Tensor product of Hilbert spaces, abstract) Let $\mathfrak{H}$ and $\mathfrak{K}$ be two Hilbert spaces. Denote the vectors of $\mathfrak{H}$ by ϕ and those of $\mathfrak{K}$ by ψ, with indices, if necessary. Define $\phi \otimes \psi$ to be a form that is linear in both factors, i.e.,

$$(a\phi + a'\phi') \otimes \psi = a\phi \otimes \psi + a'\phi' \otimes \psi$$

and

$$\phi \otimes (b\psi + b'\psi') = b\phi \otimes \psi + b'\phi \otimes \psi'.$$

Define the inner product of $\phi \otimes \psi$ and $\phi' \otimes \psi'$ to be

$$(\phi \otimes \psi, \phi' \otimes \psi') = (\phi, \phi') \cdot (\psi, \psi').$$

Consider now the set $\mathfrak{S}$ of finite linear combinations of the forms $\phi \otimes \psi$ (with complex coefficients), and define on this set the inner product

$$\left(\sum_{j=1}^{m} a_j \phi_j \otimes \psi_j, \sum_{k=1}^{n} a'_k \phi'_k \otimes \psi'_k \right) = \sum_{j,k=1}^{m,n} \bar{a}_j a'_k \, (\phi_j, \phi'_k) \cdot (\psi_j, \psi'_k). \qquad \text{(A6.5)}$$

It has, of course, to be verified that (A6.5) does indeed define an inner product. The space $\mathfrak{H} \otimes \mathfrak{K}$ is defined to be the completion of $\mathfrak{S}$ in the norm induced by the inner product (A6.5). Then $\mathfrak{H} \otimes \mathfrak{K}$ is complete by definition, but one has to verify that it is separable.

The above definition has a variant that is often encountered in the literature. The set $\mathfrak{S}$ is replaced by the set of *infinite* linear combinations (as in the concrete case), together with appropriate convergence conditions. The space so defined is complete. The details are left to the reader.

The construction may be extended easily to more than two factors, but we shall have no occasion to use them. The following result is one of the easy consequences of the definition,

$$L^2(\mathbb{R}^3) = L^2(\mathbb{R}^2) \otimes L^2(\mathbb{R}) = L^2(\mathbb{R}) \otimes L^2(\mathbb{R}) \otimes L^2(\mathbb{R}),$$

and gives an intuitive grasp of the idea of the tensor product.

A6.2 Operators on Hilbert space

A linear tansformation $T : \mathfrak{H} \to \mathfrak{H}$ is called a **linear operator**, or simply an **operator**[5] on $\mathfrak{H}$. In the metric topology on $\mathfrak{H}$, T is **continuous at** $x \in \mathfrak{H}$ if, given $\varepsilon > 0$, there exists $\delta > 0$ such that $||T(x-y)|| < \varepsilon$ for all y such that $||x-y|| < \delta$. Equivalently, T is continuous at x if, for any sequence $\{x_n\}$ in $\mathfrak{H}$ that converges to x, the sequence $\{Tx_n\}$ converges to Tx. The operator T is **continuous** iff it is continuous at every $x \in \mathfrak{H}$.

Theorem A6.5 *Let $T : \mathfrak{H} \to \mathfrak{H}$ be a linear operator. The following conditions are equivalent:*

(1) *T is continuous.*

(2) *T is continuous at the origin (i.e., at the zero vector).*

(3) *There exists a number $K \geq 0$ such that*

$$||Tx|| \leq K||x|| \text{ for all } x \in \mathfrak{H}.$$

Proof (1) $\Rightarrow$ (2) is trivial.

To prove (2) $\Rightarrow$ (3), assume that there is no such K, i.e., for any $C > 0$ one can find $x \in \mathfrak{H}$ such that $||Tx|| > C||x||$. Then for each $n > 0$ one can find x_n such that $||Tx_n|| > n||x_n||$. Define

$$y_n = \frac{1}{n}\frac{x_n}{||x_n||}.$$

Then $y_n \to 0$. But

$$||Ty_n|| = \frac{1}{n}\frac{||Tx_n||}{||x_n||} > 1,$$

so that $Ty_n \not\to 0$, i.e., T is not continuous at the origin, which contradicts (2).

To prove (3) $\Rightarrow$ (1), fix ϵ and choose $\delta = \epsilon/K$. Then $||x-y|| < \delta \Rightarrow ||T(x-y)|| < K||x-y|| < K\delta = \epsilon$. □

Definition A6.6 (Bounded operators) An operator $T : \mathfrak{H} \to \mathfrak{H}$ is called **bounded** if there exists $K \geq 0$ such that $||Tx|| \leq K||x||$ for all $x \in \mathfrak{H}$. The number

$$\inf_{x \in \mathfrak{H}} \{K : ||Tx|| \leq K||x||\} \tag{A6.6}$$

is called the **norm** of T, and is denoted by $||T||$.

[5] If A is an operator on $\mathfrak{H}$ and B an operator on $\mathfrak{K}$, then their **tensor product** is the operator (denoted by $A \otimes B$) on $\mathfrak{H} \otimes \mathfrak{K}$ which is defined by $(A \otimes B)(\phi \otimes \psi) = A\phi \otimes B\psi$.

The norm can also be defined as follows:

$$||T|| = \sup_{||x||=1} ||Tx||;$$

the supremum is taken over all unit vectors in $\mathfrak{H}$. The norm satisfies the following conditions:

(i) $||aT|| = |a|||T||$ for all $a \in \mathbb{C}$.

(ii) $||T + U|| \leq ||T|| + ||U||$.

Condition (i) is almost self-evident; condition (ii), known, unsurprisingly, as the **triangle inequality**, is fairly easy to prove, but we shall omit the details.

Remark A6.7 Observe that the definition and the basic properties of bounded operators given above involve only the norm, and not the inner product. Therefore they are equally applicable to linear transformations of Banach spaces (page 290).

Unbounded operators are not just pathologies that can be disregarded by the physicist, as the following theorem shows.

Theorem A6.8 *On a Hilbert space $\mathfrak{H}$, the canonical (Born–Jordan) commutation relations*

$$QP - PQ = \mathrm{i}I$$

cannot be satisfied by any pair of bounded operators $P, Q \in \mathfrak{H}$. Here I is the identity operator.

Proof From $QP - PQ = \mathrm{i}I$, it follows by induction that $Q^nP - PQ^n = \mathrm{i}nQ^{n-1}$. Transposing sides, taking norms, and using successively the triangle inequality and the Cauchy–Schwartz inequality, we obtain

$$\begin{aligned} n||Q||^{n-1} &= ||Q^nP - PQ^n|| \\ &\leq ||Q^nP|| + ||PQ^n|| \\ &\leq 2||Q||^n||P||. \end{aligned}$$

Since Q is not the zero operator, $||Q|| > 0$. Dividing the last inequality by $2||Q||^{n-1}$, we obtain

$$\frac{n}{2} \leq ||Q|| \cdot ||P||,$$

which contradicts the assumption that both Q and P are bounded. □

An operator which is not bounded is called **unbounded**. Theorem A6.8 shows that at least one of Q and P is unbounded. Since, from Theorem A6.5, an operator which is continuous at *a single point* is bounded, an unbounded operator is *nowhere continuous*. The actual situation is even worse; *unbounded operators are not defined everywhere on* $\mathfrak{H}$; if A is an unbounded operator, one can find a convergent sequence $\{\phi_n\} \in \mathfrak{H}$ such that $||A\phi_n|| \to \infty$, i.e., A is not defined on $\lim \phi_n = \phi \in \mathfrak{H}$. The following examples are instances of position and momentum operators in quantum mechanics.

Examples A6.9

(i) *Multiplication by* x Let $\mathfrak{H} = L^2(\mathbb{R})$. The function

$$f(x) = \frac{1}{\mathrm{i} + x}$$

is clearly square-integrable, and so belongs to $\mathfrak{H}$. Let X be the operator of multiplication by x: $X(f(x)) = xf(x)$. Then $Xf = x/(\mathrm{i}+x)$ which, equally clearly, is not square-integrable; the operator X is not bounded. The position operator for a particle constrained to lie on a straight line cannot be bounded.

There are many functions $f(x)$ in $L^2(\mathbb{R})$ such that $xf(x)$ is not square-integrable. Let $f(x) = 1/x$ for $|x| \geq 1$ and $f(x) = 0$ for $|x| < 1$. Then $|xf(x)|^2 = 1$ for $|x| \geq 1$.

Multiplication by x is a bounded operator on $L^2([a, b])$.

(ii) *Differentiation* Consider $L^2([0, 2\pi])$. The functions (more precisely, the equivalence classes of these functions)

$$f_n(x) = \exp(inx) \text{ and } f_n'(x) = in\exp(inx)$$

belong to this space. Therefore, if the operator D of differentiation with respect to x is definable on $L^2([0, 1])$, one should have $Df_n(x) = f_n'(x)$ a.e. But $||f_n(x)|| = 1$, whereas $||f_n'(x)|| = n$, i.e., $||Df_n(x)|| \to \infty$ as $n \to \infty$, which shows that D cannot be defined everywhere on $\mathfrak{H}$.

As explained in Section A5.8, most vectors in $L^2(\mathbb{R})$ or $L^2([a, b])$ will not be differentiable, i.e., will not belong to the domain of the differentiation operator; the momentum operator in quantum theory can never be bounded. The unboundedness of the differentiation operator is what makes the theory of differential equations (as compared with the theory of integral equations) into a difficult subject, and provides a perspective for the role of the theory of distributions in it.

It follows from the above that the full definition of an unbounded operator U on $\mathfrak{H}$ must include specification of its **domain**, i.e., the subset $\mathfrak{D}_U \subset \mathfrak{H}$ on which U is defined. The best one can hope for is that $\mathfrak{D}_U$ be dense in $\mathfrak{H}$. We shall return to unbounded operators in Section A6.5.

A6.3 Bounded operators

We shall now define a few subclasses of bounded operators that are of particular interest. We begin with the definition of the adjoint.

Definition A6.10 (Adjoint of a bounded operator) Let B be a bounded operator on $\mathfrak{H}$. Define an operator B^* on $\mathfrak{H}$ by[6]

$$(f, B^*g) = (Bf, g) \text{ for all } f, g \in \mathfrak{H}.$$

B^* is clearly bounded; it is called the **adjoint** of B.

The properties of adjoints on finite-dimensional vector spaces continue to hold on $\mathfrak{H}$; in particular, $(B^*)^* = B$.

Definition A6.11 (Bounded self-adjoint operators) If $B = B^*$, then B is called **self-adjoint**, or **Hermitian**, or **symmetric**.

(The last term is seldom used in the physics literature.) As B^* is defined everywhere, it has to be bounded. The reader is invited to prove that $||B^*|| = ||B||$.

Definition A6.12 (Normal operators) If $B^*B = BB^*$, i.e., if a bounded operator commutes with its adjoint, it is called **normal**.

Every self-adjoint operator is trivially normal; a normal operator need not be self-adjoint, but if B is normal, then $B + B^*$ and $\mathrm{i}(B - B^*)$ are a pair of commuting self-adjoint operators.

Definition A6.13 (Projection operators) A bounded self-adjoint operator E that satisfies the condition $E^2 = E$ is called a **projection operator**.

A projection operator E has only two eigenvalues, namely 0 and 1, and splits $\mathfrak{H}$ into two mutually orthogonal subspaces $\mathfrak{M}$ and $\mathfrak{M}^\perp$, $\mathfrak{H} = \mathfrak{M} \oplus \mathfrak{M}^\perp$, such that E acts as the identity operator on $\mathfrak{M}$ and the zero operator on $\mathfrak{M}^\perp$. It is easily seen that: (i) $E^\perp = I - E$ is also a projection operator; (ii) $E^\perp$ acts as the zero operator on $\mathfrak{M}$ and the identity operator on $\mathfrak{M}^\perp$, (iii) $EE^\perp = E^\perp E = 0$ and (iv) $E + E^\perp = I$.

Let $\mathfrak{K}$ be a subspace of $\mathfrak{H}$, and T an operator on $\mathfrak{H}$. The **restriction** of T to $\mathfrak{K}$, which we shall write as $T|_{\mathfrak{K}}$, is the linear map

$$T|_{\mathfrak{K}} : \mathfrak{K} \to \mathfrak{H}$$

[6] The existence of the operator B^* on an infinite-dimensional space is not obvious; it is assured by a theorem known as the **Riesz representation theorem**. We shall not discuss it here. The interested reader is referred to the **Riesz lemma** in (Reed and Simon, 1972) or **Riesz's theorem** in (Friedman, 1970).

which is defined by $(T|_{\mathfrak{K}})\phi = T\phi$ for all $\phi \in \mathfrak{K}$. In the cases we shall consider, $\mathfrak{K}$ will also be the range of $T|_{\mathfrak{K}}$. Then, if E is the projection operator from $\mathfrak{H}$ onto $\mathfrak{K}$, we shall have $E \cdot (T|_{\mathfrak{K}}) = (T|_{\mathfrak{K}}) \cdot E$.

Definition A6.14 (Positive operators) An operator A on $\mathfrak{H}$ is called **positive** if $(f, Af) \geq 0$ for all $f \in \mathfrak{H}$. By implication, A is bounded.

Definition A6.15 (The trace) Let A be a positive operator and $\{e_n\}, n \in \mathbf{N}$ an orthonormal basis on $\mathfrak{H}$. The sum

$$\sum_{n=1}^{\infty}(e_n, Ae_n)$$

is called the **trace** of A, and is written $\operatorname{Tr} A$.

The trace is an invariant (i.e., independent of the basis $\{e_n\}$). The trace of the product of a finite number of (positive) operators is invariant under cyclic permutations of the operators. The proofs for the finite-dimensional cases hold unchanged for the infinite-dimensional case.

The trace need not be finite; for example, the identity operator on $\mathfrak{H}$ is bounded, but its trace is infinite. The class of positive operators on $\mathfrak{H}$ with finite trace is important enough to be given a name:

Definition A6.16 (Trace class) The class of positive operators on $\mathfrak{H}$ with finite trace is called the **trace class** on $\mathfrak{H}$. (We shall denote this class by $\mathcal{T}$.) One speaks of **trace class operators** on $\mathfrak{H}$.

We now come to a class of operators which represent – as we shall soon see – the simplest generalizations of finite-dimensional matrices to infinite-dimensional Hilbert spaces.

Definition A6.17 (Compact operators) A bounded operator B on $\mathfrak{H}$ is called **compact** (or **completely continuous**) if it satisfies the following condition: for any bounded sequence $\{f_n\}$, the sequence $\{Bf_n\}$ contains a convergent subsequence.

Compactness is a very strong condition on operators. For example, the identity operator I is not compact; a sequence of vectors $\{e_n\}$ that forms a complete orthonormal set is bounded, but does not contain any convergent subsequence.

Note that if B is compact, then so is its restriction to any subspace $\mathfrak{M}$ of $\mathfrak{H}$, as one sees by taking $\{f_n\}$ to be a sequence in $\mathfrak{M}$.

The following theorem plays a crucial role in von Neumann's measurement theory.

Theorem A6.18 *A trace class operator on* $\mathfrak{H}$ *is compact.*

For a proof, see (Reed and Simon, 1972).

A6.4 Spectral theorems: bounded operators

Spectral theorems lie at the heart of operator theory. Although there are some extensions to normal operators, the chief concern of spectral theory is the structure of self-adjoint operators. We shall consider *only* self-adjoint operators, and shall often omit the adjective 'self-adjoint'. The subject may be introduced by recalling the basic structure theorem for a Hermitian operator on an n-dimensional $(0 < n < \infty)$ inner product space V over $\mathbb{C}$. Recall that (i) the eigenvalues of a Hermitian operator T are real, and (ii) the normalized eigenvectors of T form an orthonormal basis for V.

A6.4.1 The finite-dimensional case

Suppose that 0 is an eigenvalue of T, and denote by V_0 the eigenspace of V belonging to the eigenvalue 0. We may then split V into the direct sum

$$V = V_0 \oplus V_0^{\perp}, \tag{A6.7}$$

where $V_0^{\perp}$ is the orthogonal complement of V_0. Then T has only nonzero eigenvalues on $V_0^{\perp}$. Corresponding to (A6.7), we may split T into the direct sum[7]

$$T = T_0 \oplus T_0^{\perp}, \tag{A6.8}$$

where the notation is obvious. $T_0^{\perp}$ is a Hermitian operator on $V_0^{\perp}$ which has only nonzero eigenvalues. Since the splittings (A6.7) and (A6.8) are always possible, *there is no loss of generality in assuming that the operator T has only nonzero eigenvalues.*

Theorem A6.19 (Spectral theorem on finite-dimensional spaces) *Let A be a Hermitian operator on an n-dimensional vector space V over $\mathbb{C}$ which does not have zero as an eigenvalue. Let $\lambda_1, \lambda_2, \ldots, \lambda_k$ be the distinct eigenvalues of A, W_{λ_j} the eigenspace of A belonging to the eigenvalue λ_j, and E_{λ_j} the orthogonal projection of V on W_{λ_j}, i.e., $E_{\lambda_i} E_{\lambda_j} = \delta_{ij} E_{\lambda_j}$. Then*

$$V = \bigoplus_{i=1}^{k} W_{\lambda_i},$$

[7] In matrix theory and operator theory, the notation + seems to be preferred over $\oplus$ for the direct sum of operators; in the theory of group representations, it is usually the other way round. Using + to denote the direct sum of operators has the drawback that one has to work out whether the summands are defined on the whole space, or only on subspaces. We have chosen the notation $\oplus$ to avoid this ambiguity.

and

$$A = \sum_{i=1}^{k} \lambda_i E_{\lambda_i}. \tag{A6.9}$$

Note that had one of the λ_i been zero, it would have contributed the summand $0 \cdot E_0 = 0$ to the sum on the right-hand side of (A6.9); the term would simply have dropped out. Note also that

$$I = \sum_{i=1}^{k} E_{\lambda_i}. \tag{A6.10}$$

Recall now that, if p is any polynomial, then

$$p(A) = \sum_{i=1}^{k} p(\lambda_k) E_{\lambda_i}. \tag{A6.11}$$

Theorem A6.19 is purely algebraic; it is stated and proven using concepts that are based upon the binary operations of addition and multiplication by scalars. On a Hilbert space $\mathfrak{H}$, these concepts have to be supplemented by topological ones; the useful notion of a subspace is one which is closed both linearly and topologically, i.e., one that contains its limit points. The term **subspace**, applied to a Hilbert space, will mean one that is *closed both linearly and topologically.*

A6.4.2 Spectral theorem: compact operators

The analysis of linear transformations on a finite-dimensional vector space V began with posing the eigenvalue problem. The existence of solutions of the characteristic equation was ensured by the fundamental theorem of algebra. This method cannot be used in the infinite-dimensional case; to make matters worse, operators with purely continuous spectra do not have eigenvectors (which, essentially, is why von Neumann wrote his book (von Neumann, 1955, Preface). The key to unravelling the structure of compact operators is the following result.[8]

Theorem A6.20 (Hilbert) *Every compact self-adjoint operator A on $\mathfrak{H}$ has at least one non-zero eigenvalue, and an associated eigenvector.*

For any compact self-adjoint operator A, the set of eigenvectors belonging to the eigenvalue 0 forms a subspace of $\mathfrak{H}$ which we shall denote by $\mathfrak{M}_0$. Its

[8] For a proof, see (Riesz and Sz-Nagy, 1955, pp. 231–232). The reader who wishes to go into details may be better advised to make a systematic study of the foundations of modern analysis, an elegant and economic account of which is provided in the text of the same name by Friedman (Friedman, 1970). The text by Reed and Simon is somewhat longer, but may be more attuned to the needs of physicists (Reed and Simon, 1972). One should add that the results of Subsections A6.4.2–A6.4.4 were established by Hilbert in 1905.

orthogonal complement $\mathfrak{M}_0^\perp$ contains no eigenvector of A with eigenvalue 0; the operator A can be decomposed uniquely into the direct sum $A_0 \oplus A_0^\perp$. As the rest of the analysis will be carried out exclusively on $\mathfrak{M}_0^\perp$, *we shall assume that* $\mathfrak{M}_0^\perp = \mathfrak{H}$; this is equivalent to assuming that A does not have 0 as an eigenvalue; it is the exact parallel of the finite-dimensional case considered earlier, and is not a restrictive assumption.

As in the finite-dimensional case, eigenvectors belonging to different eigenvalues are orthogonal to each other. Let λ be an eigenvalue of A, and denote by $\mathfrak{M}_\lambda$ the eigenspace of A belonging to the eigenvalue λ. The restriction of A to this subspace acts like a multiple of the identity on it. If $\mathfrak{M}_\lambda$ is infinite-dimensional, then the identity operator on it will not be compact, from which it follows that A will not be compact. Therefore $\mathfrak{M}_\lambda$ must be *finite-dimensional.* It follows from this (and the separability of $\mathfrak{H}$) that *A has countably many distinct eigenvalues.*

Let λ_k, $k = 1, 2, \ldots$ be the distinct eigenvalues of A. The numbers $|\lambda_k|$ can be arranged in a decreasing sequence $\{|\lambda_n|\}$. It turns out that only a finite number of these are greater than any preassigned positive number,[9] so that $|\lambda_n| \to 0$. Furthermore, it turns out that

$$\mathfrak{H} = \mathfrak{M}_{\lambda_1} \oplus \mathfrak{M}_{\lambda_2} \oplus \cdots .$$

Since each $\mathfrak{M}_{\lambda_k}$ is finite-dimensional, this means that the normalized eigenvectors of A form a countable and complete orthonormal set in $\mathfrak{H}$.

We may summarize the discussion so far as follows:

Theorem A6.21 (Spectral theorem for compact operators) *Let A be a compact self-adjoint operator on $\mathfrak{H}$ such that 0 is not an eigenvalue of A. Then A has countably many distinct eigenvalues λ_k, $k = 1, 2, \ldots$ that have 0 as a limit point and, associated with each eigenvalue λ_k, a subspace $\mathfrak{M}_{\lambda_k}$ of $\mathfrak{H}$ such that:*

(1) $0 < \dim \mathfrak{M}_{\lambda_k} < \infty$.

(2) *If $x \in \mathfrak{M}_{\lambda_k}$, then $Ax = \lambda_k x$.*

(3) *If $j \neq k$, then $\mathfrak{M}_{\lambda_j} \perp \mathfrak{M}_{\lambda_k}$.*

(4) $\mathfrak{H} = \oplus_{n=1}^{\infty} \mathfrak{M}_{\lambda_n}$.

Consequently, A can be written as

$$A = \sum_{n=1}^{\infty} \lambda_n E_{\lambda_n}, \tag{A6.12}$$

where E_{λ_j} is the projection operator onto the subspace $\mathfrak{M}_{\lambda_j}$, with $E_{\lambda_j} E_{\lambda_k} = 0$ for $j \neq k$.

[9] For a simple proof, see (Lorch, 1962, p. 114).

Parallel to (A6.10), we have here:

$$I = \sum_{n=1}^{\infty} E_{\lambda_n}. \tag{A6.13}$$

The above results have a converse, which may be stated as follows:

Theorem A6.22 (Converse of spectral theorem) *Let $\{\lambda_n\}$, $n \in \mathbf{N}$ be a bounded sequence of nonzero real numbers that converges to zero, $\{\mathfrak{M}_n\}$ a countable set of pairwise-orthogonal finite-dimensional subspaces of $\mathfrak{H}$ such that $\mathfrak{H} = \oplus_n \mathfrak{M}_n$, and $E_{\lambda_n} : \mathfrak{H} \to \mathfrak{M}_n$ the projection operator from $\mathfrak{H}$ to $\mathfrak{M}_n$. Then there exists a compact operator A such that*

$$A = \sum_{n=1}^{\infty} \lambda_n E_{\lambda_n}.$$

The condition that 0 not be an eigenvalue of A was essential for unravelling the structure of A, i.e., for arriving at the spectral decomposition (A6.12). However, once this has been achieved, it may be dropped. Let A be a self-adjoint operator that has 0 as an eigenvalue. The eigenvectors of A with eigenvalue zero span a subspace $\mathfrak{H}_0$ of $\mathfrak{H}$ (called the **null space** of A), and $\mathfrak{H}$ can be decomposed as

$$\mathfrak{H} = \mathfrak{H}_0 \oplus \mathfrak{H}_0^{\perp},$$

where $\mathfrak{H}_0^{\perp}$ is the orthogonal complement of $\mathfrak{H}_0$. The restriction of A to $\mathfrak{H}_0^{\perp}$ has the spectral decomposition (A6.12). Denoting the projection operator on $\mathfrak{H}_0$ by E_0 and taking $\lambda_0 = 0$, we may generalize (A6.12) and (A6.13) to

$$A = \sum_{n=0}^{\infty} \lambda_n E_{\lambda_n} \tag{A6.14}$$

and

$$I = \sum_{n=0}^{\infty} E_{\lambda_n}. \tag{A6.15}$$

In these formulae, only the lower limit of summation has changed.

A6.4.3 Resolutions of the identity

Theorem A5.9 suggests that it may be possible to write (A6.12) at least formally as a Riemann–Stieltjes integral. To see what is involved, let us consider the spectral decomposition of a compact operator O which has only positive eigenvalues. We noted earlier that the set $\{\lambda_n\}$ has a maximum. We may therefore assume the eigenvalues λ_n of O to be arranged in a decreasing sequence, i.e., $\lambda_k < \lambda_{k+1}$

for all k. (They can be arranged in an increasing sequence, but then there will be no first member.)

For each positive integer k define, inductively, a subspace $V^{(k)}$ of $\mathfrak{H}$ as follows: $V^{(1)} = 0$, $V^{(k)} = \mathfrak{M}_{\lambda_1} \oplus \cdots \oplus \mathfrak{M}_{\lambda_{k-1}}$ for $k > 1$. Then $V^{(k)} \subset V^{(k+1)}$. Next, define $\mathfrak{M}^{(k)} = (V^{(k)})^{\perp}$. Then $\mathfrak{M}^{(k)} \supset \mathfrak{M}^{(k+1)}$. Each $V^{(k)}$ is finite-dimensional, but each $\mathfrak{M}^{(k)}$ is infinite-dimensional. Finally, define a function

$$\mathfrak{M} : \mathbb{R} \longrightarrow \{\text{set of subspaces of } \mathfrak{H}\}$$

as follows:

$$\mathfrak{M}_\lambda = \begin{cases} \mathfrak{M}^{(k)}, & \text{for} \quad \lambda < \lambda_k, \\ \mathfrak{M}^{(k+1)}, & \text{for} \quad \lambda \geq \lambda_k. \end{cases}$$

$\mathfrak{M}_\lambda$ so defined is a *step function* on $\mathbb{R}$, taking values in the set of subspaces of $\mathfrak{H}$, which has the following properties:[10]

(i) For $\lambda < 0$, $\mathfrak{M}_\lambda = 0$.

(ii) If $\lambda < \lambda'$, then $\mathfrak{M}_\lambda \subset \mathfrak{M}_{\lambda'}$. For $\lambda_k \leq \lambda < \lambda_{k+1}$, $\mathfrak{M}_\lambda$ is constant, and equals $\mathfrak{M}^{(k)}$. At $\lambda = \lambda_{k+1}$, $\mathfrak{M}_\lambda$ jumps from $\mathfrak{M}^{(k)}$ to $\mathfrak{M}^{(k+1)}$; it is, in some sense,[11] continuous from the right but discontinuous from the left at each λ_k.

(iii) For $\lambda > \lambda_1$, $\mathfrak{M}_\lambda = \mathfrak{H}$.

The properties of the set $\{\mathfrak{M}_\lambda\}$ can be translated into the language of projection operators. In the process, which does not make any reference to any specific operator A, we shall obtain a certain generalization.

Definition A6.23 (Bounded resolutions of the identity) Let $\lambda \in \mathbb{R}$. For each λ, let $(\mathfrak{M}_\lambda, \mathfrak{M}_\lambda^{\perp})$ be a decomposition of $\mathfrak{H}$ into a pair of orthogonal subspaces (i.e., , $\mathfrak{H} = \mathfrak{M}_\lambda \oplus \mathfrak{M}_\lambda^{\perp}$),[12] and $E_\lambda : \mathfrak{H} \to \mathfrak{M}_\lambda$ the associated projection operator. The family $\{E_\lambda : \lambda \in \mathbb{R}\}$ is called a **bounded resolution of the identity** if it satisfies the following conditions:[13]

(a) There exist $M, N > 0$ such that $E_\lambda = 1$ for $\lambda > M$, and $E_\lambda = 0$ for $\lambda < -N$.

(b) $E_\lambda E_{\lambda'} = E_{\min(\lambda, \lambda')}$.

Condition (b) is often expressed as follows: for $\lambda < \lambda'$, $E_\lambda \leq E_{\lambda'}$.

[10] This definition embodies an important mathematical observation: more often than not, the properties of a function or map depend more on the domain than on the range.

[11] This sense can be made precise by defining a topology on the set of subspaces of $\mathfrak{H}$. We shall omit the details, because in this instance intuition will not lead us astray.

[12] $\mathfrak{H}$ has enough subspaces to permit different decompositions for different values of λ.

[13] Friedman uses the term **spectral family**. It should be noted that we have glossed over some continuity conditions (the same as in footnote 11) which may be found in (von Neumann, 1955, p. 118) or (Friedman, 1970, p. 226).

The above definition does not relate to any specific operator. A resolution of the identity specific to an operator A has to satisfy the additional condition $AE_\lambda x = E_\lambda Ax$ for all $x \in \mathfrak{H}$.

Definition A6.23 is not empty; as we have seen, any compact operator that does not have zero as an eigenvalue defines a bounded resolution of the identity. The generalization we have obtained is that E_λ may depend continuously on λ, in the naive intuitive sense.

Let us now return to the compact operator A of Theorem A6.21. Let $\Delta\lambda$ be a small interval of λ. It is easily seen that the quantities $\Delta E_\lambda = E_{\lambda+\Delta\lambda} - E_\lambda$ are well-defined projection operators; $\Delta E_\lambda = 0$ if the interval $\Delta\lambda$ does not contain any eigenvalue, and $\Delta E_\lambda = E_{\lambda_k}$ if $\Delta\lambda$ contains the single jump at λ_k of λ. The reader is invited to verify that for any $x, y \in \mathfrak{H}$, the 'matrix element' (y, Ax) of A may be expressed as the Riemann–Stieltjes integral

$$(y, Ax) = \int_{-\infty}^{\infty} \lambda \, \mathrm{d}(y, E_\lambda x). \tag{A6.16}$$

Formally, we may write A as

$$A = \int_{-\infty}^{\infty} \lambda \, \mathrm{d}E_\lambda. \tag{A6.17}$$

Equation (A6.17) is, for the moment, to be interpreted as shorthand for the set of equations (A6.16) for all $x, y \in \mathfrak{H}$. It is known as the **spectral resolution of a compact operator**. There is an alternative interpretation of (A6.17) in terms of what are called **projection-valued measures**, which does not use the Riemann–Stieltjes integral, but is essentially the same. As the term is encountered rather frequently in quantum-mechanical measurement theory and in Mackey's theory of infinite-dimensional group representations, the definition is given below.

Definition A6.24 (Projection-valued measures) Let X be a topological space, $\mathcal{B}$ a Borel structure on it (page 301) and $\mathfrak{H}$ a separable Hilbert space. A **projection-valued measure** is an assignment $B \mapsto E_B$, for each $B \in \mathcal{B}$, of a projection operator E_B on $\mathfrak{H}$ to the Borel subset B of X that satisfies the following conditions:

(a) $E_X = I$.

(b) $E_{B_1 \cap B_2} = E_{B_1} E_{B_2}$ for all $B_1, B_2 \in \mathcal{B}$.

(c) For every countable subset $\{B_i\} \subset \mathcal{B}$ such that $j \neq k \Rightarrow B_j \cap B_k = \emptyset$,

$$E\left(\bigcup_{k=1}^{\infty} B_k\right) = \sum_{k=1}^{\infty} E_{B_k}.$$

In the last equation we have, exceptionally, written the argument of E on the left within brackets, rather than as a subscript, for a more agreeable visual display.

This definition should be compared with the definition A5.15 of measure on page 302. Condition A6.24(a) shows that projection-valued measures have a certain similarity with probability measures.

If we take for X the spectrum of the operator A, then, after tying up some loose ends, we are able to write

$$A = \int \lambda \mathrm{d} E_\lambda .$$

We leave the details to the reader.

Since A is compact one may, if one wishes, write (A6.16) and (A6.17) as infinite sums rather than as Stieltjes integrals. However, resolutions of the identity can be defined, with (A6.16) and (A6.17) continuing to hold, for operators more general than the compact ones. The classes of operators that remain to be considered are the bounded and the unbounded ones. Each step involves a huge increase in complexity.

A6.4.4 Spectral theorem: bounded operators

As we have just seen, the structure of compact operators can be described in terms of eigenvalues and eigenvectors. This is no longer possible when we move from compact to bounded operators, where we encounter, for the first time, the phenomenon of the continuous spectrum. Our first task is to make this notion precise.

A6.4.4.1 The spectrum of an operator

The eigenvalue problem $(A - \lambda I)x = 0$ has a nonzero solution x for given λ if and only if the operator $A - \lambda I$ is not invertible for that particular λ. This leads us to the following definitions.

Definition A6.25 (Resolvent set and spectrum) The **resolvent set** $\rho(A)$ of the operator A is the set of all $\lambda \in \mathbb{C}$ such that the operator $(A - \lambda I)^{-1}$ exists, *and is bounded.* The **spectrum** of A is the complement of its resolvent set in $\mathbb{C}$, and is denoted by $\sigma(A)$.

Let $\lambda \in \sigma(A)$. There are three possibilities:

(i) $(A - \lambda I)^{-1}$ does not exist. This is the same as saying that the equation $(A - \lambda I)\varphi_\lambda = 0$ has a nonzero solution $\varphi_\lambda \in \mathfrak{H}$. Such a λ is called an **eigenvalue** of A, and φ_λ an **eigenvector** corresponding to it. The set of eigenvalues of A is called its **discrete** or **point spectrum** and denoted by $\sigma_{\mathsf{d}}(A)$.

(ii) $(A - \lambda I)^{-1}$ exists, is unbounded, and its domain is dense in $\mathfrak{H}$. The set of such values of λ constitutes the **continuous spectrum**[14] of A, and is denoted by $\sigma_c(A)$.

(iii) $(A - \lambda I)^{-1}$ exists, is unbounded, but its domain is not dense in $\mathfrak{H}$. The set of such values of λ constitutes the **residual spectrum** of A.

Definition A6.26 (Spectral radius) The quantity

$$r(A) = \sup |\lambda|, \; \lambda \in \sigma(A)$$

is called the **spectral radius** of A.

We shall state the following result without proof.

Proposition A6.27 *If A is bounded and self-adjoint, then $r(A) = ||A||$; in words, the spectral radius of A equals its norm.*

We shall also state, without proof, the following results on self-adjoint operators. The proof of the first is fairly straightforward, but that of the second is not.

(i) The spectrum of a self-adjoint operator is a subset of $\mathbb{R}$.

(ii) The residual spectrum of a *bounded* self-adjoint operator is empty.

A6.4.4.2 Approximate eigenvectors

There is a very important result concerning the continuous spectrum of A which we shall sketch under the simplifying assumption that $\sigma_c(A)$ is an interval on $\mathbb{R}$. Although the equation $(A - \lambda I)x = 0$ cannot be satisfied by any vector $x \in \mathfrak{H}$ for $\lambda \in \sigma_c(A)$, *there exist vectors x with $||x|| = 1$ such that $||(A - \lambda I)x||$ is arbitrarily small* (we have called such vectors **approximate eigenvectors** in Chapter 9).[15] Probing a little further, one finds that for $\epsilon/N > 0$, there exists a small interval $\Delta\lambda$ around λ and a subspace $\mathfrak{M}_{\Delta\lambda}$ such that

$$||(A - \lambda I)x|| < \frac{\epsilon}{N} \quad \text{for all} \quad x \in \mathfrak{M}_{\Delta\lambda}.$$

That is, A behaves approximately like λI on $\mathfrak{M}_{\Delta\lambda}$. Then. by partitioning $\sigma_c(A)$ into small subintervals $\Delta_k\lambda$, $k = 1, \ldots, N$ such that $\lambda_k \in \Delta_k\lambda$, one can show that there exists a set of pairwise-orthogonal subspaces $\mathfrak{M}_{\lambda_k}$ of $\mathfrak{H}$ and the

[14] This definition of the continuous spectrum is used in (Lorch, 1962) and (Friedman, 1970). A slightly different definition, in which the discrete and continuous spectra are not necessarily disjoint, is used in (Reed and Simon, 1972). See the paragraphs following (A6.23), pages 339–340.

[15] The proof is not difficult, but we shall omit it.

corresponding projection operators ΔE_{λ_k} such that

$$||A - \sum_{n=1}^{N} \lambda_n \Delta E_{\lambda_n}|| < \epsilon.$$

This means that A *can be approximated arbitrarily well in the norm by linear combinations of projection operators.*

A6.4.4.3 The theorem

We are now in a position to state the spectral theorem for bounded operators.

Theorem A6.28 (Spectral theorem for bounded operators) *Let A be a bounded self-adjoint operator on $\mathfrak{H}$. Then there exists an essentially unique bounded resolution of the identity with projection operators E_λ that are continuous from the left such that A is given by the Riemann–Stieltjes integral*

$$A = \int_{-\infty}^{\infty} \lambda \, \mathrm{d}E_\lambda. \tag{A6.18}$$

Conversely, given any such bounded resolution of the identity, the operator A defined by (A6.18) *is a bounded operator.*

If an operator A is given in the form (A6.18), it is natural to ask which values of λ belong to its discrete spectrum, and which to its continuous spectrum. The answer is as follows: λ belongs to the discrete spectrum iff, for $\epsilon > 0$,

$$\lim_{\epsilon \to 0} \mathfrak{M}_{\lambda+\epsilon} \neq \mathfrak{M}_{\lambda-\epsilon}.$$

λ belongs to the resolvent set iff there exists an interval $(\lambda - \epsilon, \lambda + \epsilon)$ such that, for $\omega \in (\lambda - \epsilon, \lambda + \epsilon)$,

$$\mathfrak{M}_{\lambda-\epsilon} = \mathfrak{M}_\omega = \mathfrak{M}_{\lambda+\epsilon},$$

i.e., $\mathfrak{M}_\omega$ is constant for such ω. Finally, λ belongs to the continuous spectrum if it does not belong either to the resolvent set or to the discrete spectrum.

If the function f is continuous a.e., then $f(A)$ may be defined as

$$f(A) = \int_{-\infty}^{\infty} f(\lambda) \, \mathrm{d}E_\lambda.$$

A6.5 Unbounded operators

The notion of an unbounded operator is a little too general to grapple with effectively. To establish a measure of control, one needs a little more structure.

This 'little more' is the requirement that the domain $\mathfrak{D}_A$ of the operator A satisfy the following conditions:

(i) $\mathfrak{D}_A$ is a linear space over $\mathbb{C}$, i.e., if $f, g \in \mathfrak{H}, \alpha \in \mathbb{C}$, then $f + g \in \mathfrak{H}$ and $\alpha f \in \mathfrak{H}$.

(ii) $\mathfrak{D}_A$ is dense in $\mathfrak{H}$.

If the domain $\mathfrak{D}_A$ of A satisfies the above conditions, then A is said to be **densely defined** (on $\mathfrak{H}$). Henceforth the term *unbounded operator* will always mean one which is densely defined.

It was discovered by von Neumann that the domain of an unbounded operator need not be the same as that of its adjoint (which needs to be defined anew in the present context), and that this fact has profound consequences. We begin with the definition of the adjoint of the operator A with domain $\mathfrak{D}_A$.

Let M be the set of all pairs $\{g, h\}$, $f, g \in \mathfrak{H}$ such that

$$(Af, g) = (f, h) \text{ for all } f \in \mathfrak{D}_A. \tag{A6.19}$$

M is nonempty, because $\{0, 0\} \in M$. If M contains the pair $\{g, h\}$, it cannot contain a pair $\{g, h'\}$ with $h \neq h'$, for then $(f, h) = (f, h')$ for all $f \in \mathfrak{D}_A$, which is impossible. Finally, if $\{g, h\} \in M$ and $\{g', h'\} \in M$, then $\{g + g', h + h'\} \in M$ and $\{\alpha g, \alpha g'\} \in M$. This suggests that an operator $A^\star$ be defined as follows:

$$A^\star g = h. \tag{A6.20}$$

The domain $\mathfrak{D}_{A^\star}$ of $A^\star$ will then be the set of all g appearing in the pairs $\{f, g\} \in M$. Formally:

Definition A6.29 (Adjoint of an unbounded operator) Let A be a (densely defined) unbounded operator on $\mathfrak{H}$ with domain $\mathfrak{D}_A$. The adjoint $A^\star$ of A is a densely defined operator with domain $\mathfrak{D}_{A^\star}$ such that

$$(Af, g) = (f, A^\star g)$$

whenever $f \in \mathfrak{D}_f$ and $g \in \mathfrak{D}_{A^\star}$.

Using the adjoint, we define two generalizations of Hermitian operators of the finite-dimensional case:

Definition A6.30 (Unbounded symmetric operators) If

$$\mathfrak{D}_A \subset \mathfrak{D}_{A^\star} \text{ and } Af = A^\star f \text{ for all } f \in \mathfrak{D}_A,$$

then the operator A is called **symmetric** (von Neumann uses the term **Hermitian**).

Definition A6.31 (Unbounded self-adjoint operators) If

$$\mathfrak{D}_A = \mathfrak{D}_{A^*} \text{ and } Af = A^*f \text{ for all } f \in \mathfrak{D}_A,$$

then the operator A is called **self-adjoint** (von Neumann uses the term **Hermitian hypermaximal** or simply **hypermaximal**).

Unbounded self-adjoint operators possess two critical properties that are not shared by (unbounded) symmetric operators: (1) They can be exponentiated; if H is a self-adjoint operator, then the expression $U(t) = \exp(\mathrm{i}Ht)$ always makes sense. (2) They admit spectral decompositions (which we shall touch upon in Section A6.6). Proofs of these assertions may be found in (Reed and Simon, 1972).

A6.5.1 Families of unbounded operators

Let A and B be unbounded operators on $\mathfrak{H}$, with domains $\mathfrak{D}_A$ and $\mathfrak{D}_B$. Then it is possible for $\mathfrak{D}_A \cap \mathfrak{D}_B = \emptyset$. In physics, one often has to deal with several unbounded operators at once, and the best that one can hope for is that the intersection of their domains be dense in $\mathfrak{H}$. The reader who would like more information on the subject is referred to (Reed and Simon, 1972, Chapter VIII). An irreducible unitary representation of a locally compact noncompact Lie group is infinite-dimensional, and some representatives of its Lie algebra are unbounded (Stone's theorem). If they were not defined on a common dense domain, it would be hard to understand why day-to-day exploitation of representations of the Poincaré and Galilei Lie algebras in physics has not led to contradictions. This problem went unnoticed until around 1972, when Flato and coworkers established that the unbounded generators are indeed defined on a common dense domain (Flato *et al.*, 1972; Simon, 1972; Flato and Simon, 1973).

A6.6 The spectral theorem for unbounded operators

The fact that an unbounded operator on $\mathfrak{H}$ is not defined everywhere on $\mathfrak{H}$ is a vast complication. Von Neumann discovered (or invented) methods by which the study of unbounded self-adjoint operators could be reduced to that of bounded operators and functions of bounded operators. We shall not enter into the subject, but shall content ourselves with giving an example and stating the final result.

If A is a bounded operator, then its positive integral powers A^n are well-defined operators on $\mathfrak{H}$. Therefore, if $p_n(z)$ is a polynomial of degree n (over $\mathbb{C}$), the function $p_n(A)$ is also defined. Let now $f(z)$ be an analytic function of z, with radius of convergence R:

$$f(z) = \sum_{n=0}^{\infty} a_n z^n \text{ for } |z| < R.$$

Then, if $||A|| < R$, the expression

$$f(A)\psi = \left(\sum_{n=0}^{\infty} a_n A^n\right)\psi \tag{A6.21}$$

converges to a vector in $\mathfrak{H}$ for every $\psi \in \mathfrak{H}$ (we leave the details to the reader). If, however, $||A|| > R$, then the right-hand side of (A6.21) will not converge for every ψ; if (A6.21) defines an operator at all, it will define an unbounded operator. This makes it plausible that the function $f(A)$ be defined as follows, via the spectral resolution of A:

$$f(A) = \int f(\lambda)\mathrm{d}E_\lambda. \tag{A6.22}$$

Stone's theorem provides another example of how an unbounded operator can result from the differentiation of a unitary operator which depends on a parameter t, where $t \in \mathbb{R}$. Von Neumann constructed other examples of bounded functions of unbounded operators from which one could recover the unbounded operator. We shall stop here, and refer the interested reader to the very readable account by Lorch (Lorch, 1962, pp. 124–131).

When there is no residual spectrum, an unbounded operator has a spectral resolution of the same form as (A6.18), but using *unbounded resolutions of the identity* (for which at least one of M, N in Definition A6.23 is infinite). This will suffice for our purposes.

Spectral resolutions may be written in a form in which discrete and continuous spectra are separated. Consider an electron which may or may not be bound to an atomic nucleus. The discrete part of its energy spectrum is negative, and describes its bound states. The nonnegative values of its energy spectrum belong to the scattering states, and fill the continuum $[0, \infty)$. We may write the spectral decomposition of its Hamiltonian as

$$H = \sum_{n=0}^{\infty} \lambda_n E_{\lambda_n} + \int_0^{\infty} \lambda \mathrm{d}E_\lambda. \tag{A6.23}$$

We shall conclude this appendix with a few remarks about the spectrum. It turns out that the subdivision of the spectrum into a discrete, a continuous and a residual part, irrespective of how one defines the continuous part, is not detailed enough for physics. For example, the continuous spectrum may be resolved – via the notion of spectral measures, which we have not defined – into an absolutely continuous and a singular continuous part (see, for example, (Reed and Simon, 1972)). The existence of a singular continuous part in the spectrum of a Hamiltonian would require major revision of quantum-mechanical scattering theory. However, in 1971 Balslev and Combes succeded in establishing its absence in a large class of Schrödinger Hamiltonians (Balslev and Combes, 1971). Their results were integrated into the comprehensive treatment of the mathematical theory of scattering in quantum mechanics that was planned by Jauch and carried out after his death by Amrein and Sinha (Amrein, Jauch and Sinha, 1977). The latter monograph was quickly followed by an article by Pearson with 'examples of potentials giving rise to purely singular continuous spectra' (Pearson, 1978).

The mathematical study of Anderson localization,[16] which was initiated a little later, revealed an unexpected richness and complexity in the spectra of random Schrödinger operators. This has become a large and rapidly developing subject, about which the present author is poorly informed. A few of the highlights that he is aware of are (Goldsheid, Molchanov and Pastur, 1977; Avron and Simon, 1981; Fröhlich and Spencer, 1983; Fröhlich et al, 1985; Simon, 1995). For an account of the earlier works, the reader is referred to the monograph (Carmona and Lacroix, 1990). A more recent review, concentrating on 'exotic spectra', may be found in (Last, 2006).

[16] The Nobel prize for physics was awarded to Anderson and Mott (and van Vleck) in 1977 for the discovery of this phenomenon. In his prize lecture, Anderson said: 'It has yet to receive adequate mathematical treatment, and one has to resort to the indignity of numerical simulations to settle even the simplest questions about it. Only now, and through primarily Sir Neville Mott's efforts, is it beginning to gain general acceptance.' Apparently he was unaware of the paper by Goldsheid, Molchanov and Pastur which 'opened the floodgates', or (more surprisingly) of the paper entitled 'On Mott's problem' by Goldsheid and Molchanov which had been published a year earlier (Goldsheid and Molchanov, 1976).

A7

Conditional expectations

We shall assume that the reader is familiar with elementary probability theory. The only purpose of Section A7.1 is to recall the basic definitions and to set up the basic notation. The reader is referred to (Feller, 1970) for a detailed and authoritative treatment.

A7.1 Recapitulation of basic notions

In discrete probability theory the **sample space** Ω is a countably infinite set.[1] Its elements, called **sample points**, will be denoted by $\omega_n, n \in \mathbf{N}$. An **event** will be a subset of the sample space. The **probability** of the event A will be denoted by $\boldsymbol{P}(A)$. The function $\boldsymbol{P}$ is a countably additive nonnegative set function on the power-set $\mathcal{P}(\Omega)$ of Ω that satisfies $\boldsymbol{P}(\Omega) = 1$. It follows that it is completely determined by its values on the $\omega_n \in \Omega$, that $0 \leq \boldsymbol{P}(\omega_n) \leq 1$ for all $\omega_n \in \Omega$, and that

$$\sum_{n=1}^{\infty} \boldsymbol{P}(\{\omega_n\}) = 1. \tag{A7.1}$$

The quantity $\boldsymbol{P}(\{\omega_n\})$ is denoted, for simplicity, by p_n. Since sample points of zero probability will not interest us, *we shall exclude these cases*, i.e., stipulate that $0 < p_n < 1$ for all $n \in \mathbf{N}$.

Two events A and B are said to be **independent** if

$$\boldsymbol{P}(A \cap B) = \boldsymbol{P}(A)\boldsymbol{P}(B). \tag{A7.2}$$

A **random variable** $\boldsymbol{X}$ is a real-valued function on Ω. Let $\{x_j | j = 1, 2, \ldots\}$ be the set of distinct values it assumes. Then, for each j, the set $\{\boldsymbol{X} = x_j\} = \boldsymbol{X}^{-1}(x_j)$ is a subset (finite or infinite) of Ω; it is an event. The probability of the event $\{\boldsymbol{X} = x_j\}$ is denoted by

$$f(x_j) = \boldsymbol{P}(\boldsymbol{X} = x_j). \tag{A7.3}$$

[1] A finite sample space is a simplification which need not be considered separately.

We then have $0 < f(x_j) < 1$, and

$$\sum_j f(x_j) = 1,$$

where the upper limit of the summation is not made explicit; it may be finite or infinite, depending on $\boldsymbol{X}$. The set $\{f(x_j)\}$ is known as the **probability distribution** of the random variable $\boldsymbol{X}$.

Notations A7.1 We shall denote random variables by boldface upper-case letters such as $\boldsymbol{X}$ and $\boldsymbol{Y}$. A special class of random variables, denoted by $\mathbf{1}_S$, will be introduced in Section A7.2.

Let now $\boldsymbol{Y}$ be another random variable defined on Ω, and let $\{y_k | k = 1, 2, \ldots\}$ be the set of different values it assumes. The set of points of Ω at which both conditions $\boldsymbol{X} = x_j$ and $\boldsymbol{Y} = y_k$ are satisfied is an event, the probability of which[2] is denoted by $\boldsymbol{P}(\boldsymbol{X} = x_j, \boldsymbol{Y} = y_k)$. The function

$$f(x_j, y_k) = \boldsymbol{P}(\boldsymbol{X} = x_j, \boldsymbol{Y} = y_k), \quad j, k = 1, 2, \ldots \tag{A7.4}$$

is called the **joint probability distribution** of $\boldsymbol{X}$ and $\boldsymbol{Y}$. It satisfies the conditions

$$0 \leq f(x_j, y_k) < 1 \quad \text{and} \quad \sum_{j,k} f(x_j, y_k) = 1. \tag{A7.5}$$

Define now the functions $g : \boldsymbol{X} \to [0, 1]$ and $h : \boldsymbol{Y} \to [0, 1]$ as follows:

$$\begin{aligned} g(x_j) &= \sum_k f(x_j, y_k), \\ h(y_k) &= \sum_j f(x_j, y_k). \end{aligned} \tag{A7.6}$$

The functions g and h are called the **marginal distributions** of $\boldsymbol{X}$ and $\boldsymbol{Y}$ respectively. They satisfy the conditions

$$0 < g(x_j) = \boldsymbol{P}(\boldsymbol{X} = x_j), \quad 0 < h(y_k) = \boldsymbol{P}(\boldsymbol{Y} = y_k)$$

and

$$\sum_j g(x_j) = \sum_k h(y_k) = 1.$$

[2] It may happen that $\boldsymbol{P}(\boldsymbol{X} = x_j, \boldsymbol{Y} = y_k) = 0$ for *some* pairs (j, k), but we assume that the cases $\boldsymbol{P}(\boldsymbol{X} = x_j, \boldsymbol{Y} = y_k) = 0$ for fixed j and all k, or vice versa, are excluded, so that $g(x_j)$ and $h(y_k)$ defined by (A7.6) are never zero. It is possible to relax this last condition.

The (discrete) random variables $\boldsymbol{X}$ and $\boldsymbol{Y}$ are said to be **independent** if, for every j, k

$$\begin{aligned} \boldsymbol{P}(\boldsymbol{X} = x_j, \boldsymbol{Y} = y_k) &= f(x_i, y_j) \\ &= g(x_i)\, h(y_j) = \boldsymbol{P}(\boldsymbol{X} = x_j) \cdot \boldsymbol{P}(\boldsymbol{Y} = y_k). \end{aligned} \tag{A7.7}$$

Our main concern will be with random variables that are *not* independent.

A7.1.1 Conditioning

For any two events $A, B \subset \Omega$, the **conditional probability of** A **given** B is defined to be[3]

$$\boldsymbol{P}(A|B) = \frac{\boldsymbol{P}(A \cap B)}{\boldsymbol{P}(B)}. \tag{A7.8}$$

Note that, according to our definition, $\boldsymbol{P}(B) \neq 0$. It follows from (A7.2) that if A and B are independent, then $\boldsymbol{P}(A|B) = \boldsymbol{P}(A)$; one says that the occurrence or nonoccurrence of B does not affect the probability of A. For random variables, this translates into the following definition:

$$\boldsymbol{P}(\boldsymbol{X} = x_j | \boldsymbol{Y} = y_k) = \frac{\boldsymbol{P}(\boldsymbol{X} = x_j, \boldsymbol{Y} = y_k)}{\boldsymbol{P}(\boldsymbol{Y} = y_k)} = \frac{f(x_j, y_k)}{h(y_k)}. \tag{A7.9}$$

Comparing this definition with (A7.7), we see immediately that, if $\boldsymbol{X}$ and $\boldsymbol{Y}$ are independent, then

$$\begin{aligned} \boldsymbol{P}(\boldsymbol{X} = x_j | \boldsymbol{Y} = y_k) &= g(x_j), \\ \boldsymbol{P}(\boldsymbol{Y} = y_k | \boldsymbol{X} = x_j) &= h(y_k) \end{aligned}$$

for all y_k and x_j respectively.

The **mean**, or **expectation**, or **expected value**,[4] of the random variable $\boldsymbol{X}$ is defined to be

$$E(\boldsymbol{X}) = \sum_n \boldsymbol{X}(\omega_n)\, p_n = \sum_j x_j\, \boldsymbol{P}(\boldsymbol{X} = x_j), \tag{A7.10}$$

provided the series converges absolutely.[5] If $\boldsymbol{X}$ and $\boldsymbol{Y}$ are two random variables with a joint distribution f given by (A7.5), then the **conditional mean** of $\boldsymbol{X}$

[3] The separator | in $\boldsymbol{P}(A|B)$ signifies that the quantity to the right of it, here B, is held constant, whereas the quantity to the left of it, here A, may vary. This asymmetry is preserved in more detailed notations, as in (A7.9).

[4] Another frequently used notation for $E(\boldsymbol{X})$ is $\langle X \rangle$. The term **expectation value**, in common use in quantum mechanics, is seldom used in the mathematical literature.

[5] Absolute convergence ensures that $E(\boldsymbol{X})$ has physical meaning; it does not depend on the enumeration of the set of values of $\boldsymbol{X}$.

for given $\boldsymbol{Y}$ is the function which, when it is known that $\boldsymbol{Y} = y_k$, has the value

$$\sum_j x_j \boldsymbol{P}(\boldsymbol{X} = x_j | \boldsymbol{Y} = y_k) = \sum_j x_j \frac{f(x_j, y_k)}{h(y_k)} \tag{A7.11}$$

provided that the right-hand side converges absolutely. In this case the set of coefficients of the x_j on the right is called the **conditional distribution** of $\boldsymbol{X}$ given $\boldsymbol{Y}$.

From (A7.11) we see immediately that

$$\begin{aligned}\sum_k h(y_k)\left(\sum_j x_j \frac{f(x_j, y_k)}{h(y_k)}\right) &= \sum_{j,k} x_j f(x_j, y_k) \\ &= \sum_j x_j \sum_k f(x_j, y_k) \quad = \quad \sum_j x_j g(x_j) \\ &= E(\boldsymbol{X}).\end{aligned}$$

This suggests that a random variable $\boldsymbol{E}(\boldsymbol{X}|\boldsymbol{Y})$, called the **conditional expectation** of $\boldsymbol{X}$ given $\boldsymbol{Y}$, be defined as follows:

(a) $\boldsymbol{E}(\boldsymbol{X}|\boldsymbol{Y})(\omega) = \sum_j x_j[f(x_j, y_k)/h(y_k)]$ wherever $\boldsymbol{Y}(\omega) = y_j$, and is zero elsewhere.

(b) The distribution of $\boldsymbol{E}(\boldsymbol{X}|\boldsymbol{Y})$ is such that the following relation holds:

$$E(\boldsymbol{E}(\boldsymbol{X}|\boldsymbol{Y})) = E(\boldsymbol{X}). \tag{A7.12}$$

Set

$$r_k = \sum_j x_j \left(\frac{f(x_j, y_k)}{h(y_k)}\right).$$

If $k \neq k' \Rightarrow r_k \neq r_{k'}$, then the distribution of $\boldsymbol{E}(\boldsymbol{X}|\boldsymbol{Y})$ is exactly $\{h(y_k)\}$. However, it is entirely possible that $r_k = r_{k'}$ for $k \neq k'$, in which case the distribution $\{h(y_k)\}$ of $\boldsymbol{Y}$ has to be suitably aggregated to give a distribution for $\boldsymbol{E}(\boldsymbol{X}|\boldsymbol{Y})$ so that (A7.12) is satisfied. Details, which consist mainly of devising a transparent notation, are left to the reader.

Equation (A7.12) is known as the **law of total expectations**. The conditional expectation $\boldsymbol{E}(\cdot|\boldsymbol{Y})$ is the unique random variable that satisfies the law of total expectations. The reader is invited to express this as a commutative diagram.

A7.2 Probability: general theory

In Chapter 10 we used a notion of conditional expectation on a noncommutative algebra of operators. Elementary probability theory is not formulated in terms of algebras at all, but the general mathematical theory of probability is, and it will provide us with the required hint.

A triple $\mathcal{P} = (\Omega, \mathcal{F}, \mu)$ is called a **probability space** if Ω is a set, $\mathcal{F}$ a σ-algebra on Ω and μ a measure on $(\Omega, \mathcal{F})$ such that $\mu(\Omega) = 1$; such a measure is called a **probability measure.** When Ω is countable (finite or infinite), $\mathcal{F}$ is the family of all subsets of it, but this, as we have seen in Appendix A5, can no longer be expected to hold when Ω is uncountable.

We shall use the term **event** to denote an element of $\mathcal{F}$ in the probability space $(\Omega, \mathcal{F}, \mu)$. (The Banach–Tarski paradox (page 299) shows that it would be wholly inappropriate to call a nonmeasurable set of Ω a nonevent.) We shall denote a point in Ω by ω. The probability of the event A will be defined by

$$\boldsymbol{P}(A) = \mu(A) = \int_A \mathrm{d}\mu(\omega). \tag{A7.13}$$

The notation $\mathrm{d}\mu(\omega)$ is intended to remind the reader of the Stieltjes integral; in future we shall use[6] the simpler $\mathrm{d}\mu$. For example, the normal distribution on the real line $\mathbb{R}$, with mean a and variance σ, is defined by the measure

$$\mathrm{d}\mu(x) = \frac{1}{\sqrt{2\pi\sigma}} \mathrm{e}^{-\frac{(x-a)^2}{\sigma^2}} \mathrm{d}x.$$

A **random variable X** in $\mathcal{P}$ will be defined to be a real-valued measurable function on Ω. The characteristic function χ^S of the subset $S \subset \Omega$ was defined on page 304. This function is called the **indicator function** in probability theory,[7] where it is denoted by $\mathbf{1}_S$. The notation is justified, because an indicator function is also a random variable.[8] Using the indicator function, (A7.13) may be rewritten as

$$\boldsymbol{P}(A) = \mu(A) = \int_\Omega \mathbf{1}_A(\omega) \mathrm{d}\mu(\omega). \tag{A7.14}$$

The **distribution** (or **cumulative distribution**) of the random variable $\boldsymbol{X}$ is defined to be the function

$$F_{\boldsymbol{X}}(t) = \boldsymbol{P}\{\boldsymbol{X} \leq t\} = \boldsymbol{P}\{\omega | \boldsymbol{X}(\omega) \leq t\}. \tag{A7.15}$$

[6] A large variety of notations are in use in the literature to write the integral in (A7.13); we shall not attempt to list them. The reader who wishes to consult a book on the subject is advised to bear this in mind.

[7] The name *characteristic function* is generally used for something akin to the Fourier transform in probability theory. See (Feller, 1971, Chap. XV).

[8] At this point the reader will surely appreciate Feller's remark that the term *random function* would have been more appropriate than *random variable*.

If $\boldsymbol{X}$ is a random variable, so is $|\boldsymbol{X}|$, defined by $|\boldsymbol{X}|(\omega) = |\boldsymbol{X}(\omega)|$. In the following, *we shall only consider random variables* $\boldsymbol{X}$ *that satisfy* the condition

$$\int_{\Omega} |\boldsymbol{X}(\omega)| \mathrm{d}\mu < \infty.$$

For such $\boldsymbol{X}$, the **expectation** $E(\mathbf{X})$ of $\boldsymbol{X}$ is defined to be

$$E(\mathbf{X}) = \int_{\Omega} \boldsymbol{X}(\omega) \mathrm{d}\mu, \tag{A7.16}$$

which is a direct generalization of (A7.10).

A7.3 Conditioning

Let S be an event with nonzero probability, $\mu(S) > 0$. By analogy with (A7.8), we may write

$$\mu_S(A) = \frac{\mu(A \cap S)}{\mu(S)} \quad \text{for all } A \in \mathfrak{F}. \tag{A7.17}$$

The quantity μ_S so defined is clearly a countably additive set function on $\mathfrak{F}$ with $0 \leq \mu_S(A) \leq 1$, $\mu_S(S) = 1$. It is therefore a measure on $(\Omega, \mathfrak{F})$. The conditional probability $\boldsymbol{P}(A|S)$ of A given S may therefore be expressed as

$$\boldsymbol{P}(A|S) = \mu_S(A) = \int_{\Omega} \mathbf{1}_A(\omega) \mathrm{d}\mu_S, \tag{A7.18}$$

and one may define the conditional expectation in terms of the conditional probability, by using the measure μ_S instead of μ:

$$E(\mathbf{X}|S) = \int_{\Omega} \boldsymbol{X}(\omega) \mathrm{d}\mu_S = \frac{1}{\mu(S)} \int_{S} \boldsymbol{X} \mathrm{d}\mu. \tag{A7.19}$$

However, this method will fail if the right-hand side of (A7.17) is an indeterminate form $0/0$ for some S.

Note that when μ_S is defined by (A7.17), then

$$\mu(A) = 0 \Rightarrow \mu_S(A) = 0, \tag{A7.20}$$

i.e., μ_S is absolutely continuous with respect to μ. This suggests that conditional probabilities and expectations may be definable in terms of Radon–Nikodym derivatives (page 314).

Conditional expectations are encountered in several branches of pure and applied mathematics. There are alternative definitions tailored to specific needs, and a considerable body of mathematical theory. By contrast, our interest is limited to writing down a version of the law of total expectations that was the model for Lemma 10.7.

A random variable generates a σ-subalgebra of $\mathcal{B}$ in a fairly natural fashion. The definition is given below.

Definition A7.2 (σ-algebra generated by a random variable) Let $\boldsymbol{X} : \Omega \to \mathbb{R}$ be a random variable, $\mathcal{B}$ the Borel structure on $\mathbb{R}$ and $B \in \mathcal{B}$. The set $\sigma(\boldsymbol{X}) = \{\boldsymbol{X}^{-1}(B) | B \in \mathcal{B}\}$ is a σ-subalgebra of $\mathcal{F}$; it is called the **σ-algebra generated by $\boldsymbol{X}$**.

The algebra $\sigma(\boldsymbol{X})$ is the smallest σ-algebra with respect to which $\boldsymbol{X}$ is measurable.

We may now proceed to define conditional expectations in the general setting. We shall first give the construction, which establishes the existence and uniqueness of the required object, and then condense the result into a formal definition.

Fix a random variable $\boldsymbol{Y}$, and let μ_0 be μ restricted to $\sigma(\boldsymbol{Y})$, i.e., $\mu_0(A) = \mu(A)$ for $A \in \sigma(\boldsymbol{Y})$. Let $\boldsymbol{X}^+$ and $\boldsymbol{X}^-$ be the nonnegative and negative parts of $\boldsymbol{X}$, i.e.,

$$\boldsymbol{X}^{\pm}(\omega) = \tfrac{1}{2}\left(|\boldsymbol{X}| \pm \boldsymbol{X}\right).$$

Then $\boldsymbol{X} = \boldsymbol{X}^+ - \boldsymbol{X}^-$. For $S \in \sigma(\boldsymbol{Y})$, define

$$\begin{aligned} \rho^{\pm}(S) &= \boldsymbol{E}(\boldsymbol{X}^{\pm}\mathbf{1}_S) = \int_{\Omega} \boldsymbol{X}^{\pm}(\omega)\mathbf{1}_S(\omega)\mathrm{d}\mu \\ &= \int_{S} \boldsymbol{X}^{\pm}(\omega)\mathrm{d}\mu_0. \end{aligned} \tag{A7.21}$$

Then $\rho^{\pm}$ are measures on $\sigma(\boldsymbol{Y})$ that are absolutely continuous with respect to μ_0. Define now

$$\boldsymbol{E}(\boldsymbol{X}|\boldsymbol{Y}) = \frac{\mathrm{d}\rho^+}{\mathrm{d}\mu_0} - \frac{\mathrm{d}\rho^-}{\mathrm{d}\mu_0}, \tag{A7.22}$$

where the quantities on the right are Radon–Nikodym derivatives (A5.26). Then, using the Radon–Nikodym theorem (Theorem A5.35), we have

$$\begin{aligned} \int_{\Omega} \boldsymbol{E}(\boldsymbol{X}|\boldsymbol{Y})(\omega)\mathbf{1}_S(\omega)\mathrm{d}\mu &= \int_S \frac{\mathrm{d}\rho^+}{\mathrm{d}\mu_0}\mathrm{d}\mu_0 - \int_S \frac{\mathrm{d}\rho^-}{\mathrm{d}\mu_0}\mathrm{d}\mu_0 \\ &= \rho^+(S) - \rho^-(S) \\ &= \boldsymbol{E}(\boldsymbol{X}\mathbf{1}_S). \end{aligned}$$

We condense the above into a definition (Feller, 1971, pp. 162–165):

Definition A7.3 (Expectation conditioned on a random variable) Let $\boldsymbol{X}$ and $\boldsymbol{Y}$ be bounded random variables on the probability space $(\Omega, \mathcal{F}, \mu)$. The **conditional expectation** $\boldsymbol{E}(\boldsymbol{X}|\boldsymbol{Y})$ of $\boldsymbol{X}$ given $\boldsymbol{Y}$ is the unique $\sigma(\boldsymbol{Y})$-measurable random variable that satisfies

$$E(\boldsymbol{E}(\boldsymbol{X}|\boldsymbol{Y})\mathbf{1}_S) = E(\boldsymbol{X}\mathbf{1}_S)$$

for all $S \in \sigma(\boldsymbol{Y})$. The conditional expectation $\boldsymbol{E}(\boldsymbol{X}|\boldsymbol{Y})$ is unique up to a set of measure zero.

It follows from the uniqueness of $\boldsymbol{E}(\boldsymbol{X}|\boldsymbol{Y})$ that, if $\boldsymbol{X}$ itself is $\sigma(\boldsymbol{Y})$-measurable, then

$$\boldsymbol{E}(\boldsymbol{X}|\boldsymbol{Y}) = \boldsymbol{X}.$$

Definition A7.4 (Expectation conditioned on a σ-subalgebra) Let X be a random variable with $E(|\boldsymbol{X}|) < \infty$, and $\mathcal{G}$ a σ-subalgebra of $\mathcal{F}$. The **conditional expectation of $\boldsymbol{X}$ given** $\mathcal{G}$ is the unique $\mathcal{G}$-measurable random variable $\boldsymbol{E}(\boldsymbol{X}|\mathcal{G})$ that satisfies

$$E(\boldsymbol{E}(\boldsymbol{X}|\mathcal{G})\boldsymbol{Z}) = E(\boldsymbol{X}\boldsymbol{Z}) \tag{A7.23}$$

for all bounded and $\mathcal{G}$-measurable random variables $\boldsymbol{Z}$. The conditional expectation $\boldsymbol{E}(\boldsymbol{X}|\mathcal{G})$ is unique up to a set of measure zero.

When $\mathcal{G} = \sigma(\boldsymbol{Y})$, Definition A7.4 is equivalent to Definition A7.3. The proof is straightforward, and is omitted.

It is obvious that $\boldsymbol{E}(\cdot|\mathcal{G})$ is linear; it is positive, i.e., if $\boldsymbol{X} \geq 0$ then $\boldsymbol{E}(\boldsymbol{X}|\mathcal{G}) \geq 0$. Moreover, $\boldsymbol{E}(\cdot|\mathcal{G})$ is a **projection**, i.e.,

$$\boldsymbol{E}(\boldsymbol{E}(\boldsymbol{X}|\mathcal{G})|\mathcal{G}) = \boldsymbol{E}(\boldsymbol{X}|\mathcal{G}).$$

These properties can be translated into the language of operators on the Hilbert space $L^2(\Omega, \mathcal{F}, \mu)$. We refer the reader to (Streater, 2000; Rédei and Summers, 2007) for further information, both on classical and quantum probability theory. These articles have been written from the perspective of mathematical physics rather than pure mathematics. The classic text, in English, on classical probability theory is the two-volume work (Feller, 1970, 1971).

A8

Fibre bundles, differentiable manifolds, Lie groups and Lie algebras

The final result of Part I of this book, Theorem 5.10, was stated using the term *differentiable manifold.* Some complexities of the relation between Lie groups and Lie algebras were encountered in Sections 7.4.3 and 7.5.2. One needs precise definitions of these mathematical objects to set these questions in their geometrical perspective. In Chapter 12 we made use of Banach and Hilbert bundles. These are special cases of *fibre bundles.* Fibre bundles also provide deep insights into Lie groups and Lie algebras. The purpose of this appendix is to provide the basic definitions, and brief introductions to these subjects that would suffice for our needs. We begin with the topological concept of a fibre bundle and continue with that of a differentiable manifold. The two are brought together through the notion of tangent spaces to lead us to the desired relation between Lie groups and Lie algebras.

A8.1 Fibre bundles

The topological spaces called fibre bundles are generalizations of the topological product; they have the product form *locally,* but not globally. The simplest example is the Möbius strip. The rectangular strip of paper *abcd* shown in Fig. A8.1 may be formed into a cylinder by glueing the short edges together; a is joined to b, and c to d. However, if one gives the strip a twist and joins a to c and b to d, one obtains a Möbius strip. The cylinder is homeomorphic to the product $S_1 \times I$,

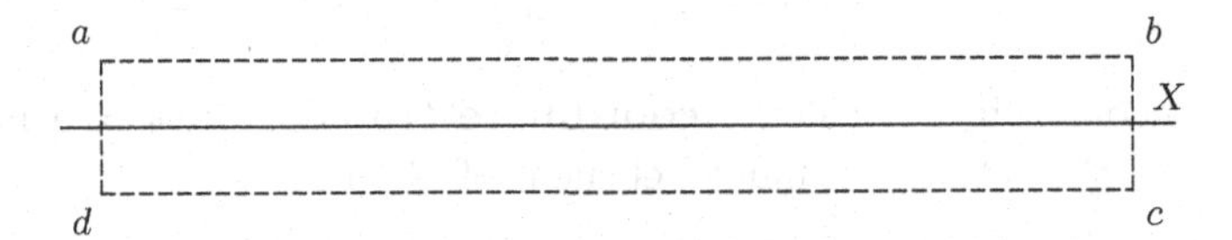

Fig. A8.1. Forming the cylinder and the Möbius strip

where S_1 is the circle (1-sphere) and I the unit interval; both the cylinder and the Möbius strip are *locally* homeomorphic to $(\alpha, \beta) \times I$, where (α, β) is any open interval on the real line, but the latter is not globally homeomorphic to $S_1 \times I$. The twisting, which introduces the difference, is accomplished mathematically by the action of a two-element group, one element of which flips the Y-axis. The definition of a fibre bundle is a generalization of the above, but, unlike the Möbius strip, a fibre bundle does not have to be nonorientable.

The material of this section is taken from Steenrod's book (Steenrod, 1972). References to the original sources will be found in this book.

A8.1.1 Definition of fibre bundles

The definition of a **fibre bundle** is as follows:

(a) It is a quintuple $\mathcal{B} = (B, X, \pi, G, Y)$, where

  (i) B, X, Y are topological spaces. B is known as the **total space**, X the **base space** and Y the **fibre** of the bundle,

  (ii) $\pi : B \to X$ is a continuous map called the **projection**, and

  (iii) G is a topological group of transformations of Y which acts effectively on Y.[1] It is called the **group of the bundle**.

(b) The inverse image $\pi^{-1}(x)$ is called the **fibre over** x. The fibres $\pi^{-1}(x)$ are homeomorphic to Y for each x.

(c) There is an open cover $\{U_j | j \in J\}$ of X such that, for each $j \in J$, there exists a homeomorphism

$$\varphi_j : U_j \times Y \longrightarrow \pi^{-1}(U_j) \tag{A8.1}$$

which satisfies

$$\pi \circ \varphi_j(x, y) = x \text{ for } x \in U_j, \quad y \in Y. \tag{A8.2}$$

The maps φ_j are called **local trivializations.**[2]

(d) For each j and each $x \in U_j$, the homeomorphism φ_j defines a map $\varphi_{j,x} : Y \to \pi^{-1}(x)$ by

$$\varphi_{j,x}(y) = \varphi_j(x, y);$$

and for each pair $j, k \in J$ and each $x \in U_i \cap U_j$, the map

$$g_{kj}(x) = \varphi_{k,x}^{-1} \circ \varphi_{j,x} : Y \longrightarrow Y \tag{A8.3}$$

is a homeomorphism (called a **coordinate transformation**) which coincides with the action of a unique element of G on Y.

(e) For each pair $j, k \in J$, the map

$$g_{kj} : U_k \cap U_j \longrightarrow G$$

is continuous.

[1] The action of a group G on a topological space Y is said to be **effective** if $gy = y$ for all y implies that $g = e$, where e is the identity of G.

[2] They are called **coordinate functions** in (Steenrod, 1972), and elsewhere. As this term is also used in the definition of differentiable manifolds, we prefer to avoid using it here, but the reader should note that such terminological distinctions may be sacrificed if it makes a phrase less cumbersome.

Any local trivialization ϕ_α (more precisely, a pair (ϕ_α, U_α) satisfying (A8.2)) that is compatible with the **glueing condition** (A8.3) may be adjoined to the existing ones. We shall assume that the set of local trivializations is maximal in this respect. This assumption has the effect of eliminating the dependence on local trivializations, and will be used again in the definition of differentiable manifolds in Section A8.2.

A map

$$s : X \to B \tag{A8.4}$$

that satisfies the condition

$$\pi \circ s(x) = x \text{ for all } x \in X \tag{A8.5}$$

is called a **section**, or **cross-section** (of X in B).[3] *Continuous* cross-sections do not always exist in fibre bundles; however, every bundle has continuous **local cross-sections**, which are defined not over the entire bundle but only over the local trivializations. Choose a fixed point $y_0 \in Y$ and define $\sigma : U_i \to \pi^{-1}(U_i)$ by $\sigma(x) = \phi_i(x, y_0)$. Then $\pi(\sigma(x)) = x$, and σ is continuous. In the following, *local cross-sections will always be assumed to be continuous.*

If the total space B is homeomorphic to $X \times Y$, then the bundle is called **trivializable, equivalent to the product**, or simply a **product bundle**.

A8.1.2 Principal bundles; the bundle structure theorem

A bundle $\mathcal{B} = (B, X, \pi, G, G)$ in which the fibre is the same as the (underlying topological space of) its group and the group acts upon it by left-translations is called a **principal bundle**. If Y is a topological space on which G has an effective action, then, using the coordinate transformations of $\mathcal{B}$, one can define a new bundle $\mathcal{A}$ with base X, fibre Y, group G, and the same coordinate transformations as in $\mathcal{B}$ but with a new total space A (and, obviously, a new projection p). The bundle $\mathcal{A}$ is said to be **associated** with $\mathcal{B}$. It is an important result in the theory of fibre bundles that *a bundle $\mathcal{A}$ is equivalent to the product if and only if the associated principal bundle has a* continuous *cross-section.*

Let G be a topological group, and H a closed subgroup of it. Denote the space of left-cosets of H in G, furnished with the quotient topology (page 264) by G/H. Then there is a natural projection $\pi : G \to G/H$ which sends every element of G into its left-coset in G/H, and G has a natural action on G/H. The following result is central to the representation theory of topological groups in physics, where it is often used without being recognized:

[3] There is some nonuniformity in terminology at this point. Some authors, for example Steenrod (who uses the term cross-section), require sections to be continuous, while others do not.

Theorem A8.1 (The bundle structure theorem) *If G is a topological group and H a closed subgroup of it such that G/H has a local cross-section in G, then G is a principal fibre bundle $(G, G/H, \pi, H, H)$ with group and fibre H, with $\pi : G \to G/H$ defined as above.*

Local cross-sections (of G/H in G) do not always exist. But, if G is a Lie group and H a closed subgroup of it, then G/H always has a local cross-section in G. Therefore *the bundle structure theorem holds unrestrictedly for Lie groups and their closed subgroups.*

A8.1.3 Hilbert bundles

A bundle $\{B, X, \pi, U, \mathfrak{H}\}$, in which the fibre is an infinite-dimensional separable Hilbert space $\mathfrak{H}$ over the complex numbers and U the unitary group of $\mathfrak{H}$ is called a **Hilbert bundle**. If the base space X of a Hilbert bundle is a topological manifold (defined on page 354), then the bundle is equivalent to the product $X \times \mathfrak{H}$. Our base spaces will always be topological manifolds, and therefore all our Hilbert bundles will be equivalent to the product. The result we have quoted is given in (Steenrod, 1972); some exceptional cases are covered by a theorem of Kuiper (Kuiper, 1965).

A8.1.4 Bundle maps

Let (B, X, π, G, Y) and (B', X', π', G, Y) be two bundles with the same fibre and group. A continuous map $\theta : B \to B'$ is called a **bundle map** if it maps a fibre Y_x of B onto a fibre $Y_{x'}$ of X', thus inducing a continuous **base map** map $\bar{\theta} : X \to X'$ such that

$$\pi' \circ \theta = \bar{\theta} \circ \pi. \tag{A8.6}$$

Equation (A8.6) may be represented by the commutative diagram of Fig. A8.2.

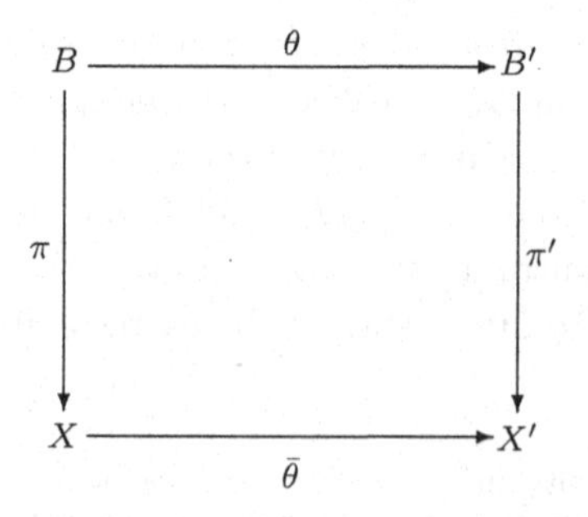

Fig. A8.2. The induced base map

Remark A8.2 The map θ is required to satisfy some other (fairly mild) conditions, involving the group of the bundle. These conditions will always be satisfied in cases of interest to us, and we shall not dwell on them. The reader is referred to (Steenrod, 1972, p. 9) for details. A bundle map is often called a **fibre-preserving map** in the literature.

A8.2 Differentiable manifolds

We begin by describing a hierarchy of functions. Let O be a connected open set in $\mathbb{R}^n$. The set of real-valued functions on O with continuous partial derivatives of order k is denoted by $C^k(O)$; here $k = 1, 2, \ldots$ The set of continuous functions on O is denoted by $C^0(O)$; that of infinitely differentiable (called **smooth**) functions, by $C^\infty(O)$. The set of **analytic** functions, i.e., functions that have convergent Taylor series expansions around every point in O, is denoted by $C^\omega(O)$. One has

$$C^0(O) \supsetneqq \cdots \supsetneqq C^k(O) \supsetneqq C^{k+1}(O) \supsetneqq \cdots \supsetneqq C^\infty(O) \supsetneqq C^\omega(O)$$

for $k = 1, 2, \ldots$ The inclusion $C^0 \supsetneqq C^1$ is decidedly nontrivial; Weierstrass caused a sensation in 1872 when he announced his everywhere-continuous but nowhere-differentiable function.[4] The inclusions $C^k \supsetneqq C^{k+1}$ are straightforward. The function $f : \mathbb{R} \to \mathbb{R}$ defined by

$$f(x) = \begin{cases} \exp(-1/x) & \text{for} \quad x > 0, \\ 0, & \text{for} \quad x \leq 0 \end{cases} \tag{A8.7}$$

is smooth, but not analytic. This example shows that $C^\infty(\mathbb{R}) \supsetneqq C^\omega(\mathbb{R})$.

A8.2.1 Definition of a differentiable manifold

In elementary differential geometry, curves are studied by embedding them in $\mathbb{R}^2$ or $\mathbb{R}^3$, and surfaces by embedding them in $\mathbb{R}^3$. In 1858, Riemann showed that embeddings were unnecessary; curvature could be studied *intrinsically*, via the metric and tensor that now bear his name. In 1917, Levi-Civita defined the notion of *parallel displacement* and forged a tool which became the key: the covariant derivative, familiar to every physicist (Levi-Civita, 1917). But the objects of study of differential geometry, *differentiable manifolds*, were defined in their generality only later, by Whitney in the 1930s. This definition is given below.[5] It should be noted that a differentiable manifold *can* be embedded in a higher-dimensional Euclidean space; this is a major theorem, first proved by Levi-Civita

[4] This example may be found in (Goursat, 1959, pp. 423–425).

[5] A different definition, 'inspired by the definition of a Riemann surface given by Hermann Weyl', was used by Chevalley in his systematization of the global theory of Lie groups (Chevalley, 1946). The two definitions are equivalent.

and then refined by Whitney. The result, that an n-dimensional differentiable manifold can be embedded in $\mathbb{R}^{2n+1}$, is known as the *Whitney embedding theorem.*

A **topological manifold** is a *second-countable* Hausdorff space X such that *every* point of X has a neighbourhood which is homeomorphic to an open set in $\mathbb{R}^n$ (n is fixed).[6]

Let X be a topological manifold, U_α, U_β open subsets of X that are homeomorphic to open sets in $\mathbb{R}^n$, and $\varphi_\alpha : U_\alpha \to \mathbb{R}^n$, $\varphi_\beta : U_\beta \to \mathbb{R}^n$ maps that are homeomorphisms onto their ranges. Suppose now that the intersection $U_{\alpha,\beta} = U_\alpha \cap U_\beta$ is nonempty, and set

$$\varphi_\alpha(U_{\alpha,\beta}) = W_\alpha, \quad \varphi_\beta(U_{\alpha,\beta}) = W_\beta,$$

as shown in Fig. A8.3. We shall denote the restrictions of φ_α and φ_β to $U_{\alpha,\beta}$ by the same symbols.

We now have a homeomorphism of $U_{\alpha,\beta}$ with W_α and another with W_β. These sets are shown by the shaded regions in Fig. A8.3. The map

$$W_\alpha \xrightarrow{\varphi_\alpha^{-1}} U_{\alpha,\beta} \xrightarrow{\varphi_\beta} W_\beta$$

shown in Fig. A8.3, being the composition of two continuous maps, is continuous.

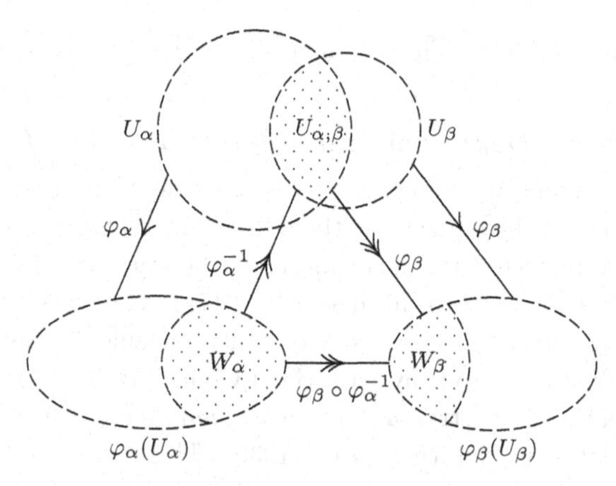

Fig. A8.3. Transition maps

[6] Some authors do not require the second-countability condition in the initial definition of a manifold, mainly to give counterexamples showing that it is not required for defining a differentiable structure. It has to be introduced at some stage, because many important theorems on differentiable manifolds depend essentially upon the second-countability axiom.

It is, in fact, a homeomorphism, as the arrows in Fig. A8.3 can be reversed, with the corresponding maps being replaced by their inverses.

The maps $\varphi_\alpha \circ \varphi_\beta^{-1} : W_\alpha \to W_\beta$ are from one open subset of $\mathbb{R}^n$ to another, as are their inverses. It therefore makes sense to ask whether these maps are differentiable. They do not have to be; the textbook example is the map $f : \mathbb{R} \to \mathbb{R}$ defined by $y = f(x) = x^3$; this map is invertible, its inverse being $x = y^{1/3}$, which is continuous. While f is differentiable, its inverse fails to be differentiable at the origin. We *assume* that the maps $\varphi_\beta \circ \varphi_\alpha^{-1}$ and their inverses are smooth for all α, β for which $U_\alpha \cap U_\beta \neq \emptyset$. They are called **transition maps**.

Remark A8.3 This assumption is not as restrictive as it looks. We could have chosen the transition maps and their inverses to be of class $C^k, k \geq 1$; this would have led to the definition of C^k-manifolds. However, it can be shown that every C^k-atlas contains a C^{k+1}-subatlas, and therefore a C^∞-subatlas. See, for example, (Hirsch, 1976). Hirsch even defines a smooth manifold to be a manifold of class $C^k, k \geq 1$. One sometimes says that the C^k-structure is **subordinate to** the C^∞-structure.

Now comes a mass of terminology. A pair $(U_\alpha, \varphi_\alpha)$ is a **local coordinate system**, or a **chart**. The collection of charts $\{(U_\alpha, \varphi_\alpha) | \alpha \in A\}$ is an **atlas** for X. The maps $\varphi_\alpha \circ \varphi_\beta^{-1}$ are **transition maps**.[7] The quantities $(x_1, \ldots, x_n) = \varphi_\alpha(x)$ are called **local coordinates** of x in the chart $(U_\alpha, \varphi_\alpha)$. This chart is often referred to simply as the 'chart' U_α.

Clearly, one can adjoin to the atlas any chart (U, φ) such that the coordinate transformations it defines are infinitely differentiable. It is therefore assumed that the atlas $\mathcal{A}$ is maximal in this sense, i.e., it contains every chart that is compatible with the others. The atlas $\mathcal{A}$ defines a (smooth) **differentiable structure** on X. The pair $(X, \mathcal{A})$ is called a (smooth) **differentiable manifold** or a **differential manifold**. All our manifolds will be smooth. Note that there are oddities among topological manifolds that cannot carry a differentiable structure.[8]

Let M, M' be homeomorphic manifolds. If the homeomorphism ψ and its inverse ψ^{-1} are smooth, then ψ is called a **diffeomorphism**.[9] The question arises: if two *smooth* manifolds are homeomorphic, are they necessarily diffeomorphic? The answer is *no*! This counterintuitive phenomenon was first discovered by Milnor for the seven-dimensional sphere S_7.

[7] These maps are also called **coordinate transformations** in the literature. However, we have already used this term in the definition of fibre bundles, following (A8.3), and shall not use it here. See also footnote 2 on page 350.

[8] This statement is a little too loose for comfort, but we lack the machinery to make it more precise. An excellent introductory account may be found in (Hilton, 1968).

[9] The definition of differentiability for the map $\psi : M \to M'$ is given in Section A8.4.

The spaces $\mathbb{R}^n$ admit only *one* differentiable structure each, *except* for $n = 4$. The manifold $\mathbb{R}^4$ has infinitely many inequivalent differential structures; all but the standard one are termed **exotic.**[10]

Let us consider a few examples of one-dimensional manifolds embedded in the two-dimensional plane. The first is the X-axis itself. This is a one-dimensional manifold that can be covered by a single chart. Next, consider the graph of the function $y = \sin x$. This is also a one-dimensional manifold which is homeomorphic with the X-axis, but it is 'not flat'. To make this idea precise, one needs to define (for example) the notion of curvature, which will give precise meaning to the intuitive statements that the graph of $y = \sin x$ is curved, whereas the X-axis is flat. This also shows that the notion of curvature cannot be a topological invariant, and the same is true of the notion of torsion. Our last example is the circle S_1, which cannot be covered by a single chart. It is impossible to make a flat space out of the circle without topological violence.

There is one further point that deserves mention. Recall that in the definition of a topological manifold, every point is assumed to have a neighbourhood homeomorphic to an open set in $\mathbb{R}^n$. Consider now the disc $D_r(O) = \{(x, y)|(x^2 + y^2)^{1/2} \leq r\}$ in $\mathbb{R}^2$, with the usual metric topology. Every point in the interior of the disc has a neighbourhood that is open in $\mathbb{R}^2$, but that is not true for any point on its perimeter. The discs D_r are examples of **manifolds with boundary**, which have to be defined in a slightly different manner. The term differential manifold usually applies to manifolds *without* boundary, and we shall be concerned only with these.

The definition of Lie groups uses the notion of the **product** $M_1 \times M_2$ **of two manifolds** M_1 and M_2. The definition is analogous to that of the topological product of two topological spaces. Details may be found in (Matsushima, 1972, pp. 32–33) or (Auslander and MacKenzie, 1963, pp. 92–94). The result $\mathbb{R}^m \times \mathbb{R}^n = \mathbb{R}^{m+n}$ also holds for differentiable manifolds.

A8.2.2 Functions on manifolds

We shall only be concerned with real-valued functions on manifolds. The continuity and differentiability properties of functions defined on manifolds can be examined in local coordinates. As we are dealing with smooth manifolds, the transition maps are smooth, and therefore the differentiability properties of functions at a point will

[10] The subject of *exotic differentiable structures* on $\mathbb{R}^4$ developed from a theorem proven by Donaldson using gauge theory (Donaldson, 1983). Exotic $\mathbb{R}^4$'s were first constructed in (Gompf, 1983). A technical review of the early works using gauge theory will be found in (Lawson, 1985). A short, semipopular account was given by Freedman (Freedman, 1984). Relativity theorists may be interested in the more recent results of Brans (Brans, 1994). The book (Asselmeyer-Maluga and Brans, 2007) is aimed largely at physicists. The geometrization of gauge theory was initiated by Lubkin (Lubkin, 1963) and Mayer (Mayer, 1977), well before it was taken up by mathematicians.

not change from one local coordinate system to another. It therefore makes sense to speak of the sets of functions $C^k(M)$, where $k \in \mathbb{N}$ or $k = \infty$.

A8.3 Tangent vectors and the tangent bundle

From elementary geometry, we are used to visualizing tangent vectors as jutting out into the embedding space. How do we define tangent vectors when the embedding space is no longer there?

Consider the quantities that define tangent vectors. For a plane curve φ, the tangent at a point $p \in \varphi$ is determined by the derivative $\varphi'(p)$. The tangent vectors at a point p on a smooth surface $\varphi(x, y) = \text{const}$ form a plane. This plane is determined by the partial derivatives

$$\frac{\partial \varphi}{\partial x}(p), \ \frac{\partial \varphi}{\partial y}(p)$$

at that point. The mathematical entities that determine the tangent vectors in two dimensions are the two partial differential operators $\partial/\partial x$ and $\partial/\partial y$. The coordinate system (x, y) is arbitrary; the way to eliminate this arbitrariness is to admit all possible coordinate systems on equal footing. This leads to a two-dimensional vector space over $\mathbb{R}$ which has, as a base, the pair of partial differential operators $\partial/\partial x, \partial/\partial y$. It is this two-dimensional vector space that is realized, geometrically, as a plane when the partial derivatives are evaluated at a given point. The extension to n dimensions is obvious. We proceed to the formal definitions.

Definition A8.4 (Tangent vectors) Let M be a smooth n-manifold and $p \in M$. Denote by $\mathfrak{F}(p)$ the set of real-valued functions on M that are infinitely differentiable at p. Then $\mathfrak{F}(p)$ is a linear space over $\mathbb{R}$. A map (think of it as an operator!)

$$v_p : \mathfrak{F}(p) \longrightarrow \mathbb{R}$$

is a **tangent vector** at p if it satisfies the conditions

$$\begin{aligned} v_p(\lambda f + \mu g) &= \lambda v_p(f) + \mu v_p(g), \\ v_p(fg) &= v_p(f)g + f v_p(g) \end{aligned} \tag{A8.8}$$

for all $f, g \in \mathfrak{F}$, $\lambda, \mu \in \mathbb{R}$. It follows from the above that $v_p(c) = 0$ for all $c \in \mathbb{R}$.

Let v_p, v'_p be tangent vectors to M at p and $\lambda \in \mathbb{R}$. Define λv_p and $v_p + v'_p$ by

$$(\lambda v_p)f = \lambda(v_p f), \ \ (v_p + v'_p)f = v_p(f) + v'_p(f)$$

for all $f \in \mathfrak{F}(p)$. It is obvious that λv_p and $v_p + v'_p$ are also tangent vectors at p. The linear space of tangent vectors at p is called the **tangent space** at p, and denoted by T_pM. We have:

Theorem A8.5 *Let M be a smooth n-manifold. Then T_pM is n-dimensional for every $p \in M$, and is homeomorphic with $\mathbb{R}^n$.*

This theorem looks obvious but requires proof. We omit the proof, which may be found, for example, in (Matsushima, 1972, pp. 41–42).

In local coordinates, a tangent vector at p has the form

$$v_p = \sum_{i=1}^{n} a_i \left(\frac{\partial}{\partial x^i}\right)_p.$$

In the above, the coefficients a_i are constants and the subscript p on the right indicates that the partial derivatives have to be evaluated at p.

A8.3.1 The tangent bundle

It can be shown that the collection $\{(p, T_pM)|p \in M\}$ is a *fibre bundle* over the base space M with fibre $\mathbb{R}^n$ and group $GL(n, \mathbb{R})$, the general linear group over the reals in n dimensions. This bundle is known as the **tangent bundle** of M. In this case the coordinate transformations (A8.3) and the action of the group on the fibre are not merely continuous; they are smooth, and moreover if M is n-dimensional then the tangent bundle is a $2n$-dimensional differentiable manifold. The explicit verification of these facts may be found in (Steenrod, 1972, pp. 20–24). We shall denote the tangent bundle of M by $\{TM, M, \pi, GL(n, \mathbb{R}), \mathbb{R}^n\}$, where π is the obvious projection $\pi : TM \to M$. The bundle itself and its total space are both denoted by TM; this will cause no confusion. The tangent space T_pM is the fibre over p.

It is clear that a cross-section of TM is an assignment of a tangent vector to each point $p \in M$. As remarked in Section A8.1, continuous cross-sections may not always exist in the topological category. However, they exist trivially in the tangent bundle, e.g., the *zero section* $\sigma(p) = 0$ for every $p \in M$. Cross-sections can be added, and multiplied by constants; the space of cross-sections is a linear space over $\mathbb{R}$ which is n-dimensional at every $p \in M$. The question arises: does the space of cross-sections contain n linearly independent continuous cross-sections? The answer is: only in those rare instances in which TM is equivalent to the product $M \times \mathbb{R}^n$. We shall not stop to explain the origins of this result; the reader is referred to (Steenrod, 1972).

A8.4 Maps of manifolds

Let M and M' be smooth manifolds of dimensions m and n, respectively, and $\varphi : M \to M'$ a continuous map. We wish to define the notion of differentiability for the map φ. One possibility would be to express φ in terms of local coordinates on M and M' and require these vector-valued functions to be differentiable, but we would like to achieve the same result with greater economy of effort.

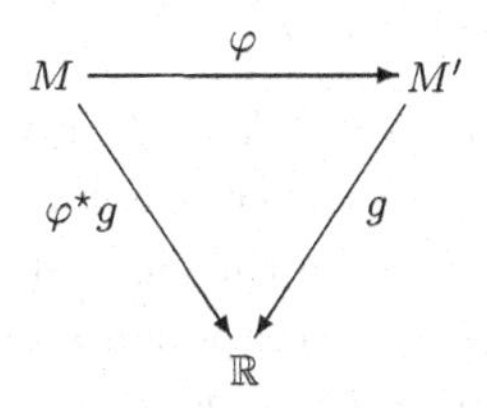

Fig. A8.4. Differentiable maps

If g is any smooth function on M', then φ defines a continuous map $\varphi^\star g : M \to \mathbb{R}$ by $\varphi^\star g = g \circ \varphi$, or $(\varphi^\star g)(q) = g(\varphi(q))$, as shown by the commutative diagram of Fig. A8.4. The map φ is **smooth** if $\varphi^\star g$ is smooth for every smooth function g on M'.

A smooth map $\varphi : M \to M'$ induces, as one might expect, a fibre-preserving map from TM to TM'. This map is denoted by $\varphi_\star$ and is called the **differential** of φ.

For $v_p \in T_pM$, the vector $\varphi_\star(v_p)$ is defined by

$$\varphi_\star(v_p)g = v_p(\varphi^\star g) \tag{A8.9}$$

for every $g \in \mathfrak{F}'(p')$, where $p' = \varphi(p)$ and $\mathfrak{F}'(p')$ is the set of smooth functions at $p' \in M'$. It may be verified directly, from (A8.9), that $\varphi_\star(v_p)$ has the properties (A8.8) of tangent vectors.

We could also look at the situation the other way round: $\varphi_\star : TM \to TM'$ is a bundle map which induces the base map $\varphi : M \to M'$, i.e.,

$$\pi' \circ \varphi_\star = \varphi \circ \pi,$$

where π, π' are the projections in the bundles TM, TM', respectively. The procedure we have followed, namely to start with a base map as given and to construct the bundle map out of it, may be easier to grasp intuitively.

A8.5 Vector fields and derivations

Let $p \in M$, and X_p a tangent vector at p. The set $\{X_p | p \in M\}$ is called a **vector field** on M, and is denoted by X.

In words, a vector field is an assignment of a tangent vector to each point of M – it is a cross-section of TM – but vector field is the term that is commonly used.

Let $(x_1, \ldots x_n)$ be the local coordinates of $x \in M$ in a chart (U, φ). In these local coordinates, a vector field X has a unique expression

$$X = \sum_{i=1}^{n} \xi_i \frac{\partial}{\partial x_i}. \tag{A8.10}$$

The functions ξ_i are the *components* of X in the given chart. By transforming to a different chart, one sees easily that the continuity and differentiability properties of the coefficients ξ_i do not depend on the chart. It therefore makes sense to talk about **smooth vector fields**, the components of which are smooth functions in every chart. Taking a cue from calculus, a map $D : \mathfrak{A} \to \mathfrak{A}$ of an associative algebra over the field K into itself is called a **derivation** (**polarization**, by Hermann Weyl (Weyl, 1946)) if it satisfies the conditions[11]

$$\begin{aligned} D(\lambda a + \mu b) &= \lambda D a + \mu D b, \\ D(ab) &= D(a)b + aD(b) \end{aligned} \tag{A8.11}$$

for all $\lambda, \mu \in K$ and all $a, b \in \mathfrak{A}$.

Let D_1 and D_2 be two derivations of an algebra $\mathfrak{A}$. Define their **commutator product** $[D_1, D_2] : \mathfrak{A} \to \mathfrak{A}$ by

$$[D_1, D_2] = D_1 D_2 - D_2 D_1.$$

This product is clearly antisymmetric, $[D_1, D_2] = -[D_2, D_1]$, and it satisfies the **Jacobi identity**

$$[D_1, [D_2, D_3]] + [D_2, [D_3, D_1]] = [D_3, [D_1, D_2]] = 0.$$

It may be verified by direct computation that, for any $a, b \in \mathfrak{A}$,

$$[D_1, D_2](ab) = ([D_1, D_2]a)b + a([D_1, D_2]b),$$

i.e., $[D_1, D_2]$ is also a derivation.

According to the definitions of vector fields and derivations, a vector field X on M is a derivation D_X on $C^\infty(M)$. The converse is also true.

Theorem A8.6 *If D is a derivation of the associative algebra $C^\infty(M)$, then there is a unique vector field X on M such that $D = D_X$.*

Now let X, Y be two vector fields on M, and D_X, D_Y the corresponding derivations on $C^\infty(M)$. Then their commutator product $[D_X, D_Y]$ is also a derivation

[11] The algebraic notion of a field was defined on page 17.

on $C^\infty(M)$. It now follows from Theorem A8.6 that there is a unique vector field Z on $C^\infty(M)$ that corresponds to the derivation $[D_X, D_Y]$. This vector field is called the **commutator product** of X and Y, and is denoted by $[X, Y]$. We then have

$$[D_X, D_Y] = D_{[X,Y]}. \tag{A8.12}$$

This commutator product is also antisymmetric and satisfies the Jacobi identity.

Under the commutator product, the set of derivations of an algebra $\mathfrak{A}$ forms a *nonassociative* algebra over K. This algebra satisfies the Jacobi identity, and is called a **Lie algebra**. We may sum up the discussion as follows:

Theorem A8.7 *The set of smooth vector fields on a smooth manifold M forms a Lie algebra under the commutator product. This Lie algebra is in* 1–1 *correspondence with the Lie algebra of derivations on $C^\infty(M)$.*

A8.6 One-parameter groups of transformations

Let $\Gamma = (a, b) \subset \mathbb{R}$. A continuous map $\gamma : \Gamma \to M$ is called a **curve** in M. (Common parlance usually associates the term curve with the image set $\{\gamma(t) | t \in \Gamma\}$; there is something to be said for it.) Clearly Γ is a smooth manifold, and we shall assume that the map γ is also smooth. It then induces a fibre-preserving map $\gamma_\star$ from $T\Gamma$ to TM. The tangent spaces $T_t\Gamma$ are one-dimensional, and a tangent vector at $t \in \Gamma$ is $v_t = (\mathrm{d}/\mathrm{d}t)_{t=t}$. The image of v_t under $\gamma_\star$ is a tangent vector at $\gamma(t) \in M$.

If there is a vector field X on M such that $\gamma_\star(v_t) = X_{\gamma(t)}$, then the curve γ is called an **integral curve** of X.[12]

The following theorem is a restatement of the fundamental existence and uniqueness theorem for a system of linear first-order differential equations:[13]

Theorem A8.8 (Uniqueness theorem for integral curves) *Let X be a smooth vector field on M. There is exactly one integral curve of X that passes through any given point $p \in M$.*

Note that this is a *local* result, because, as $\Gamma = (a, b)$, the curve γ is generally defined only on an open set in M.

Our next concern is with one-parameter groups of transformations of M. The definition is a trivial modification of the one used for Stone's theorem (Subsection 7.5.3). A parametrized family of maps $\{\varphi_t : M \to M | t \in \mathbb{R}\}$ is called a **one-parameter group of transformations of** M if it satisfies the following conditions:

[12] Not every smooth curve is an integral curve of a vector field; a simple example is provided by a curve which intersects itself.

[13] A proof will be found as the lemma in (Matsushima, 1972, pp. 78–79).

(a) φ_t is a diffeomorphism for each t.

(b) $\varphi_s \circ \varphi_t = \varphi_{s+t}$ for all $s, t \in \mathbb{R}$.

(c) The map $\mathbb{R} \times M \to M$ defined by $(t, p) \mapsto \varphi_t(p)$ is differentiable.

Let $f \in C^\infty(M)$. Since $\varphi_t(p) \in M$, the expression $f(\varphi_t(p))$ defines a function of t. Let now

$$X_p f = \left[\frac{\mathrm{d}f(\varphi_t(p))}{\mathrm{d}t} \right]_{t=0} . \tag{A8.13}$$

It is easily seen that X_p, so defined, is a smooth vector field on M. It is called the **infinitesimal generator**, or simply the **generator**, of the one-parameter group of transformations $\{\varphi_t\}$ of M.

A vector field X on M which generates a one-parameter group of transformations of M is said to be **complete**. Formula (A8.13) shows that a one-parameter group of transformations defines a unique complete vector field. The following theorem, which is basically a global version of Theorem A8.8, shows that the converse is also true.

Theorem A8.9 (Uniqueness theorem for one-parameter groups) *A complete vector field on M defines a unique one-parameter group of transformations of M.*

The difference between Theorems A8.8 and A8.9 is that the first is local, whereas the second is global. Incomplete vector fields exist, but they do not define one-parameter groups of transformations of M. We shall see below that the geometrical picture of integral curves is of great help in appreciating the difference between the local and global situations.

A8.7 Lie groups and Lie algebras

A Lie group is a group, a topological space and an analytic manifold. Recall that in a topological group the operations of multiplication and taking the inverse are continuous. In a Lie group, they are analytic. More precisely:

Definition A8.10 (Lie group) A real **Lie group** is a group G which is also a real-analytic manifold, such that the maps

$$G \times G \longrightarrow G, \quad G \longrightarrow G$$

defined by $g_1 g_2 \mapsto g$ and $g \mapsto g^{-1}$, respectively, are analytic.[14]

It is a remarkable fact that the following theorem holds.

[14] We are restricting ourselves to real Lie groups here; *complex Lie groups* will be considered briefly in Section A8.7.4.

Theorem A8.11 (Gleason, Montgomery and Zippin) *Let G be a group which is also a topological manifold such that the group operations of multiplication and taking the inverse are continuous.*[15] *Then G is a Lie group.*

Theorem A8.11 constitutes the affirmative solution of *Hilbert's fifth problem.*[16] As every physicist knows, a Lie group may have several disconnected components. In the following, *we shall restrict ourselves to the connected component which contains the identity of the group.*

A8.7.1 Left-invariant vector fields and Lie algebras

Let G be a *connected* Lie group. Denote by L_g the **left-translation** of G by $g \in G$, i.e., the map[17] $L_g : G \to G$ defined by $L_g(g') = g \cdot g'$ for all $g' \in G$. It is clear that

$$L_g \cdot L_h = L_{gh} \quad \text{and} \quad (L_g)^{-1} = L_{g^{-1}}.$$

Furthermore, L_g is a diffeomorphism of G onto itself. Therefore, for each $g \in G$, one can define a bundle map $(L_g)_\star : TG \to TG$. Let now X be a vector field on G. Then so is $(L_g)_\star X$. A vector field X on G is **left-invariant** if it satisfies the condition

$$(L_g)_\star X = X \tag{A8.14}$$

for all $g \in G$.

Left-invariant vector fields exist, and are smooth; this is not hard to verify. The following results, which we state without proof, are deeper.

Theorem A8.12 *A left-invariant vector field on G is complete.*

Theorem A8.13 *If G is n-dimensional,*[18] *then there are exactly n linearly independent left-invariant vector fields on G.*

Theorem A8.13 implies that the tangent bundle TG of G is homeomorphic to the product $G \times \mathbb{R}^n$, which is certainly not true for tangent bundles in general.

Since vector fields are derivations, it makes sense to talk about the Lie algebra of left-invariant vector fields on G. In view of Theorem A8.13, one may choose a basis $\{X_1, X_2, \ldots, X_n\}$ in the space of smooth left-invariant vector fields on G. Combining Theorem A8.13 with formula (A8.12), we conclude that there exists

[15] It is not enough to require that G be a topological group. See the last paragraph of Appendix A3, page 276. A very clear statement may be found in (Kaplansky, 1974, p. 87).

[16] An English translation of the relevant part of it will be found in (Montgomery and Zippin, 1955, pp. 67–69).

[17] Right-translations can be defined analogously, but we shall not use them.

[18] What we have called an n-dimensional Lie group is also known as an n-**parameter Lie group**, especially in the older literature.

a set of constants c^k_{ij} such that

$$[X_i, X_j] = \sum_{j=1}^{n} c^k_{ij} X_k. \tag{A8.15}$$

The Lie algebra defined by (A8.15) is known as the **Lie algebra of the group** G, and the c^k_{ij} its **structure constants.**

A8.7.2 The exponential map

We know from Theorem A8.9 that a complete vector field on a manifold determines a unique one-parameter group of transformations of it. If the manifold is a (connected) Lie group G, then an element X of its Lie algebra $\mathfrak{g}$, being a left-invariant vector field, determines a one-parameter subgroup of G. This subgroup $\{\varphi_t | t \in \mathbb{R}\}$ has to satisfy the conditions $\varphi_s \varphi_t = \varphi_{s+t}, \varphi_0 = e$. Were φ_t to be a function of a real variable, it could only be the exponential function, $\varphi_t = \exp t$. By imitation, the one-parameter subgroup of G determined by $X \in \mathfrak{g}$ is written as

$$\{\exp(tX) | t \in \mathbb{R}\}. \tag{A8.16}$$

It is easily seen that the function $\exp(tX)$ has all the formal properties of the real exponential function, and that

$$X = \left(\frac{\mathrm{d}}{\mathrm{d}t} \exp(tX) \right)_{t=0}.$$

The map $\mathfrak{g} \to G$ defined by $X \mapsto \exp X$ is called the **exponential map**. It is a map from $\mathbb{R}^n$ into the n-dimensional Lie group G.

If $G = GL(n, \mathbb{R})$, then the elements of the group are defined as invertible $n \times n$ matrices with real entries. Then the elements $X \in \mathfrak{g}$ of its Lie algebra are also $n \times n$ matrices, and $\exp(tX)$ is the exponential of the matrix tX, which is well defined for all $t \in \mathbb{R}$. In the general case, the exponential map may be written in local coordinates as follows.[19] Assume that G is n-dimensional, let X_k be an element of $\mathfrak{g}$, and choose a chart (U, φ) with local coordinates $\varphi(p) = (x_1, \ldots, x_k, \ldots, x_n)$ such that

$$X_k = \frac{\partial}{\partial x_k}.$$

[19] The account given above is heuristic. A proper mathematical account may be found in (Matsushima, 1972, pp. 203–207; Auslander and MacKenzie, 1963, Chap. 7; Varadarajan, 1974, pp. 84–92).

Then, if $(x_1, \ldots, x_k + a, \ldots, x_n) \in U$ and f is a function with a Taylor series expansion at p which converges in U,

$$\begin{aligned} \exp(aX_k)f(\cdot x_k \cdot) &= \sum_{m=1}^{n} \left(\frac{1}{m!} \frac{\partial^m}{(\partial x_k)^m} f(\cdot x_k \cdot) \right) \\ &= f(x_1, \ldots, x_k + a, \ldots, x_n). \end{aligned} \tag{A8.17}$$

In the first line of (A8.17), $f(\cdot x_k \cdot)$ is short for $f(x_1, \ldots, x_k, \ldots, x_n)$. On an analytic manifold, a map $f(p) \mapsto f(p')$ is the same as the map $p \mapsto p'$.

A8.7.3 Some examples and counterexamples

Consider the group of translations of $\mathbb{R}^2$. The space $\mathbb{R}^2$ can be covered by a single chart, and the Lie algebra of the group written as

$$X = \frac{\partial}{\partial x}, \; Y = \frac{\partial}{\partial y}. \tag{A8.18}$$

The group is Abelian.

Let $f(x, y)$ be an analytic function of (x, y). Then

$$\exp(aX)f(x) = f(x + a, y) \text{ and } \exp(bY) = f(x, y + b)$$

for any $a, b \in \mathbb{R}$. It follows that the vector fields X and Y are complete, and generate the translations

$$(x, y) \mapsto (x + a, y) \text{ and } (x, y) \mapsto (x, y + b).$$

The integral curves of X and Y are the lines parallel to the horizontal and vertical axes, respectively.

Consider now the punctured plane $\mathbb{R}^2 \setminus \{(0, 0)\}$. This space can also be covered by a single chart, and X and Y defined by (A8.18) are vector fields on it. But these vector fields are *no longer complete*. Geometrically, this is obvious; the punctured plane is not invariant under translations. Analytically, let $f(x, y)$ be an analytic function on $\mathbb{R}^2$. Then

$$\exp(-aX)f(a, 0) = f(0, 0), \tag{A8.19}$$

which is well defined on $\mathbb{R}^2$, is not well defined on the punctured plane; the point $(0, 0)$ does not belong to the latter, and therefore the right-hand side of (A8.19) is undefined. The punctured plane is, however, invariant under rotations around the origin; the complete vector field which generates the rotations is

$$x\frac{\partial}{\partial y} - y\frac{\partial}{\partial k}.$$

It is known that if G is a Lie group and Δ a discrete normal subgroup of it, then G/Δ is a Lie group and the Lie algebras of G and G/Δ are isomorphic. However, given a finite-dimensional Lie algebra $\mathfrak{g}$, there is a unique *simply connected* Lie group G that has $\mathfrak{g}$ as its Lie algebra. The notion of an infinite-dimensional Lie algebra makes sense, but there may be no 'infinite-dimensional Lie group' that has a given infinite-dimensional Lie algebra as its Lie algebra. The present author is not knowledgeable in these subjects, and the interested reader is advised to consult a mathematician friend who will lead him or her to a specialist.

A8.7.4 Complex Lie groups and algebras; complexification

Complex manifolds can be defined almost exactly as real manifolds; one replaces $\mathbb{R}^n$ in the definition by $\mathbb{C}^n$ and requires the transition maps to be holomorphic.[20] This leads quite naturally to the definition of **complex Lie groups** and **complex Lie algebras** (Matsushima, 1972; Varadarajan, 1974). However, every complex Lie group (algebra) in n dimensions over $\mathbb{C}$ turns out to be a real Lie group (algebra) in $2n$ dimensions, and this will suffice for our purposes.

The rotation group in three dimensions $O^+(3, \mathbb{R})$ is real, as is its Lie algebra. Yet the latter is invariably written in physics texts as

$$[L_i, L_j] = \mathrm{i}\epsilon_{ijk}L_k, \quad i, j, k = 1, 2, 3,$$

where ϵ_{ijk} is the completely antisymmetric tensor in three dimensions, and Einstein's summation convention has been employed. The reason is that the quantities L_i, $i = 1, 2, 3$ are generally represented as Hermitian operators on a complex vector space. What is needed, therefore, is an extension of the notion of a Lie algebra over the reals to a Lie algebra over the complex numbers *that does not change the dimension of the algebra.* This process is known as the **complexification** of a real Lie algebra. The complexification of a real Lie algebra is *not* a complex Lie algebra. A formal definition of the process of complexification is that it is the tensor product of a real vector space with the complex numbers. As all physicists are familiar with the process (though perhaps not with the name), we shall not reproduce the details here. They may be found, for instance, in (Matsushima, 1972, Sec. **16 B**, pp. 100–102) in concrete form (without using the term tensor product), or in (Varadarajan, 1974, p. 47), in abstract form.

[20] As every student of quantum field theory will know, there are important differences between the theories of functions of one and several complex variables. A concise account of the elementary theory of functions of several complex variables may be found in (Narasimhan, 1971).

List of symbols for Part I

Symbol	Page	Description
${}^{l}\!>$, $<^{l}$		total order on light rays
on l	23	
on $\check{l}$	70	
on ℓ	77	
$<^{ll}$		nonreflexive, nonsymmetric order on light rays
on l	23	
on $\check{l}$	70	
on ℓ	77	
$>, <$		causal order on the entire space
on M	29	
on $\check{M}$	72	
$\ll$, $\not\ll$		nonreflexive timelike order and its negation
on M	31	
on $\check{M}$	72	
$\gg$, $\not\gg$		nonreflexive timelike order and its negation
on M	31	
on $\check{M}$	72	
$\bowtie$	25	concatenation of light-ray segments
$\check{\partial}$	71	mantle operator
C_x^+	28	forward (future) cone at x
C_x^-	28	backward (past) cone at x
C_x	28	cone at x
$\beta C_x^\pm$	29	β-boundaries of cones at x
$\tau C_x^\pm$	28	τ-interiors of cones at x
$C_{a;D}^\pm$	70	local cones at $a \in M$
$\beta C_{a;D}^\pm$	70	β-boundaries of local cones at a

$C^{\pm}_{a;\check{D}}$	70	$C^{\pm}_{a;D}$, considered as a subset of $\check{D}$
$\widetilde{C}^{\pm}_{a;\check{D}}$	70	uniform completion of $C^{\pm}_{a;\check{D}}$
$\check{C}^{\pm}_{a;\check{D}}$	70	the intersection $\widetilde{C}^{\pm}_{a;\check{D}} \cap \check{D}$
$_{T}\check{C}^{\pm}_{a;\check{D}}$	70	interior of $\check{C}^{\pm}_{a;\check{D}}$
D	34–35	D-set
$\widetilde{D}$	68	uniform completion of D
$\check{D}$	68	interior of $\widetilde{D}$
$\mathcal{E}_U$	67	order uniformity on U
$F[a, b; p]$	85	2-cell in D-interval
$I(a, b)$	33	open order interval in M
$I[a, b]$	33	closed order interval in M
$\check{I}(x, y)$	71	open order interval in $\check{M}$
$\check{I}[x, y]$	71	closed order interval in $\check{M}$
M	23	point-set
$\widetilde{M}$	68	the union $\cup_{\alpha \in A} \widetilde{D}^{\alpha}$
$\check{M}$	66, 68	order completion of M
$P(x_0, x_1, \ldots, x_n)$	25	l-polygon connecting x_0 with x_n
$P^{\uparrow}(x_0, x_1, \ldots, x_n)$	27	ascending l-polygon
$P^{\downarrow}(x_0, x_1, \ldots, x_n)$	27	descending l-polygon
$S(p, q)$	53	spacelike hypersphere in M
$\check{S}(x, y)$	71	spacelike hypersphere in $\check{M}$
l	23	light ray in M
$\check{l}$	70	order completion of the light ray l
ℓ	74	new light ray in $\check{M}$
l_x	23	light ray through x
$l_{x,y}$	23	light ray through x, y
l^{+}_{x}, l^{-}_{x}	28	forward and backward rays from x
$l^{+}_{b,c}$	27	forward ray from b passing through c
$l^{-}_{b,a}$	27	backward ray from b passing through a
$l(a, b)$	25	open light ray segment from a to b
$l[a, b]$	25	closed light ray segment from a to b

References

Aleksandrov, A D (1959). Filosofskoe soderzhanie i znachenie teorii otnositel'nosti, *Voprosy Filosofii* **1**, 67–84.

Allahverdyan, A E, Balian, R and Nieuwenhuizen, Th M (2003). Curie–Weiss model of the quantum measurement process, *Eur. Phys. Lett.* **61**, 452–458.

Amrein, W O, Jauch, J M and Sinha, K B (1977). *Scattering Theory in Quantum Mechanics.* New York: W A Benjamin.

Apostol, T M (1974). *Mathematical Analysis*, 2nd edn. Addison-Wesley World Student Series. Reading, MA: Addison-Wesley.

Araki, H and Woods, E J (1963). Representations of the canonical commutation relations describing a nonrelativistic infinite free Bose gas, *J. Math. Phys.* **4**, 637–662.

Araki, H and Yanase, M M (1960). *Phys. Rev.* **120**, 622–626. Reprinted in (Wheeler and Zurek, 1983, pp. 707–711).

Asselmeyer-Maluga, T and Brans, C H (2007). *Exotic Smoothness and Physics: Differential Topology and Spacetime Models.* Singapore: World-Scientific.

Auslander, L and MacKenzie, R E (1963). *Introduction to Differentiable Manifolds.* New York: McGraw-Hill.

Avron, J and Simon, B (1981). Transient and recurrent spectrum, *J. Funct. Anal.* **43**, 1–31.

Bahr, B (2006). The hot bang state of massless fermions, *Lett. Math. Phys.* **78**, 39–54.

Balslev, E and Combes, J-M (1971). Spectral properties of many-body Schrödinger operators with dilatation-analytic interactions, *Commun. Math. Phys.* **22**, 280–294.

Bargmann, V (1954). On unitary ray representations of continuous groups, *Ann. Math.* **59**, 1–46.

Bargmann, V (1964). Note on Wigner's theorem on symmetry operations, *J. Math. Phys.* **5**, 862–868.

Barton, G (1963). *Introduction to Advanced Field Theory*, Interscience Tracts in Physics and Astronomy. New York: Interscience Publishers.

Bell, J S (1975). On wave packet reduction in the Coleman-Hepp model, *Helv. Phys. Acta* **48**, 93–98.

Bell, J S (2004). *Speakable and Unspeakable in Quantum Mechanics*, 2nd edn., with an Introduction by Alain Aspect. Cambridge: Cambridge University Press.

BenDaniel, D J (1998a). The unreasonable effectiveness of mathematics in physics, in: *Causality and Locality in Modern Physics*, Eds. G Hunter, S Jeffers and J-P Vigier, Fundamental Theories of Physics. Dordrecht: Kluwer.

BenDaniel, D J (1998b). The definability of fields, *Chaos, Solitons and Fractals*, special issue, ed. K Svozil, **10**, 975–979.

BenDaniel, D J (2006). Constructibility and quantum mechanics, based on a talk given at the 26th Annual Colloquium in Group-Theoretical Methods in Physics, New York. arXiv:0806.1365.

Blanchard, Ph and Jadczyk, A (1993). On the interaction between classical and quantum systems, *Phys. Lett.* **A 175**, 157–164.

Blanchard, Ph and Jadczyk, A (1995). Theory of events, in: *Proceedings of the Conference on Nonlinear, Deformed and Irreversible Quantum Systems, Clausthal*, 1994, pp. 265–272. Singapore: World-Scientific.

Bohm, D (1951). *Quantum Theory.* Englewood Cliffs, NJ: Prentice-Hall. Reprinted (1989), New York: Dover Publications.

Bohr, N (1935). Can quantum-mechanical description of physical reality be considered complete? *Phys. Rev.* **48**, 696–702. Reprinted in (Wheeler and Zurek, 1983, pp. 145–151).

Bohr, N, Kramers, H R and Slater, J C (1924). The quantum theory of radiation, *Phil. Mag.* Ser. 6, **47**, 785–802. Reprinted in (van der Waerden, 1967, pp. 159–176).

Bohr, N and Rosenfeld, L (1933). Zur Frage der Messbarkeit der elektromagnetischen Feldgrössen, *Mat.-fys. Medd. Dan. Vid. Selsk.* **12**, No 8. English translation, 'On the question of measurability of electromagnetic field quantities', in (Rosenfeld, 1979, pp. 357–400). Reprinted in (Wheeler and Zurek, 1983, pp. 479–522).

Bohr, N and Rosenfeld, L (1950). Field and charge measurements in quantum electrodynamics, *Phys. Rev.* **78**, 794–98. Reprinted in (Rosenfeld, 1979, pp. 401–412) and in (Wheeler and Zurek, 1983, pp. 523–534).

Borchers, H-J and Hegerfeldt, G C (1972a). The structure of space-time transformations, *Commun. Math. Phys.* **28**, 259–266.

Borchers, H-J and Hegerfeldt, G C (1972b) Über ein Problem der Relativitätstheorie: Wann sind Punktabbildungen des $\mathbb{R}^n$ linear? *Nachr. Akad. Wiss. Göttingen*, 205–229.

Borchers, H-J and Sen, R N (1975). Relativity groups in the presence of matter, *Commun. Math. Phys.* **42**, 101–126.

Borchers, H-J and Sen, R N (2006). *Mathematical Implications of Einstein–Weyl Causality*, Lecture Notes in Physics, Vol. 709, Berlin: Springer.

Born, G V R (2002). The wide-ranging family history of Max Born, *Notes Rec. Roy. Soc. London* **56**(2), 219–262.

Born, H and Born, M, (1969). *Der Luxus des Gewissens*, Ed. A Hermann. München: Nymphenburger Verlagshandlung.

Born, M (1926). Zur Quantenmechanik der Stossvorgänge, *Zeits. f. Phys.* **37**, 863–867. English translation, On the quantum mechanics of collisions, in (Wheeler and Zurek, 1983, pp. 52–55).

Born, M (1949). *The Natural Philosophy of Cause and Chance.* Oxford: Clarendon Press.

Born, M (1954). The statistical interpretation of quantum mechanics, Nobel lecture. In *Nobel lectures: Physics 1942–1962*, pp. 256–267, published for the Nobel Foundation in 1964. Amsterdam. Elsevier Publishing Company.

Born, M (1955). Continuity, determinism and reality, *Dan. Mat.-fys. Medd.* **30**, No. 2, Dedicated to Professor Niels Bohr on the Occasion of his 70th Birthday, Copenhagen: Ejnar Munksgaard.

Born, M (1961). Bemerkungen zur statistischen Deutung der Quantenmechanik, in: *Werner Heisenberg und die Physik unserer Zeit*, Ed. F Bopp, pp. 103–118. Braunschweig: F Vieweg.

Born, M (1962). *Zur statistischen Deutung der Quantentheorie.* Stuttgart: Ernst Battenberg.

Bose, S K (1995a). The galilean group in $2 + 1$ space-time and its central extension, *Comm. Math. Phys.* **169**, 385–395.

Bose, S K (1995b). Representations of the (2+1)-dimensional galilean group, *J. Math. Phys.* **36**, 875–890.

Brans, C H (1994) Localized exotic smoothness, *Class. Quantum Grav.* **11**, 1785–1792.

Bratteli, O and Robinson, D W (1979). *Operator Algebras and Quantum Statistical Mechanics I. $C^\star$- and $W^\star$-algebras, Symmetry Groups, Decomposition of States.* New York: Springer-Verlag.

Bratteli, O and Robinson, D W (1981). *Operator Algebras and Quantum Statistical Mechanics II. Equilibrium States, Models in Quantum Statistical Mechanics.* New York: Springer-Verlag.

Brown, H R (1986). The insolubility proof of the quantum mechanical measurement problem, *Found. Phys.* **16**, 857–870.

Buchholz, D, Ojima, I and Roos, H (2002). Thermodynamic properties of non-equilibrium states in quantum field theory, *Ann. of Phys.* **297**, 219–242.

Busch, P and Shimony, A (1996). Insolubility of the quantum measurement problem for unsharp observables, *Stud. Hist. Phil. Mod. Phys.* **27**, 397–404.

Carmona, R and Lacroix, J (1990). *Spectral Theory of Random Schrödinger Operators*, Probability and its Applications. Boston: Birkhäuser.

Cartier, P (2001). A mad day's work: from Grothendieck to Connes and Kontsevich. The evolution of concepts of space and symmetry. *Bull. Am. Math. Soc.* **38**, 389–408.

Chevalley, C (1946). *The Theory of Lie Groups*, I. Princeton: Princeton University Press.

Compton, A H and Simon, A W (1925). Directed quanta of scattered X-rays, *Phys. Rev.* **26**, 289–299.

Courant, R and Hilbert, D (1953). *Methods of Mathematical Physics*, Vol. I. New York: Interscience Publishers. Translated (with additions and modifications) from the German original, *Methoden der mathematischen Physik*, 2nd edn. (1930). Berlin: Julius Springer.

Courant, R and Hilbert, D (1962). *Methods of Mathematical Physics*, Vol. II, *Partial Differential Equations* (by R Courant). New York: Interscience Publishers.

Dauben, J W (1990). *Georg Cantor: His Mathematics and Philosophy of the Infinite*. Princeton: Princeton University Press. Originally published (1979), Cambridge, MA: Harvard University Press.

Davies, E B (1976). *Quantum Theory of Open Systems*. London: Academic Press.

D'Espagnat, B (1971). *Conceptual Foundations of Quantum Mechanics*. Menlo Park, CA: W A Benjamin.

Derfel, G and Sen, R N (1997). Singular continua, fractals and functional equations, *Open Sys. and Information Dyn.* **4**, 125–139.

Dirac, P A M (1930). *Principles of Quantum Mechanics*. Oxford: Clarendon Press.

Dirac, P A M (1938–39). The relation between mathematics and physics, *Proc. Roy. Soc. (Edinburgh)*, **59**, part II, 122–129.

Donaldson, S K (1983). An application of gauge theory to 4-dimensional topology, *J. Differential Geom.* **18**, 279–315.

Dyson, F J (1952). Divergence of perturbation theory in quantum electrodynamics, *Phys. Rev.* **85**, 631–632.

Dyson, F J (1966). *Symmetry Groups in Nuclear and Particle Physics*: A Lecture-Note and Reprint Volume. New York: W A Benjamin.

Einstein, A (1953). Elementare Überlegungen zur Interpretation der Grundlagen der Quanten-Mechanik, in: *Scientific Papers Presented to Max Born*, Ed. E V Appleton, pp. 33–40. New York: W A Benjamin.

Einstein, A, Podolsky, B and Rosen, N (1935). Can quantum-mechanical description of physical reality be considered complete? *Phys. Rev.* **47**, 777–780. Reprinted in (Wheeler and Zurek, 1983, pp. 138–141).

Einstein, A (1971). *The Born–Einstein Letters*: Correspondence between Albert Einstein and Max and Hedwig Born from 1916 to 1955, with Commentaries by Max Born. London: MacMillan. (Sometimes catalogued under Born, M.)

Eisenhart, L P (1964). *Riemannian Geometry*, 5th printing. Princeton: Princeton University Press.

Ellis, R S (1985). *Entropy, Large Deviations and Statistical Mechanics*. New York: Springer-Verlag.

Emch, G G (1964). Coarse-graining in Liouville space and master equation, *Helv. Phys. Acta* **37**, 532–544.

Emch, G G (1972). *Algebraic Methods in Statistical Mechanics and Quantum Field Theory*. New York: Wiley-Interscience.

Emch, G G and Piron, C (1963). Symmetry in quantum theory, *J. Math. Phys.* **4**, 469–473.

Fehrs, M and Shimony, A (1974). Approximate measurements in quantum mechanics, I, *Phys. Rev.* **D9**, 2317–2320. Reprinted with a comment added, in (Shimony, 1993, pp. 34–41).

Feller, W (1970). *An Introduction to Probability Theory and its Applications*, Vol. 1, 3rd edn., Revised Printing. New York: John Wiley, Wiley International Edition.

Feller, W (1971). *An Introduction to Probability Theory and its Applications*, Vol. 2, 2nd edn., New York: John Wiley.

Fine, A (1969). On the general quantum theory of measurement, *Proc. Camb. Phil. Soc.* **65**, 111–121.

Flato, M and Simon, J (1973). Separate and joint analyticity in Lie group representations, *J. Funct. Anal.* **13**, 268–276.

Flato, M, Simon, J, Snellman, H and Sternheimer, D (1972). Simple facts about analytic vectors and integrability, *Ann. Scient. de l'École Norm. Sup.* **5**, 423–434.

Fraenkel, A A (1953). *Abstract Set Theory*, Studies in Logic and the Foundations of Mathematics. Amsterdam: North-Holland.

Fraenkel, A A, Bar-Hillel, Y and Levy, A (2001). *Foundations of Set Theory*, 2nd revised edn, Studies in Logic and the Foundations of Mathematics. Amsterdam: North-Holland.

Freedman, M H (1984). There is no room to spare in four-dimensional space, *Notices Amer. Math. Soc.* **31**, 3–6.

Friedman, A (1970). *Foundations of Modern Analysis*. New York: Holt, Rinehart and Winston. Corrected edn. (1982), New York: Dover Publications.

Fröhlich, J and Spencer, T (1983). Absence of diffusion in the Anderson tight binding model for large disorder or low energy, *Commun. Math. Phys.* **88**, 151–184.

Fröhlich, J, Martinelli, F, Scoppola, E and Spencer, T (1985). A constructive proof of localization in Anderson tight binding model, *Commun. Math. Phys.* **101**, 21–46.

Fuglede, B (1967). On the relation $PQ - QP = -\mathrm{i}I$, *Math. Scand.* **20**, 79–88.

Goldsheid, I and Molchanov, S A (1976). On Mott's problem, *Soviet Math. Dokl.* **17**, 1369–1373.

Goldsheid, I, Molchanov, S A and Pastur, L A (1977). A pure point spectrum of the stochastic one-dimensional Schrödinger equation, *Funct. Anal. and Appl.* **11**, 1–10.

Gompf, R (1983). Three exotic $\mathbb{R}^4$'s and other anomalies, *J. Differential Geom.* **18**, 317–328.

Goursat, E (1959). *A Course in Mathematical Analysis*, Vol. I. New York: Dover Press. Originally published (1904), Boston: Ginn & Company. Translated from the French *Cours d'analyse mathématique* (1902), Paris: Gauthier-Villars.

Grad, H (1969). Principles of the kinetic theory of gases, in: *Handbuch der Physik*, Vol. XII, Ed. S. Flügge, pp. 205–294. Berlin: Springer-Verlag.

Haag, R (1955). On quantum field theories, *Kon. Dan. Vid. Selsk. Mat.-fys. Medd.* **29**, no. 12. Reprinted in *Dispersion Relations and the Abstract Approach to Field Theory* (1961), Ed. L Klein, pp. 55–89. New York: Gordon and Breach.

Haag, R (1993). *Local Quantum Theory*. Berlin: Springer-Verlag.

Haag, R, Hugenholtz, N M and Winnink, M (1967). On the equilibrium states in quantum statistical mechanics, *Commun. Math. Phys.* **5**, 215–236.

Halmos, P (1950). *Measure Theory*. Princeton: Van Nostrand.

Hamermesh, M (1962). *Group Theory and its Applications to Physical Problems*. Reading, MA: Addison-Wesley.

Hausdorff, F (1957). *Set Theory*. New York: Chelsea. Translated from the 3rd German edn. (1935) of *Mengenlehre*. Berlin: de Gruyter.

Hegerfeldt, G C (1974). Remark on causality and particle localization, *Phys. Rev.* **D 10**, 3320–3321.

Heisenberg, W (1925). Über quantentheoretische Umdeutung kinematischer und mechanischer Bezeihungen, *Zeits. f. Phys.* **33**, 879–893. English translation: Quantum-theoretical re-interpretation of kinematic and mechanical relations, in (van der Waerden, 1967, pp. 261–276).

Heisenberg, W (1927). Über den anschaulichen Inhalt der quantentheoretischen Kinematik und Mechanik, *Zeits. f. Phys.* **43**, 172–98. English translation: The physical content of quantum kinematics and mechanics, in (Wheeler and Zurek, 1983, pp. 62–84).

Heisenberg, W (1928). Discussione sulla communicazione Bohr, *Congresso Internazionale dei Fisici*, 11–20 Settembre 1927, Como-Pavia-Roma, Volume Secondo, p. 593. Bologna: Nicola Zanichelli. Reprinted in: Niels Bohr, *Collected Works*, Vol. 6, p. 141. Amsterdam: North-Holland.

Heisenberg, W (1930). *The Physical Principles of Quantum Theory.* Chicago: University of Chicago Press. English translation of *Die physikalischen Prinzipien der Quantenmechanik*, (1930). Leipzig: Hirzel-Verlag.

Heisenberg, W (1975). The great tradition, *Encounter* **44**, 52–58.

Hepp, K (1972). Quantum theory of measurement and macroscopic observables, *Helv. Phys. Acta* **45**, 237–248.

Hewitt, E and Ross, K A (1963). *Abstract Harmonic Analysis*, Vol. I, Vol. 115 in Die Grundlehren der mathematischen Wissenschaften. Berlin: Springer-Verlag.

Hilbert, D (1912). Begrundung der kinetischen Gastheorie, *Math. Ann.* **71**, 562–577.

Hille, E (1965). What is a semi-group? in: *MAA Studies in Mathematics*, Vol. 3: *Studies in Real and Complex Analysis*, Ed. I I Hirschman, Jr, pp. 55–66. Englewood Cliffs, NJ: Prentice-Hall.

Hilton, P J (1968). Introduction: Modern Topology, in: *MAA Studies in Mathematics,* Vol. 5*: Studies in Modern Topology*, Ed. P J Hilton, pp. 1–22. Englewood Cliffs, NJ: Prentice-Hall.

Hirsch, M W (1976). *Differential Topology.* New York: Springer-Verlag.

Huang, K (1963). *Statistical Mechanics.* New York: John Wiley.

Inönü, E and Wigner, E P (1952). Representations of the Galilei group, *Nuovo Cimento* **IX**, 706–718.

Jacobson, N (1974). *Basic Algebra I.* San Francisco: W H Freeman and Company.

James, I M (1999). *Topologies and Uniformities.* London: Springer-Verlag.

Jauch, J-M (1968). *Foundations of Quantum Mechanics.* Reading, MA: Addison-Wesley.

Joos, H, Zeh, D, Kiefer, C, Giulini, D, Kupsch, J and Stamatescu, I-O (2003). *Decoherence and the Appearance of a Classical World in Quantum Theory*, 2nd edn. Berlin: Springer-Verlag.

Kaplansky, I (1974). *Lie Algebras and Locally Compact Groups.* Chicago: University of Chicago Press.

Kelley, J L (1955). *General Topology.* New York: Van Nostrand.

Khalatnikov, I M (1965). *An Introduction to the Theory of Superfluidity.* New York: W A Benjamin. Translated from the Russian.

Kitchen, J W (1968). *Calculus of One Variable*, Addison-Wesley Series in Mathematics, Reading, MA: Addison-Wesley.

Kuhn, T S (1957). *The Copernican Revolution.* Cambridge, MA: Harvard University Press. Reprinted (2003), Harvard paperback edition, 24th printing.

Kuiper, N H (1965). The homotopy type of the unitary group of Hilbert space, *Topology* **3**, 19–30.

Landau, L D (1927). Das Dämpfungsproblem in der Wellenmechanik, *Zeits. f. Physik* **45**, 430–441.

Landau, L D and Peierls, R E (1931). Erweiterung des Unbestimmheitsprinzips für die relativistische Quantenfeldtheorie, *Zeits. f. Phys.* **69**, 56–69. English translation: Extension of the uncertainty principle to relativistic quantum theory, in *Collected Papers of L D Landau* (1965), Ed. D ter Haar, pp. 40–51. New York: Gordon and Breach. Reprinted in (Wheeler and Zurek, 1983, pp. 465–476).

Landau, L D and Lifshits, E M (1959). *Statistical Physics*, Vol. 5 of the Course of Theoretical Physics. Oxford: Pergamon Press. Translated from the Russian.

Lanford, O and Ruelle, D (1969). Observables at infinity and states with short-range correlations in statistical mechanics, *Commun. Math. Phys.* **13**, 194–215.

Last, Y (2006). Exotic spectra: a review of Barry Simon's central contributions, in: *Spectral Theory and Mathematical Physics: A Festschrift in Honor of Barry Simon's 60th Birthday*, Eds. F Gesztesy (Managing Ed.), P Deift, C Galvez, and W Schlag, Proceedings of the Symposia on Pure Mathematics, Vol. 76, 2006, pp. 697–712. Providence, RI: American Mathematical Society.

Lawson, H B (1985). *The Theory of Gauge Fields in Four Dimensions*, Conference Board of the Mathematical Sciences, Regional Conference Series in Mathematics, No. 58. Providence, RI: American Mathematical Society.

Levi-Civita, T (1917). Nozione di parallelismo in una varietà qualunque e consequente specificazione geometrica della curvatura Riemanniana, *Rendiconti di Palermo*, **42**, 173–205.

Lévy-Leblond, J-M (1972). Galilei group and galilean invariance, in: *Group Theory and its Applications*, Ed. E M Loebl, pp. 221–299. New York: Academic Press.

Li, Y S, Zeng, B, Liu, X S and Long, G L (2001). Entanglement in a two-identical-particle system, *Phys. Rev.* **A 64**, 054303, 1–4.

London, F (1964). *Superfluids,* Vol. II, *Macroscopic Theory of Superfluid Helium.* New York: Dover Press.

London, F and Bauer, E (1939). *La théorie de l'observation en mécanique quantique*, No. 775 of Actualités scientifiques et industrielle: Exposé de physique générale. Paris: Hermann. English translation: *The Theory of Observation in Quantum Mechanics*, in (Wheeler and Zurek, 1983, pp. 217–259).

Lorch, E (1962). The spectral theorem, in: *MAA Studies in Mathematics,* Vol. 1*: Studies in Modern Analysis*, Ed. R C Buck, pp. 88–137. Englewood Cliffs, NJ: Prentice-Hall.

Lorenz, E N (1963). Deterministic nonperiodic flow, *J. Atmospheric Sciences* **20**, 130–141.

Lubkin, E (1963). Geometric definition of gauge invariance, *Ann. of Phys.* **23**, 233–283.

Mackey, G W (1968). *Induced Representations of Groups and Quantum Mechanics*, New York–Torino: W. A. Benjamin–Editore Boringhieri.

Mackey, G W (1976). *The Theory of Unitary Group Representations.* Chicago: The University of Chicago Press.

Mackey, G W (1978). *Unitary Group Representations in Physics, Probability and Number Theory.* Reading, MA: Benjamin/Cummings.

MacLane, S and Moerdijk, I (1994). *Sheaves in Geometry and Logic: A First Introduction to Topos Theory.* New York: Springer.

Marsden, J E (1973). *Basic Complex Analysis.* San Francisco: W H Freeman.

Martin, G E (1975). *The Foundations of Geometry and the Non-Euclidean Plane*, Undergraduate Texts in Mathematics. New York: Springer-Verlag.

Matsushima, Y (1972). *Differential Manifolds.* New York: Marcel Dekker.

Mayer, M E (1977). Introduction to the fibre-bundle approach to gauge theories, in: W Drechsler and M E Mayer, *Fiber Bundle Techniques in Gauge Theories*, Lecture Notes in Physics, vol. 67, pp. 1–143. Berlin: Springer-Verlag.

Michel, L (1964). Invariance in quantum mechanics and group extension, in: *Group-Theoretical Concepts and Methods in Elementary Particle Physics*, Lectures of the Istanbul Summer School of Theoretical Physics, 1962, Ed. F Gürsey, pp. 135–200. New York: Gordon and Breach.

Montgomery, D and Zippin, L (1955). *Topological Transformation Groups.* New York: Intersciece Publishers.

Morinaga, M, Yasuda, M, Kishimoto, T, Shimizu, F, Fujita, J and Matsui, S (1996). Holographic manipulation of cold atomic beam, *Phys. Rev. Lett.* **77**, 802–805.

Munkres, J R (1975). *Topology: A First Course.* Englewood Cliffs, NJ: Prentice-Hall.

Murray, F J and von Neumann, J (1936). Rings of operators, *Ann. Math.* **37**, 116–229.

Namiki, M, Pascazio, S and Nakazato, H (1999). *Decoherence and Quantum Measurements.* Singapore: World Scientific.

Narasimhan, R (1971). *Several Complex Variables.* Chicago: University of Chicago Press.

Omnès, R (1999). *Understanding Quantum Mechanics.* Princeton: Princeton University Press.

Pauli, W (1928). Über das *H*-theorem vom Anwachsen der Entropie vom Standpunkt der neuen Quantenmechanik, in: *Probleme der modernen Physik: Arnold Sommerfeld zum 60. Geburtstage gewidmet von seinen Schülern*, Ed. P J W Debye, pp. 30–45. Leipzig: S Hirzel.

Pauli, W (1985). *Scientific Correspondence with Bohr, Einstein, Heisenberg, and Others*, Vol. II, Ed. K von Meyenn, Berlin: Springer.

Pearson, D B (1978). Singular continuous measures in scattering theory, *Commun. Math. Phys.* **60**, 13–36.

Peres, A (1980). Can we undo quantum measurements? *Phys. Rev.* **D 22**, 879–883. Reprinted in (Wheeler and Zurek 1983, pp. 692–696).

Peres, A (1986). When is a quantum measurement, *Am. J. Phys.* **54**, 688–692.

Peshkin, M and Tonomura, A (1989). *The Aharonov-Bohm Effect*, Lecture Notes in Physics, Vol. 340. Berlin: Springer-Verlag.

Pontrjagin, L (1939). *Topological groups.* Princeton: Princeton University Press.

Rauch, H and Werner, S A (2000). *Neutron Interferometry: Lessons in Experimental Quantum Mechanics.* Oxford: Clarendon Press.

Rédei, M and Summers, S J (2007). Quantum probability theory, *Studies in History and Philosophy of Modern Physics* **38**, 390–417.

Reed, M and Simon, B (1972). *Methods of Modern Mathematical Physics,* Vol. I. *Functional Analysis.* New York: NY, Academic Press.

Reed, M and Simon, B (1978). *Methods of Modern Mathematical Physics,* Vol. IV. *Analysis of Operators.* New York: Academic Press.

Reeh, H (1988). A remark concerning canonical commutation relations. *J. Math. Phys.* **98**, 1535–1536.

Reeh, H and Schlieder, S (1961). Bemerkungen zur Unitäräquivalenz von Lorentzinvarianten Feldern, *Il Nuovo Cimento* **22**, 1051–1068.

Riesz, F and Sz-Nagy, B (1955). *Functional Analysis.* New York: Frederick Ungar. Translated from the 2nd French edition.

Robertson, H P (1929). The uncertainty principle. *Phys. Rev.* **34**, 163–164. Reprinted in (Wheeler and Zurek 1983, pp. 127–128).

Robinson, A (1966). *Non-Standard Analysis*, Studies in Logic and the Foundations of Mathematics. Amsterdam: North-Holland.

Roos, H (1995). Defining local thermodynamic equilibrium: in *Mathematical Physics Towards the 21st Century*, Eds. R N Sen and A Gersten, pp. 99–108. Beer-Sheva: Ben-Gurion University of the Negev Press.

Roos, H and Sen, R N (1994). Is microphysics incomplete? *Open Sys. and Information Dyn.* **2**, 245–264.

Rosenfeld, L (1955). On quantum electrodynamics: in *Niels Bohr and the Development of Physics*, Essays dedicated to Niels Bohr on the occasion of his seventieth birthday, Ed. W Pauli, pp. 70–95. Oxford: Pergamon Press.

Rosenfeld, L (1979). *Selected Papers of Léon Rosenfeld*, Eds. R S Cohen and J J Stachel. Dordrecht: Reidel.

Rudin, W (1974). *Functional Analysis.* T M H edition. New Delhi: Tata McGraw-Hill.

Rudin, W (1976). *Principles of Mathematical Analysis*, 3rd edn. New York: McGraw-Hill.

Ruelle, D (1969). *Statistical Mechanics: Rigorous Results.* New York: W A Benjamin.

Schliemann, J, Cirac, J I, Kuś, M, Lewenstein and M, Loss, D (2001). Quantum correlations in two-fermion systems, *Phys. Rev.* **A 64**, 022303, 1–9.

Schmidt, E (1907). Zur Theorie der linearen und nichtlinearen Integralgleichungen. I. Entwicklung willkürlicher Funktionen nach Systemen vorgeschriebener, *Math. Ann.* **63**, 433–476.

Schmüdgen, K (1983). On the Heisenberg commutation relation II, *Publ. Res. Inst. Math. Sci., Kyoto* **19**, 601–671.

Schrödinger, E (1935). Die gegenwärtige Situation in der Quantenmechanik, *Naturwiss.* **23**, 807–812, 823–828, 844–849. English translation: 'The present situation in quantum mechanics: A translation of Schrödinger's "cat paradox" paper', *Proc. Am. Phil. Soc.* **124**, 323–338 (1980). Reprinted in (Wheeler and Zurek, 1983, pp. 152–167).

Schroeder, M (1991). *Fractals, Chaos, Power Laws: Minutes from an Infinite Paradise.* San Francisco: W H Freeman.

Schwartz, L (1957). *Théorie des distributions*, Vol. I. Paris: Hermann.

Schwartz, L (1959). *Théorie des distributions*, Vol. II. Paris: Hermann.

Schwarzschild, K (1906). Über das Gleichgewicht der Sonnenatmosphäre, *Göttinger Nachrichten, Math.-Phys. Klasse*, 41–53.

Schweber, S S (1961). *An Introduction to Relativistic Quantum Field Theory.* New York: Harper and Row.

Schweber, S S (1994). *QED and the Men Who Made It: Dyson, Feynman, Schwinger, and Tomonaga.* Princeton: Princeton University Press.

Segal, I E (1963). *Mathematical Problems of Relativistic Physics.* Providence, RI: American Mathematical Society.

Sen, R N (1972). The Galilei group and Landau excitations, in: *Statistical Mechanics and Field Theory*, Eds. R N Sen and C Weil, pp. 169–186. Jerusalem: Israel Universities Press.

Sen, R N (1978). Theory of symmetry in the quantum mechanics of infinite systems, *Physica* **94A**, I. The state space and the group action, 39–54; II. Isotropic representations of the Galilei group and its central extensions, 55–70.

Sen, R N (1986). The representation of Lie groups by bundle maps, *J. Math. Phys.* **8**, 2002–2008.

Sen, R N (1999). Why is the Euclidean line the same as the real line?, *Found. Phys. Lett.* **12**, 325–345.

Sen, R N (2008) Physics and the measurement of continuous variables, *Found. Phys.* **38**, 301–316.

Sen, R N and Sewell, G L (2002). Fiber bundles and quantum physics, *J. Math. Phys.* **43**, 1323–1339.

Sewell, G L (1989). *Quantum Theory of Collective Phenomena*, paperback edn., with corrections, Monographs on the Physics and Chemistry of Materials. Oxford: Clarendon Press.

Sewell, G L (2002). *Quantum Mechanics and Its Emergent Macrophysics.* Princeton: Princeton University Press.

Sewell, G L (2005). On the mathematical structure of the quantum measurement problem, *Rep. Math. Phys.* **56**, 271–290.

Sewell, G L (2006). Can the quantum measurement problem be resolved within the framework of Schrödinger dynamics? *Markov Processes and Related Fields*, **13**, 425–440.

Sewell, G L (2007). Can the quantum measurement problem be resolved within the framework of Schrödinger dynamics and quantum probability? in: *Quantum Theory: Reconsideration of Foundations* – 4, AIP Conference Proceedings Vol. 962, Eds. G Adenier, A Yu Khrennikov, P Lahti, V Man'ko and Th H Nieuwenhuizen, pp. 215–222. Melville, NY: American Institute of Physics.

Shimony, A (1963). Role of the observer in quantum theory, *Am. J. Phys.* **31**, 755–73. Reprinted with a comment added, in (Shimony, 1993, pp. 3–33).

Shimony, A (1974). Approximate measurements in quantum mechanics, II, *Phys. Rev.* **D9**, 2321–2323. Reprinted with a comment added, in (Shimony, 1993), pp. 41–47.

Shimony, A (1993). *Search for a Naturalistic World View*, Vol. II, *Natural Science and Metaphysics.* Cambridge: Cambridge University Press.

Shimony, A (2002). Wigner's contributions to the quantum theory of measurement, in: *Proceedings of the Wigner Centennial Conference, Pécs, Hungary*, 2002, Article 51. (Distributed on CD.)

Simmons, G F (1963). *Introduction to Topology and Modern Analysis.* New York, NY: McGraw-Hill.

Simon, B (1995). Operators with singular continuous spectra, I. General operators, *Ann. of Math.* **141**, 131–145.

Simon, J (1972). On the integrability of finite-dimensional Lie algebras, *Commun. Math. Phys.* **28**, 39–42.

Sinha, K B (1994). On the collapse postulate of quantum mechanics, in: *Mathematical Physics Towards the 21st Century*, Eds. R N Sen and A Gersten, pp. 344–350. Beer-Sheva: Ben-Gurion University of the Negev Press.

Sinha, K B and Goswami, D (2007). *Quantum Stochastic Processes and Non-Commutative Geometry*, Cambridge Tracts in Mathematics, No. 169. Cambridge: Cambridge University Press.

Stebbing, L S (1944). *Philosophy and the Physicists.* Harmondsworth, Penguin Books.

Steenrod, N (1972). *The Topology of Fiber Bundles*, eighth printing. Princeton: Princeton University Press.

Stein, H (1997). Maximal extension of an impossibility theorem concerning quantum measurement, in: *Potentiality, Entanglement and Passion-at-a-Distance. Quantum-Mechanical Studies for Abner Shimony*, Eds. R S Cohen, M Horne and J Stachel, pp. 231–243. Dordrecht: Kluwer Academic Publishers.

Stein, H and Shimony, A (1971). Limitations on measurement, in: *Foundations of Quantum Mechanics, Proceedings of the Enrico Fermi International Summer School, Course II*, Ed. B d'Espagnat, pp. 56–76. New York: Academic Press.

Streater, R F (2000). Classical and quantum probability, *J. Math. Phys.* **41**, 3556–3603.

Streater, R F and Wightman, A S (1964). *PCT, Spin and Statistics, and All That.* New York: W A Benjamin. Reprinted, with corrections (2000). Princeton: Princeton University Press.

Stueckelberg, E G C (1960). Quantum theory in real Hilbert space, *Helv. Phys. Acta* **33**, 725–752.

Stueckelberg, E G C and Guenin, M (1961). Quantum theory in real Hilbert space, *Helv. Phys. Acta* **34**, 621–628.

Stueckelberg, E G C and Guenin, M (1962). Quantum theory in real Hilbert space, *Helv. Phys. Acta* **35**, 673–695.

Stueckelberg, E G C, Guenin, M, Piron, C and Ruegg, H (1961). Quantum theory in real Hilbert space, *Helv. Phys. Acta* **34**, 675–698.

Summers, S J (2001). On the Stone-von Neumann uniqueness theorem and its ramifications, in: *John von Neumann and the Foundations of Quantum Physics*, Institut Wiener Kreis, Yearbook 8, 2000, Eds. M Rédei and M Stölzner, pp. 135–152. Dordrecht: Kluwer.

Tilgner, H (1970). A class of solvable Lie groups and their relation to the canonical formalism, *Ann. Inst. Henri Poincaré* **XIII**, 103–127.

van der Waerden, B L (1967). *Sources of Quantum Mechanics.* Amsterdam, North-Holland. Reprinted (2007), New York: Dover Publications.

van Kampen, N (1954). Quantum statistics of irreversible processes, *Physica* **XX**, 603–622.

van Kampen, N (1962). Fundamental problems in the statistical mechanics of irreversible processes, in *Fundamental Problems in Statistical Mechanics*, Ed. E D G Cohen, pp. 173–203. Amsterdam: North-Holland.

van Kampen, N (1988). Ten theorems about quantum mechanical measurements, *Physica* **A 153**, 97–113.

Varadarajan, V S (1974). *Lie Groups, Lie Algebras, and Their Representations.* Englewood Cliffs, NJ: Prentice-Hall.

Varadhan, S R S (1966). Asymptotic probabilities and differential equations, *Comm. Pure Appl. Math.* **19**, 261–286.

von Neumann, J (1931). Die Eindeutigkeit der Schrödingerschen Operatoren, *Math. Ann.* **104**, 570–578.

von Neumann, J (1931). Über Funktionen von Funktionaloperatoren, *Ann. Math.* **31**, 191–226.

von Neumann, J (1955). *Mathematical Foundations of Quantum Mechanics.* Princeton: Princeton University Press. English translation (revised by the author) of *Mathematische Grundlagen der Quantenmechanik* (1932). Berlin: Julius Springer.

Weinberg, S (1999). *The Quantum Theory of Fields*, Vol. I, *Foundations.* Cambridge: Cambridge University Press.

Weyl, H (1931) *The Theory of Groups and Quantum Mechanics.* London: Methuen. English translation of *Gruppentheorie und Quantenmechanik* (1931), 2nd edn. Reprinted by Dover Publications, New York.

Weyl, H (1946). *The Classical Groups*, 2nd edn. Princeton: Princeton University Press.

Weyl, H (1950?). *Space-Time-Matter*, New York: Dover Publications. English translation of *Raum-Zeit-Materie* (1921), 4th edn. Berlin: Julius Springer, with a preface (1950) by the author. Translation originally published (1922). London: Methuen.

Wheeler, J R and Zurek, W H (1983). (Editors) *Quantum Theory and Measurement.* Princeton: Princeton University Press.

Whitney, H (1936). Differentiable manifolds, *Annals of Math.* **37**, 645–680.

Whitten-Wolfe, B and Emch, G G (1976). A mechanical quantum measuring process, *Helv. Phys. Acta* **49**, 45–55.

Wick, G C, Wightman, A S and Wigner, E P (1952). The intrinsic parity of elementary particles, *Phys. Rev.* **88**, 101–105.

Wightman, A S (1977). Hilbert's sixth problem: mathematical treatment of the axioms of physics, in *Mathematical Developments Arising from Hilbert Problems*, Proceedings of Symposia in Pure Mathematics, Vol. XXVIII, Part I, Ed. F E Browder, pp. 147–240. Providence, RI: American Mathematical Society.

Wigner, E P (1939). On unitary representations of the inhomogeneous Lorentz group, *Ann. of Math.* **40**, 149–204. Reprinted in *Symmetry Groups in Nuclear and Particle Physics*, Ed. F J Dyson, pp. 39–94, (1966). New York: W A Benjamin.

Wigner, E P (1952). Die Messung quantenmechanischer Operatoren, *Zetis. f. Phys.* **133**, 101–108.

Wigner, E P (1959). *Group Theory and Its Application to the Quantum Mechanics of Atomic Spectra*, New York: Academic Press. Revised and expanded translation of *Gruppentheorie und ihre Anwendung auf die Quantenmechanik der Atomspektren*, (1931). Braunschweig: F Vieweg.

Wigner, E P (1960). The unreasonable effectiveness of mathematics in the natural sciences, *Comm. Pure Appl. Math.* **13**, 1–14. Reprinted in (Wigner, 1970, pp. 222–237).

Wigner, E P (1963). The problem of measurement, *Amer. J. Phys.* **31**, 6–15. Reprinted in (Wigner, 1970, pp. 153–170) and (Wheeler and Zurek, 1983 pp. 324–341).

Wigner, E P (1964). Two kinds of reality, *The Monist*, **48**, 248–264. Reprinted in (Wigner, 1970, pp. 185–199).

Wigner, E P (1970). *Symmetries and Reflections: Scientific Essays of Eugene P Wigner*, Eds. W J Moore and M Scriven, paperback. Cambridge, MA: The MIT Press. Originally published (1967), Bloomington, IN: Indiana University Press.

Wigner, E P (1983). *Interpretation of Quantum Mechanics*, in (Wheeler and Zurek, 1983, pp. 260–341).

Willard, S (1970). *General Topology*. Reading, MA: Addison-Wesley.

Yanase, M M (1961). Optimal measuring apparatus, *Phys. Rev.* **123**, 666–68. Reprinted in (Wheeler and Zurek, 1983, pp. 712–714).

Zeeman, E C (1964). Causality implies the Lorentz group, *J. Math. Phys.* **5**, 490–493.

Index

Printed in the United States
By Bookmasters